George Thomas Jackson

A Practical Treatise on the Diseases of the Hair and Scalp

George Thomas Jackson

A Practical Treatise on the Diseases of the Hair and Scalp

ISBN/EAN: 9783337088934

Printed in Europe, USA, Canada, Australia, Japan

Cover: Foto ©berggeist007 / pixelio.de

More available books at **www.hansebooks.com**

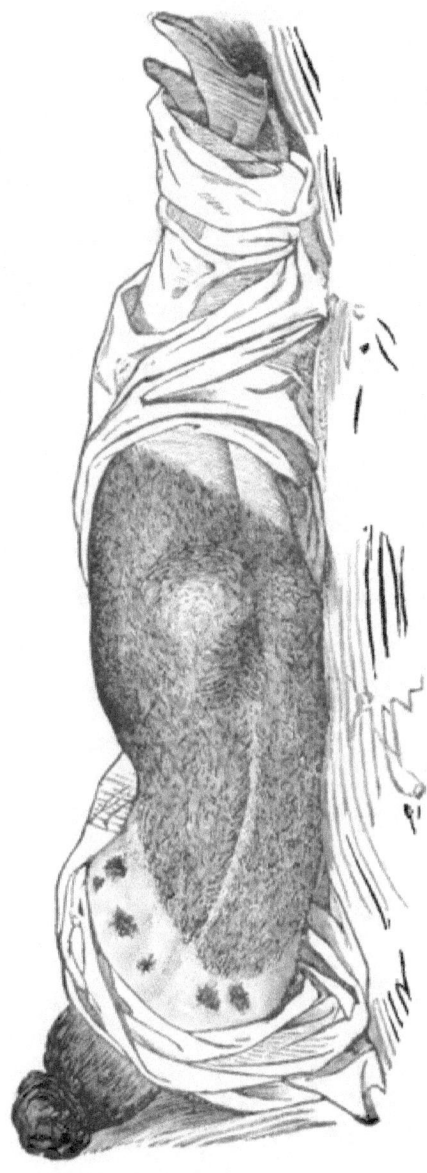

Nævus Pilosus—Hairy Mole.

[This illustration represents the case of a Hebrew maiden, who lived in Königsberg, Prussia. The hair growth was thick and soft, looking and feeling like fur. It was on the abdomen as well as on the back, and besides the large patch, there were numerous small hairy nævi scattered over the body.]

A PRACTICAL TREATISE

on

The Diseases of the Hair and Scalp

BY

GEORGE THOMAS JACKSON, M.D.

Professor of Dermatology, Woman's Medical College, N. Y. Infirmary; Chief of Clinic and Instructor in Dermatology, College of Physicians and Surgeons; Consulting Dermatologist, Presbyterian Hospital; Visiting Dermatologist, Randall's Island Hospital; Member of the American Dermatological Association, etc.

NEW, REVISED, AND ENLARGED EDITION

NEW YORK:
E. B. TREAT & COMPANY,
241-243 WEST 23D STREET.
CHICAGO: R. C. TREAT, 199 CLARK STREET.
1898.

PRESS OF STETTINER, LAMBERT & CO., 22, 24 & 26 READE ST., NEW YORK.

PREFACE.

In this edition of this book it is hoped that the reader will find all the knowledge about the hair that has been gained during the years that have gone since the appearance of the first edition of this book in March, 1887. During this time alopecia areata, the parasitic diseases, and seborrhœa have been studied with great care by many investigators.

Every page of the old edition has been revised and corrected; new articles upon folliculitis decalvans, lepothrix, and aplasia pilorum propria, and many new sections to the old chapters, have been added. The bibliography has been brought down to January, 1893, and nine new illustrations have been inserted in the text.

The author hopes that the present edition of the book will meet with as good a reception as the first edition did, and that it may prove useful to his professional brethren.

14 East 31st street, New York.

PREFACE TO FIRST EDITION.

FIVE years ago, the author of this work was in need of some complete treatise upon the diseases of the hair; and finding none of recent date, excepting such as were rather of a popular than a scientific character, he began those studies which have resulted in the present volume.

The aim of this book is to present to the medical profession a concise statement of what is known of the diseases of the hair and scalp; special attention being given to their diagnosis and treatment. To this end a great number of medical journals have been consulted, and all available books upon the hair have been read.

The chapter upon the anatomy of the hair is drawn chiefly from WALDEYER'S "*Atlas der Menschlichen und Tierischen Haare,*" Lahr, 1884; and from UNNA'S article upon the anatomy of the skin in the last number of Ziemssen's "*Handbuch der speciellen Pathologie und Therapie.*" Leipzig, 1883. Those diseases of the scalp which occur as part of a general disease of the integument have been described briefly, since all text books on dermatology treat fully of them.

In the Journal Literature there will be found but few references to papers which have appeared prior to

1860. In the last twenty-five years our knowledge of the diseases of the hair has so advanced that many of the views held by writers of an earlier date are no longer of practical value. Where experience has shown that the older writers were right, their principles and practice have been absorbed into the general sum of our knowledge, and are to be found in all systematic treatises upon the skin. For these reasons the year 1860 has been chosen arbitrarily as a dividing line, and the literature of the following years only has been consulted. The Bibliography contains the titles of books which deal solely with the hair, and also of general treatises upon the skin which have been consulted. Such references as have not been accessible to the author he has indicated by grouping them together and having them printed in smaller type.

The author takes great pleasure in acknowledging in this place his obligations and expressing his thanks to his friends Dr. George Henry Fox and Dr. Edward B. Bronson, and to his brother Rev. S. M. Jackson, for valuable suggestions and kindly criticisms during the composition of this book. The greater number of the illustrations which add so much to the value of the work are from the large collection of photographs belonging to Dr. George Henry Fox, who placed them most generously at the author's disposal

No. 14 East 31st St., New York,
  March 1st, 1887.

CONTENTS.

PART I.

GENERAL CONSIDERATIONS.

CHAP.		PAGE
I.	Anatomy of the Hair,	21
II.	Physiology of the Hair,	32
III.	Hygiene of the Scalp and Hair,	51

PART II.

ESSENTIAL DISEASES OF THE HAIR.

IV.	Canities,	63
V.	Changes in the Color of the Hair other than turning Gray,	74
VI.	Alopecia,	80
VII.	Alopecia Areata,	115
VIII.	Atrophia Pilorum Propria,	140
IX.	Hypertrophia Pilorum, or Hypertrichosis,	158
X.	Trichiasis and Distichiasis,	180
XI.	Sycosis,	182

PART III.

PARASITIC DISEASES OF THE HAIR.

CHAP.		PAGE
XII.	Trichophytosis Capitis,	205
XIII.	Kerion,	239
XIV.	Trichophytosis Barbae,	245
XV.	Favus,	256
XVI.	Pediculosis Capitis,	275
XVII.	" Pubis,	283
XVIII.	Beigel's Disease, and other unusual parasitic Diseases.	290

PART IV.

DISEASES OF THE HAIR SECONDARY TO DISEASES OF THE SKIN.

XIX.	Dandruff,	299
XX.	Keratosis Pilaris,	310
XXI.	Eczema Capitis et Barbae,	314
XXII.	Plica Polonica,	339
XXIII.	Dermatitis Papillaris Capillitii,	344
XXIV.	Nævus Pilosus,	350
XXV.	Syphilis. Lupus. Vitiligo.	354

Bibliography and Journal Literature, . . 365
Index, ... 409

CHAPTER I.

ANATOMY OF THE HAIR.

BEFORE we can understand the diseases which affect the hair, it is necessary for us to have some knowledge of its anatomy and physiology. The close connection of the sebaceous glands with the hair follicles, and the fact that disease of the former is very commonly associated with that of the latter, render it essential for us to devote a little time to the anatomy and functions of these glands. It is true that histologists are not yet in perfect accord in regard to some points in the microscopical anatomy of the hair, and that there are yet some unanswered questions in the phenomena of its development, growth, fall and regeneration; but, nevertheless, we know enough to aid us materially in our study of its diseases.

GENERAL DESCRIPTION.—The hair is an epidermic structure consisting of a *root*, which is seated in the skin and expanded below to form the *bulb;* and of a *shaft*, which projects beyond the surface of the skin and terminates in a *point*. Its form may be described as spindle-shaped, or as a slender cone gradually tapering to its apex. Its contour is circular, oval or flattened; and it is either straight, or more or less curled. It presents three main varieties: 1. Long, soft hair, such as is met with on the head, in the beard, on the pubis and in the axillae. 2. Short, stiff hair, such as is found in the eyebrows and eyelashes. 3. Lanugo, or soft, downy, colorless hair, such as is scattered all over the surface of the body, where the other varieties are absent. Each hair grows from a small nipple-shaped connective-tissue projection, the *hair papilla*, situated at the bottom of a deep slender pocket or sac-like depression in the skin which is called the *hair follicle*. To each hair follicle there is attached one or more sebaceous glands, which empty by their ducts into its upper third.

THE HAIR.—The hair is composed of three layers, which from within outwards are: 1. The medulla. 2. The cortex. 3. The cuticle. These are distinguishable even in the deepest part of the hair root, and become yet more distinct as we proceed upwards. The hairs, excepting those called lanugo, are hollow cylinders, the central cavity being filled, in fully formed healthy hair, with the medulla, and called the medullary canal. This canal begins below at the papilla, and extends to within a short distance of the point of uncut hairs. In the lanugo hairs it is generally wanting. The upper extremity or tip of the hair is pointed, if the hair has not been cut; if it has been cut it is flattened, or more or less rounded, depending upon the length of time that has intervened between the time of cutting and of examination; if a sufficient time has elapsed it may even become again pointed. The lower extremity of the hair, the root, is expanded to form the bulb, which is hollowed out so as to fit accurately like a cap over the nipple-shaped papilla upon which it rests.

THE MEDULLA.—The medulla consists of a column of superimposed cells which occupies the cavity of the medullary canal. It begins immediately upon the upper rounded top of the hair papilla as a layer of irregular cubical epithelial cells, each cell containing one or two masses of keratohyalin, which appear as dark round drops. WALDEYER (83) thinks it probable that keratohyalin, the "eleidin" of RANVIER, is identical with the "hyalin" of VON RECKLINGHAUSEN, which is found in many different kinds of cells. As we trace the medulla higher up in the hair we see that the cells, which were dispersed at first irregularly in layers, form themselves into a stratified column, like a roll of coins, with three or four cells in each horizontal layer. According to WALDEYER (83) the cells are held together by means of delicate projections, and when isolated appear like the prickle cells of the skin. The lower cells alone have nuclei. As the medulla ascends in the hair the cells become more and more flattened, the keratohyalin melts into the cell plasma, and the nucleus shrivels up and disappears. Then the cells themselves shrivel and leave spaces between them, which in the middle follicle region are filled with air, so that the cells are surrounded by a system of air canals. The air does not penetrate the cells. Towards the point of the hair the medulla is reduced to a column of single cells laid one upon the other. Then the column becomes broken up, greater or smaller spaces forming between the cells, till finally it ends. The medulla cells sometimes contain pigment. The delicate fœtal hairs are without a medulla, as are also most of the lanugo hairs. As a rule, the greater the diameter of the hair, the greater will be the diameter of the medulla. Nevertheless strong hairs will often be found with proportionately thin medullæ, as is frequently the case in the hairs of the human beard. Still more common is it to find thin hairs with stout medullæ. Towards the close of the life limit of a hair no more medulla cells are produced, and in such hairs there is a wide space between the bottom of the medulla and the end of the root.

THE CORTEX.—The second layer of the hair is the cortical substance. This is the substance proper of the hair, and consists of long spindle-shaped epithelial cells which are flattened out into fine bands, and run in the long axis of the hair. They are completely cornified and contain a shrunken nucleus, which appears as if pulled out lengthwise; it is wanting entirely towards the point of the hair, and is plainly seen only in the root. The first cells of the cortex in the neighborhood of the papilla are cube-shaped, but under pressure from without inwards they become as described above. The cortical cells are, like those of the medulla, provided with prickles. In the cortex are found pigment and air. The pigment occurs either in the form of granules, or diffused. The granular pigment is found heaped up in the cells of the cortex, especially in those lying peripherally, and sometimes the granules are crowded so closely together as to render the individual ones indistinguishable. In the upper part of the cortex pigment granules may be found lying between the cells. The diffused pigment is the essential coloring matter of the hair. The air gains entrance to the cortex on account of a separation taking place between its cells, and is found either in the form of rounded, discrete air-globules, or in groups of them, or in long streaks.

THE CUTICLE.—The third and last layer of the hair is the cuticle. It corresponds to the epidermis of the skin in location and function. It consists of flattened, non-nucleated, fully cornified cells which cover the hair like scales and are arranged like shingles on a roof with their free ends directed towards the point of the hair. In the lower part of the hair root these cells are cylindrical and contain nuclei; but they gradually become flattened and lose their nuclei. In the deepest part of the shaft of old hairs the cuticle is wanting.

HAIR-ROOT.—The root of the hair, with its bulb or expanded part, contains all the elements of the hair. As the hair shaft descends towards the lower part of the hair-follicle it widens more rapidly, and then swells out to form the bulb which covers like a cap the papilla, excepting at its narrowest part. It results from this

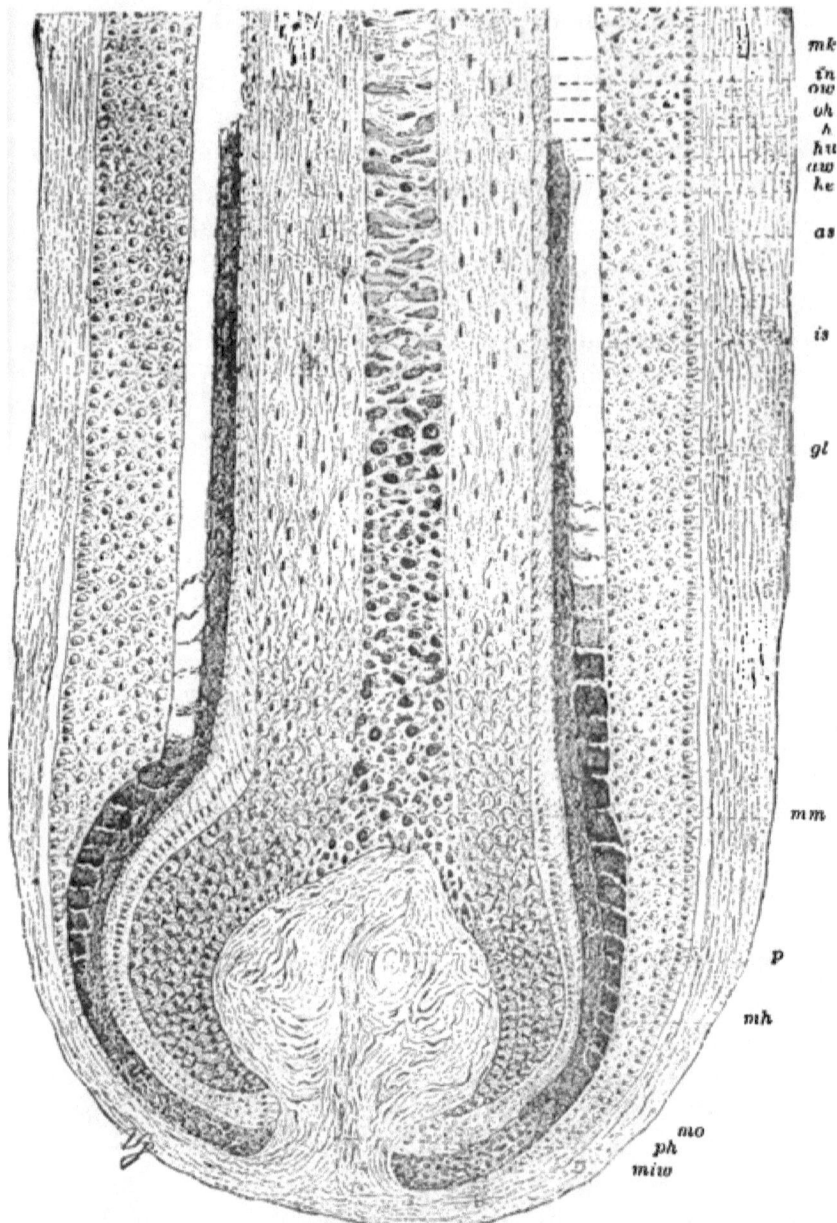

Longitudinal Section of Hair Root from Beard.

as — External sheath of hair-follicle. *is* — Internal sheath of hair-follicle. *gl* — Vitreous membrane. *aw* — External root-sheath. *iw* — Internal root-sheath. *he* — Henle's layer. *hu* — Huxley's layer. *ow* — Cuticle of root-sheath. *oh* — Cuticle of hair. *h* — Cortex of hair. *mk* — Medulla. *p* — Papilla. *miw*, *mo*, *mh*, *mm* — Matrices of *iw*, *aw*, *oh*, *mk*, and *h*. *ph* — Neck of papilla.

that the bulb is hollow below and indented like the bottom of a champagne bottle. The lower part of the bulb contains the matrices of the three layers of the hair. The matrix of the medulla occupies the part corresponding to the top of the papilla, and consists of irregular cubic cells. That of the cortical substance begins at the side of the body of the papilla, in a layer of cylindrical cells; while that of the cuticle springs from the neck of the papilla, and is likewise composed of cylindrical cells. The matrix cells soon begin to take on the characteristics of the cells of the different layers as we have already learned to know them. The sheaths which surround the hair at its root are two in number, the hair-follicle and the root-sheath. The hair-follicle is derived from the cutis, while the root-sheath is formed from the epidermis. We may represent the arrangement of the hair and its sheaths by picturing to ourselves the effect of thrusting, let us say, a dull needle into the skin. First there will take place a depression of the whole skin, and then the epidermis will be punctured at the lower part of the depression and the point will come in contact with the cutis below. The cutis will be to the outside and represent the hair-follicle, the epidermis will be in the middle and represent the root-sheath, while the blunt point of the needle will represent the hair-root.

HAIR-FOLLICLE.—The hair-follicle is always placed at an angle to the skin excepting in the eyelashes, where it is perpendicular to the tarsal edge. It is found in the upper half of the cutis when it is connected with lanugo hairs; deeper placed with stronger hairs; and in the subcutaneous connective tissue in connection with some very coarse hairs; and is from one-twelfth to one-fourth of an inch deep. It is a permanent structure, and does not leave its place when a hair is plucked from the skin. It begins above at the opening of the sebaceous gland, passes down alongside of the hair, and surrounds its lower end forming a closed pouch, while some of its fibres enter the hair-root from below to form the papilla. It consists of three layers; 1. An outer longitudinal fibrous layer of ordinary connective tissue. 2. A middle layer of circular fibres which are richly studded with long nuclei resembling those of smooth muscular fibres; but no muscular fibres are present. This layer is the true foundation of the follicle, and is the only one that enters into the formation of the papilla. 3. The vitreous membrane which is inside of all. It is a thin, smooth and homogeneous membrane, which, according to UNNA (82 a), is merely a thickening of the inner coat of the middle layer of the follicle, and is only found in its lower one third. The older the hair is, the more prominent does this membrane become. The inside of this membrane, especially in old hairs, is thrown into circular or semi-circular projections or ridges, which protrude into the prickle-cell layer of the outer root sheath in the form of variously sized, dull or sharp-pointed teeth. It does not reach to the papilla.

ROOT-SHEATH.—The root-sheath consists, as usually described, of two parts, namely: 1. An external root-sheath. 2. An internal root-sheath. UNNA (82 a) teaches that the external root-sheath should be designated as the prickle-cell layer of the hair-follicle, as it is genetically different from the internal root-sheath, being continuous with the epidermis; while the internal root-sheath springs from the hair papilla. The *external root-sheath* is continuous above with the epidermis. As it reaches the mouth of the sebaceous gland the granular and corneous layers of the epidermis cease, and the prickle and cylindrical cell layers proceed in full width to near the papilla, where they suddenly grow smaller; and finally at the neck of the papilla they form a narrow stratum. It is composed of three layers: 1. An external cylindrical cell layer which is continuous with the cylindrical cell layer of the epidermis, and like it is composed of a single row of cells, which present their narrow ends to the vitreous membrane of the hair-follicle. 2. A middle prickle cell layer, a continuation of the same layer of the epidermis. This is the thickest of all the layers of the root-sheath. 3. A single layer of flat cells lying next to the outer layer of the internal root-sheath. The *internal root-sheath* begins below at the neck of the papilla, and passing upwards ends abruptly, as if cut off, at the neck of the hair-follicle, where the sebaceous gland empties into it. It consists of three layers, according to most anatomists: 1. An external or Henle's layer. 2. A middle or Huxley's layer. 3. An internal or cuticular layer. The external or Henle's layer is composed of only a single row of flat cells. The middle or Huxley's layer is usually formed of a single row of short cylindrical cells; but when connected with the thick hairs of the beard it at times has a second row of cells. The internal or cuticular layer is similar in formation to that of the hair cuticle, being composed of a single layer of flat cells which rest one above the other like scales. In the upper regions of the hair-root they are cornified. The cuticle of the sheath differs from that of the hair, in its cells standing oblique to the long axis of the hair and pointing downwards; while those of the hair have their long diameter parallel with it and point upwards. There thus results an interlocking between them, and a fast union, especially in the middle third of the follicle, so that when a hair is plucked from the skin it usually brings with it a part of the internal root-sheath.

The internal root-sheath can readily be divided into two divisions, a lower and an upper. The lower one corresponds to the region of the papilla. All its cells

ANATOMY OF THE HAIR.

are nucleated. In the upper region the cells of the cuticle are cornified, and those of the Henle layer have lost their nuclei. The cells of Huxley's layer preserve their nuclei longer, but they become shrunken, and finally disappear. UNNA (82 a) proposes to do away with the terms Henle's and Huxley's layer, as he regards them as artificially prepared layers, and to speak of the whole as the root-sheath. As noted already he has dropped the term "outer root-sheath," regarding it simply as a part of the hair-follicle.

THE PAPILLA.—The hair-papilla is situated at the bottom of the hair-follicle in connection with its circular fibrous layer. It is a wart or nipple-shaped connective-tissue projection, which penetrates the hair bulb from below. It has a narrow neck, a

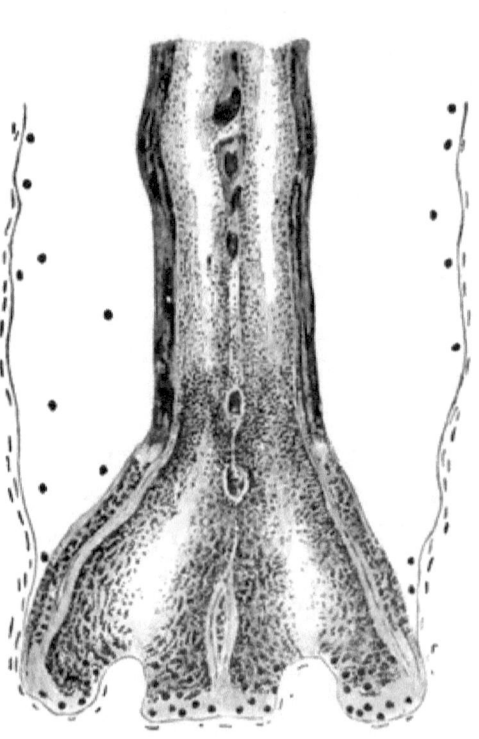

Twin Hair after Giovannini.

strongly developed, wide middle part or body, and a fine point. Upon it the hair rests; and it contains the blood-vessels for the chief supply of its nutriment.

TWIN HAIRS.—GIOVANNINI (86 ap.) has described a curious anomaly of hair-growth, where two hairs grow inside of a single follicle, surrounded by a common internal root-sheath. At the bottom of the follicle there were two separate papillæ, one for each hair. At their bases by transverse section they were of irregular shape, with their greater diameter in the same direction. The hairs were of a more or less regular oval form. Above the root they were somewhat kidney-shaped, and only occasionally oval or round. In the region of the matrix the hairs were about parallel; at the neck of the follicle they were divergent; in the shaft they were sometimes divergent, sometimes parallel. In the neighborhood of the follicle-ground the matrices of the hair were entirely separated. About the middle of the neck they were joined, to become separated and again joined higher up. The juncture was made sometimes by a melting together of the cuticle, sometimes by a bridge formed partly of cuticle, partly of cortex. Sometimes they were joined at the front, sometimes at the side.

BLOOD-VESSELS.—The blood supply of the hair is derived from arterial branches which surround the hair follicles in the form of a capillary net-work between their middle and external layers; and from small arteries which penetrate the papilla from below. These arterial branches are derived from the sub-papillary arterial anastomosis of the skin. According to UNNA (82 a) the arteries of the follicles alone are derived from this source and enter the follicle in its middle region, while the arteries of the papillæ spring from the deeper circulation of the cutis. The veins follow the same direction.

NERVES AND LYMPHATICS.—Nerves have been traced to the hair-follicles, but not to the hair-papillæ in the human subject. Some histologists have followed them to the vitreous membrane of the follicle; and UNNA (82 a) has demonstrated that the prickle-cell layer of the external root-sheath is supplied with sensitive nerves. He describes the nerve trunks as entering the follicles immediately beneath the sebaceous glands, losing their medulla when they reach the vitreous membrane through which they pass, and then breaking up into branches, which end within the cells with double-ended extremities. No lymphatics have yet been demonstrated in connection with the hair.

ARRECTORES PILORUM.—The muscles of the hair are composed of smooth muscular fibres and are named "arrectores pilorum." They arise from the lower part of the papillary layer, pass downwards beneath the sebaceous glands, and are attached with many ends to the middle part of many neighboring hair-follicles upon the side which makes an acute angle with the skin.

SEBACEOUS GLANDS.—The sebaceous glands are of the racemose variety, and are found in close connection with the hairs of the body, from two to six glands emptying into each hair-follicle in its upper third. In structure they consist of a number of acini which empty by a common duct. They are composed of a delicate, structureless capsule, the *membrana propria*, which continues into the duct, and then merges into the vitreous membrane of the hair-follicle; and of lining cells, which are large, though short, cubical or cylindrical epithelial cells, arranged in one or two layers. These pass through the duct, and are continuous with the cylindrical cells of the outer root-sheath and of the skin. The interior of the gland is filled with the fatty secretion. The size of the glands varies in different regions, the largest of those connected with the hair, being on the nose, measure 2 mm. in diameter in some places, and may have as many as twenty acini. Some glands are but $\frac{1}{505}$ of an inch in diameter. Their length varies in different regions, the longest being on the nose. As these glands are offshoots of the hair-follicles, the external layer of the hair-follicles passes around them and supports their membrana propria.

CHAPTER II.

PHYSIOLOGY.

We have now to consider the phenomena of the development, degeneration, fall and regeneration of the hair; its peculiarities of form and constitution; and the part it plays in the life history of the human being whom it clothes.

Development.—Fœtal Hair.—The germs of the future hair appear first upon the face (forehead) and eyebrows in the tenth to the twelfth week of fœtal life; on the lips in the fourteenth week; on the rest of the head in the sixteenth week; soon after upon the trunk; a few weeks later upon the arms and legs; and at last, in the seventh month, upon the backs of the hands and feet. The hair itself forms about one month later and follows the same order, first coming on the eyebrows, and so on. These germs of the future hair are short peg-like elevations of the outer embryonal skin. Reissner and Götte considered them swellings of the corium; but Waldeyer (83) and Unna (82 a) are in accord in teaching that only the stratum Malpighii of the epidermis takes part in their formation, the corneous layer passing smoothly over them. These germs are at first separated only by a fine seam of connective tissue. An increase and heaping up of round and spindle-shaped cells then slowly takes place in the cutis connective-tissue at the fundus of the germs. At first these cells surround each germ like a basket; but soon a prominence forms by the cells crowding together at one part, and this indents the bottom of the hair germ.

Thus are formed the primitive hair-follicle and the primitive hair-papilla. According to WALDEYER (83), while these changes are taking place in the connective tissue, the hair-germ sends a slightly knob-shaped projection down into the underlying cutis, and the original elevation disappears The outer cells of the hair-germ now become cylindrical, place themselves crosswise, and appear as a continuation of the basal cells of the stratum Malpighii of the skin, while those in the long axis of the hair-germ place themselves vertically, grow longer and appear, if the papilla has already penetrated from below, as a delicate cone-shaped body resting on the top of the papilla. This is the *primitive hair-cone (Haarkegel)* of UNNA (82 a).

The cornification of the cells of the primitive hair-cone begins at its point, and proceeds rapidly downwards on its external layer till to the point of the hair papilla. The lower end of the hair-cone embraces the papilla more and more as the upper end grows nearer to the outer surface of the skin. Where the hairs stand almost perpendicularly to the skin, the growing hair-cone drives the corneous layer of the skin before it, forming a rounded elevation. This at last gives way, and the point of the hair protrudes. Where the hair is placed more at an angle to the skin, it sometimes runs for some distance under the epidermis in a spiral manner before piercing it. The middle part of the hair-cone grows more rapidly than the outer part, and soon breaks through the protecting cell layer of the primitive hair-cone as a formed hair, though only consisting of cortical substance. The outer layer of cells becomes the inner root-sheath and crumbles gradually away to end at the mouth of the sebaceous gland. The remaining cells of the hair-germ, which are still to the outside of the external layer of the primitive hair-cone, become the outer root-sheath, which is continuous above with the prickle-cell layer of the skin, and ends below in a point in the neighborhood of the papilla. This is the formation of the primary or fœtal hairs. The vibrissæ, and the hairs of the eyelids, outer nose and lips, are placed almost perpendicularly to the surface of the skin. In other situations the hairs stand at an oblique angle to the surface of the skin, the obliquity increasing with the size of the hair, so as to afford more space for the lodgement of the hair-follicle. The more obliquely the hairs stand, the more plainly do the arrectores pili muscles appear, serving to bridge over the obtuse angle which the hairs form with the surface, and to sustain them in position. Where the hairs are perpendicular to the skin these muscles do not exist. The body of the fœtus at the seventh month is almost entirely covered with hair, and sometimes this condition persists till birth.

EMBRYONAL HAIR CHANGE.—It is no unusual thing to see a child born with long, colored hair upon the head, and long, light, or colorless hair upon the rest of the body as well as on the face. Normally, however, in the sixth month of fœtal life, or at the beginning of the seventh, the embryonal hair change takes place, commencing on the lip. The primitive hair is raised from the papilla, and its root end becomes knob-shaped instead of cap-shaped. It mounts up in the follicle till it reaches its middle third, where it remains for a time, and continues growing, gaining its nourishment from the epithelial cells of the part. For a time the lower part of the follicle remains open, but slowly its epithelial lining disappears, it shortens, and the papilla atrophies and vanishes. Now a new epithelial process is sent out from the lower part of the old hair, passes downwards, enters the lower part of the follicle, making it pervious again, and becomes indented from below by a new papilla. Then there is a new growth of hair, upon the same principle as in the embryonal hair. This process takes place at about the eighth month, and is coincident with the fall of the primary hair. The fall of the old hair, and the growth of the new, follows in the same order as in the first appearance of the hair. Even if the child is born with a good deal of hair it falls soon, showing that the primary hairs were already loosened from their papillæ. This change from embryonal to permanent hair is a change in type, for the new hair is soon furnished with a medulla. After this there is no change in type, excepting in the development of new hair at puberty, though the old hairs constantly fall out and are replaced by new hairs.* As the child grows older, its head, which at birth is usually sparsely supplied with fine hairs, becomes well covered with long, colored, dark hairs, and its eyebrows and eyelashes become more pronounced.

CHANGE OF HAIR AT PUBERTY.—At puberty another change in the hair growth takes place, consisting in the appearance of hair upon the pubes and in the axillæ of both sexes, and about the anus and on the face of males. Still later strong hairs grow in the nostrils and in the ears, though this growth may be long deferred. It is notable that there are usually no hairs about the anus of a woman. In women, also, the axillary hairs are generally less developed than in men, and often they are entirely wanting. By some observers, indeed, it is stated that the absence of axillary hairs in women is the rule. As far as my observation extends they have almost always been present.

VARIETIES OF HAIR GROWTH.—In fully developed adults most of the body, excepting in the regions just mentioned, is supplied only with colorless, fine lanugo hairs, but in most men and in a few women a more or less luxuriant growth of hair will be found upon the chest and extremities. In the male the pubic hair is often continuous above with a pyramidal growth of hair upon the middle line of the abdomen. In women this is never the case, the pubic hair ending by a sharp horizontal line. Hair is always absent from the glans penis and the prepuce; from the vermilion border of the lips; from the labia minora; from the last phalanges of the fingers and toes; and from the palms and soles. The hair of women grows to a greater length than does that of men, even if the latter is uncut. The hair of the scalp grows in groups, two or three hairs in each group, seldom as many as four. The hairs of each group are, according to Pincus (71), not all of the same length, because they are not all of the same age.

Hair Centres.—The hairs are not only placed at an angle to the skin, but they grow from a number of well-defined centres. Wilson (84) has studied these and carefully described them as follows: "The hairs of the head radiate from the crown with a gentle sweep behind towards the left and in front towards the right. The centre for the forehead is a median vertical line from which the hair passes to the right and left, the lower border of the growth forming the upper half of the eyebrows. This centre is distinctly visible in its whole length in many newborn children. At the inner angle of each eye is another centre from which the hair radiates, the upper and inner rays ascending to the line between the eyebrows, where they often meet those from the opposite side and form with them a line across the root of the nose; and the upper and outer rays curve along the brow and form the lower half of the eyebrow. The lower and outer rays with those of the nose, mouth and chin make a sweep over the cheek and side of the face. On the upper lip the hair grows from the nostrils outwards, and forms the mustache; on the lower lip there is a middle line for a centre. The beard is formed by the convergence of two side currents which meet at the middle line. On the trunk there is a centre of radiation from each armpit and two lines of divergence, one of the latter proceeding from this point horizontally to the middle of the front of the chest, the other vertically along the side of the trunk, across the front of the hip, and down the inside of the thigh to the bend of the knee. From the armpit centre, and from the upper side of the horizontal line, a broad and curved current sweeps upward over the upper part of the front of the chest and around the neck to the back. From the lower side of the horizontal line and from the vertical line the set of the current is downwards and inwards, with a gentle undulation to the middle line in front, and backwards to the spine.

"From the armpit centre there proceeds another line of divergence which encircles the arm like a bracelet immediately below the shoulder. From the upper margin of this line the direction of the current is upwards over the shoulder, and then backwards to the mid-line of the body. Another line commences at this ring on the front of the arm, and runs in a pretty straight course to the cleft between the index finger and thumb on the back of the hand. This is the line of divergence of the arm. From it and from the ring, the stream sets at first with a curve forwards, and then with a curve backwards to the point of the elbow. In the forearm the currents sweep downwards in front, and upwards behind towards the point of the elbow, which is thus the centre of convergence. On the back of the hand and fingers there is an outward sweep with the concavity upwards. On the lower limb there are two vertical lines of divergence, the one being the continuation of that of the side of the trunk, proceeding around the inner side of the thigh to the bend of the knee; the other, an undulating line, beginning at about the middle of the hip, and running down the outer side of the thigh and leg, and across the instep to the cleft between the great and second toes. A short oblique line connects the two vertical lines at the bend of the knee On the front of the thigh the streams from the two lines converge and descend towards the knee. On the back they converge also at the middle line, but ascend toward the trunk of the body On the leg, where there is but one line, the diverging currents sweep round the limb, and meet upon the shin, while on the foot they diverge with a sweep, as upon the back of the hand. The hair centres are called whorls.

SHEDDING OF HAIR.—At certain times of the year animals "shed their coat," that is, a rapid fall of hair takes place and the animal's coat is thinned. At the same time with the fall of the old hair there is a growth of new hair and soon the coat is as thick as ever. In the human species there is, instead of a periodic shedding, a constant fall and new growth of hair, though at certain seasons it may proceed more rapidly than at other seasons. This is accomplished in the same way as we learned when describing the embryonal hair change, namely: the hair loosens from its papilla and mounts up to the middle follicle region, where it remains for a time attached to the prickle-cell layer of the follicle, and grows there. The lower part of the follicle collapses, and the

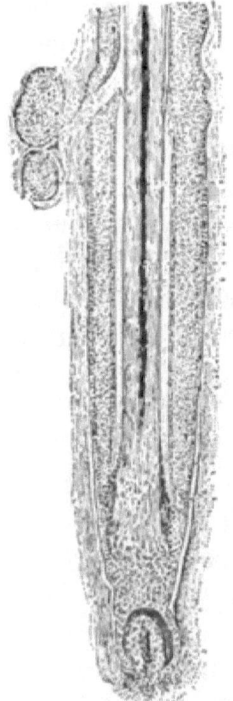

Section of hair follicle showing beginning separation of hair from papilla (Unna.)

papilla grows smaller. The lower end of the old hair becomes broom-like and knob-shaped. This appearance is due to new cornified cells being constantly attached to the root as it mounts up in the follicle.

BED-HAIR.—The hair in this position is called by UNNA (82 a) a "*bed-hair*" (Beethaar), to distinguish it from the hair seated upon its papilla, or "*papillary hair*." The part of the follicle from which it grows, he names the "Haarbeet," or "hair-bed." The "bed-hair" is always secondary to a papillary hair, and all hairs go through this stage before they fall They are distinguished from papillary hairs by absence of root-sheath, want of cuticle and medulla in their roots, and by having their pigment distributed in stripes and heaps.

The cause of the loosening and shedding of hair, whether at the close of its normal length of life, or on account of some disease, as eczema, typhus fever and the like, is to be sought for, according to UNNA (82 a), in the character of the blood supply of the hair. As we have learned, the middle region of the hair follicle is supplied by arterial twigs from the sub-papillary circulation, and the papilla of the hair is nourished from little arteries from the deeper arterial circulation of the cutis. Therefore the middle follicle region must participate in the disturbances of the sub-papillary circulation, while the papilla is relatively free from this influence. Thus any lessening of the nutritive supply to the papilla, or any increase in that from the sub-papillary circulation to the middle follicle region must cause a preponderance of nutrition in the middle over the lower follicle region. This would necessarily cause an increased growth of the prickle-cell layer of the middle follicle region, and an increased pressure upon the hair. Then either the circumference of the hair cylinder would be lessened, or the hair itself would be pressed out of the follicle and raised from the papilla. The latter is what actually takes place. The bed-hair is shoved higher and higher up, so long as it remains in the middle third of the follicle, but when it reaches the unproductive part of the follicle, that is, just below the mouth of the sebaceous gland, the circulation becomes again equalized, and the hair ceasing to grow, falls out. The final fall of the hair is hastened both by the pressure of the new hair from below, and by traction exerted upon it from above by brushing, combing, and the like. The loosening of the hair from its papilla begins at the cuticle, and proceeds from without inwards.

According to GIOVANNINI (87 ap.), when a coarse hair is plucked from its follicle, the latter shrinks up and its cavity completely disappears. The papilla diminishes slightly in volume, and approaches to the lower edge of the derma, while from below it gives off a connective-tissue pellicle containing blood-vessels. In the intra-dermic portion of the follicle the shrinkage goes on more rapidly in the upper and lower than in the middle portion of the follicle. The widest portion is where the arrectores pilorum muscles are attached. A new hair begins to form in from forty-one to seventy-two days after epilation, and is fully formed in from thirty to sixty days afterwards.

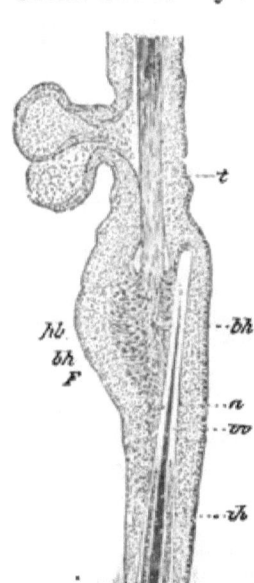

Section of hair follicle showing hair in "beethaar" stage, and growth of new hair from old papilla. *bh* = beethaar; *hb* = hair bed, from which *bh* grows (Unna.)

REGENERATION OF HAIR.—While the bed-hair is still in the middle follicle region the formation of the new hair is under way. From the lower end of the old hair down to the papilla there stretches an epithelial process composed of the old, collapsed, and greatly atrophied root-sheath. This enlarges, its cells increase, it grows downwards into the old hair-follicle, which again becomes open, and shoves the old papilla before it. Out of this a new hair is formed, seated upon the old papilla which takes on new activity. The whole process is analogous to that which occurs in the fœtus. The new hair mounts up in the old follicle and grows sometimes alongside the bed-hair, but more often it pushes it out ahead of it. This new hair lives its appointed time and then undergoes the same fate as its predecessor, and thus is constantly repeated the regular normal fall and regeneration of the hair. It is possible that new papillæ may be formed, producing new hairs, but the above is the usual course.

APPEARANCES OF HAIR.—The microscopical appearances of the hair will vary according to whether it is a lanugo hair, bed-hair, or papillary hair; or whether it has been plucked from the head, or has fallen of itself. The lanugo hairs are very fine and downy, colorless or very slightly colored, and contain no medulla. Their roots are small. The bed-hair has neither root-sheath, cuticle nor medulla in its root, and the root itself, instead of being hollow, is rounded off and shaped somewhat like an old-fashioned broom of twigs, or, it may be likened to a fir tree in shape, as UNNA suggests. Sometimes the medulla is entirely absent throughout its length. These appearances are those commonly seen in fallen hairs. This we should expect from what we have already learned of their nature. The papillary hair has a long, pliable, soft root-end, which is bulb-shaped and hollowed out for the reception of the papilla. All the component parts of the hair are present in it, and the medulla can be followed from its root to near its point. When such a hair is plucked, it presents for observation in addition to the above, an attached piece of the root-sheath which makes its lower part look swollen. These are the typical forms of hair. If the microscope is focused upon the surface of a hair, the edges of the cuticle cells will be seen as dark lines like the edges of the slates in a slate roof. If it is focused carefully upon the edge of the hair, the latter will look like the edge of a fine saw blade, as the edges of the cuticle cells overlap one another. If now the tube of the microscope be lowered, the cortical substance will come into view. It appears like a solid body marked by short stripes of a darker color, which gives it the appearance of being composed of fibres. The fibres are in reality long spindle-formed cells, as is learned by tracing the hair toward the root, when their cell form will be more apparent. Fine black granules are scattered about or gathered in heaps in its substance. These are the pigment granules. Lastly we meet with long, oval, irregularly shaped, or small, round, dark bodies lying between the fibres. These are air-globules. The small ones may be grouped and mistaken for pigment, a mistake easily rectified by altering the focus, when they will present the changes of color dependent upon their optical properties, becoming lighter as we raise the tube of the microscope. The medulla runs through the centre of the hair as a dark streak. In human hair, when examined in water alone, the separate cells of which the medulla is composed, are not visible, as a rule. This is owing to the presence of a large amount of pigment and air between the cells, both of which have the effect of rendering the medulla dark colored by transmitted light. If the hair be laid in glycerine, the air-globules will be driven out, and the cells will become more apparent. The form of the medulla cells has been given in the chapter on the anatomy of the hair.

TECHNIQUE.—A word must be said as to the proper methods of examining hair. For most examinations of the hair, an ordinary microscope, magnifying, say, 250 diameters, is sufficient. If you have higher powers so much the better. The hair should be examined at first under a covering glass alone. Then water should be added and the hair again examined. To render the hair transparent, it should be placed in liquor potassæ alone, or in liquor potassæ and glycerine. The peroxide of hydrogen acts even better than liquor potassæ as a bleaching agent, as it bleaches the pigment and does not destroy the hair. Unfortunately it soon loses its virtues by keeping. WALDEYER (83) recommends a twenty per cent. solution of nitric acid for rendering the hair transparent. If there be much foreign matter adherent to the hair, place the latter in a small test tube containing ether and shake it for some time, or put it in a corked bottle with ether and let it stand. These methods are sufficient for the proper examination of both healthy and diseased hair.

COLOR.—The color of the hair depends upon four factors, namely: 1. Diffused pigment. 2. Granular pigment. 3. Air contents; and 4. The superficial character of the hair. The cortex plays the chief part in determining the color of the hair. 1. The diffused pigment or essential color of the hair gives it a light-brown to a dark-red hue, according to its intensity. 2. The granular pigment lies in, (WALDEYER, 83) or between (UNNA, 82 a) the cells or fibres of the cortex, and, in some cases, in the cells of the medulla. It is found chiefly in the peripheral portion of the cortex, and occurs either scattered or grouped. Sometimes it is heaped up so thickly that the individual granules are indistinguishable. Its color is a shade of brown, anywhere from a light-brown to ebony. The combinations of the diffused and granular pigment makes the various shades of color met with. The darker the hair is the more granular pigment it contains, but even the lightest of blonde hair will be found to contain some granular pigment. 3. The air-globules are generally in the outer layers of the cortex. Viewed by direct light they appear under the microscope as brilliant points; by transmitted light they appear black. 4. By the superficial character of the hair is meant whether it is smooth or rough. These two last factors influence the color of the hair on account of a law in optics, namely: Every body appears white in white daylight if it reflects the white light to all sides. If the surface of the hair be uneven, and there are many little particles of air in the cortex and medulla, the light will be thus reflected and the color of the hair will be, therefore, more or less white, the tone being modified by the amount of pigment present in the cortex. A hair containing pigment never appears quite white, even if air be present, but some shade of gray. If there be no air present, or the pigment is in excess, the hair will be more or less purely of the color of the pigment. The most universal color of hair is dark brown or black.

PIGMENT.—The source of the pigment is not yet satisfactorily settled. It is derived, without doubt, from the coloring matter of the blood, as is all the pigment of the body. EHRMANN (100 and 101) of Vienna, published, in the years 1884, 1885 and 1886, some exceedingly interesting and valuable observations upon the formation of pigment in the skin and hair. His studies were made in the beginning upon frogs and salamanders; later upon the skin of dogs and men. He has found that the pigment is produced only in the corium, in parts immediately surrounding the blood-vessels, and that cell activity is absolutely essential thereto. From the corium it reaches the epidermis through protoplasmic movement. The pigment cells in men are round or oval, with short or few branches or prolongations. They are in the basal cells of the epidermis and send their prolongations downwards into the corium, to connect with the pigment-carrying cells of that part. The latter are not a peculiar species of cells, but true connective-tissue cells.

But, we are interested now more especially with the pigment of the hair. Pigment-carrying cells are found in the hair papilla, which are large in its neck and lower parts, and small in its top. As yet no branches have been observed to these cells. The pigment cells proper lie for the most part in the bases of the matrices, entirely included in them, and touch the boundary of the papilla only with one side of their circumference. Their branches pass only upwards between the cells of the matrix of the cortex up to the point where these begin to undergo cornification. The cells themselves of the cortex are devoid of pigment. Further up in the hair matrix, the branches of the pigment cells form a net-work, and the cortical cells of the third or fourth row are in close connection with this network, and themselves contain pigment.

LIFE PHENOMENA.—The length of life of the hair varies with the age, sex, character of hair, and individual peculiarity. Each hair has its determined length of life, and this is not the same for every hair of the same sort. What may be the circumstances that determine the period of its existence is not known. The lifetime of the eyelashes has been determined by MAHLY as one hundred and thirty five days. PINCUS (71) says that the period of hair growth on the human head is from two to six years. Hair is said to grow faster by day than by night, and in the warm weather rather than in the cold. Shaving and cutting the hair certainly make it coarser, and, may be, stimulate its growth. The average length of the hair of the head in women of the Anglo-Saxon race is from eighteen to twenty-four inches, when left uncut. Exceptionally it may grow to thirty-six or even fifty inches or more in length. The hair of men of the same race has an average length of six to eight inches; but custom demanding that it be cut from time to time, it is rarely seen of this length. The hair of each individual has its own determinate length, and the hair of men, even if left uncut, will not grow as long as that of women. The rate of growth, specially in young women, is from 2 to 5 mm. during each ten days after first piercing the skin. When it reaches a length of ten to fourteen inches its rate of growth is reduced one half; and later towards the end of its normal life its increase is hardly perceptible. The short, stiff hairs, as of the eyebrows, are from $\frac{1}{4}$ to 1 inch long.

Hairs, without their papillae, have been transplanted and become fixed on granulating wound surfaces and on the iris. I do not know that they have actually taken root and grown there.

The average *number* of hairs to the square inch is given by WILSON (84) as, on the scalp, 1,000. WITHOF (83) found on a man in one quarter of a square inch of the crown of the head, 293 hairs; occiput, 225; anterior part of the head, 211; chin, 39; pubes, 34; forearm, 23; back of the hand, 19; anterior surface of the thigh, 13. WILSON (84) calculates that there are 120,000 hairs upon the head of an adult. As a rule the finer the hairs, the thicker they will stand on the head.

The *diameter* of the hair varies with its color and location, and with the age and sex of the individual. WILSON (84) found that flaxen hair was the finest, and black hair the coarsest; that the hairs of the beard and whiskers were coarser than those of the breast and eyebrows, and then in order of decreasing diameter came those of the eyelashes and armpit, of the head, of the thigh, and of the leg, the latter being finest. The finest hairs of the scalp in the Anglo-Saxon race are from $\frac{1}{1500}$ to $\frac{1}{800}$ in., while the coarsest are from $\frac{1}{400}$ to $\frac{1}{170}$ in. in diameter. The head hair of a woman is somewhat coarser than that of a man, the diameter of the former being from $\frac{1}{500}$ to $\frac{1}{250}$ in.; that of the latter from $\frac{1}{525}$ to $\frac{1}{300}$ in. As a general rule, the hair of children is finer than that of adults, ranging from $\frac{1}{850}$ to $\frac{1}{700}$ in. in diameter. Even on the same head there is a great diversity in the diameter of different hairs, and individual hairs are not of the same thickness throughout. As stated in the first chapter (p. 21) the contour of the hair is circular, oval or flattened. Whether a hair is to be curly or straight is largely dependent upon its contour; the more oval or flattened it is, the more it will be curled. The curliness is influenced, also, by the condition of the atmosphere; naturally curled hair becomes more curled when the air is surcharged with moisture, and less so in dry weather. But artificially curled hair always loses its curl in damp weather.

RACE DIFFERENCES.—Ethnographical classifications have been founded upon the evident differences existing in the hair of different races of men. We are all familiar with differences of color, as between the black hair of the Negro, and the flaxen hair of the Saxon race; and of form, as between the close curled hair of the Negro, and the long straight hair of the American Indian. As marked differences between races exist in the contour of the hair, in the manner of its grouping, etc. But a discussion of these matters of classification is foreign to our purpose, and we will content ourselves with the mere statement of the fact.

PHYSICAL PECULIARITIES.—It has already been stated that the curliness of the hair is influenced by the moisture of the atmosphere. This is because the hair is hygroscopic, absorbs moisture very readily from the atmosphere, and becomes lengthened, as well as more rounded. Hair is also elastic, and is capable of being stretched from one-fifth to one-third of its normal length. When the tension is removed it will retract to nearly but not quite its original length. It possesses a considerable amount of strength, a good healthy adult hair being capable of sustaining a weight of from two to four ounces without breaking. The qualities of elasticity and resistance are mainly located in the fibrous portion or cortex of the hair. Hair is strongly electric by friction, particularly in cold and dry weather. Its electricity may readily be shown by passing a rubber comb through the long hair of a woman, when a slight crackling sound will be heard. In peculiarly susceptible individuals the hair will stand almost straight out from the head under this electrical stimulus. The electricity of the hair is negative.

CHEMICAL CONSTITUTION.—The chemical constitution of the hair, as given by WALDEYER (83) from several recent analyses, is as follows: One hundred parts of dry hair contain from $\frac{5}{10}$ to $\frac{5}{10}$ part of incombustible material. This contains 23 per cent. of alkaline sulphates, 2 to 10 per cent. oxide of iron, and 40 per cent. silica. Dark hair contains somewhat more iron. The analysis of the hair substance shows it to be composed of carbon, 50; hydrogen, 6.36; nitrogen, 17.14; oxygen, 20.85; sulphur, 5. The hair is said to contain a certain proportion of an oily substance, the color of which varies with that of the hair. The proportions of the chemical constituents vary with the color of the hair. Thus, fair hair contains least carbon and hydrogen and most oxygen and sulphur; brown hair gives the largest proportion of carbon and the smallest quantity of oxygen and sulphur; white hair of the aged contains a considerable amount of bone earth or phosphate of lime. The quantity of nitrogen remains the same in all.

Uses of Hair.—The uses of the hair are fourfold: 1. As a preservative of heat. 2 As a protective agency. 3. As an organ of touch. 4. As a promoter of beauty. (1.) As the hair is a bad conductor it serves to preserve the heat of the body. When one is exposed either to excessive heat or cold the hair is apt to grow more luxuriantly. (2.) The hair of the head forms a thick elastic cushion, and is thus an admirable defence to the skull against blows and falls. The pubic hair in like manner acts as a cushion during coitus. The eyebrows are a defence to the eye against blows, and turn the perspiration to the outside of the eye-socket, thus preventing its gaining access to the eye. The eyelashes catch flying particles of dust; the hairs of the nostrils and ears prevent insects from crawling into the cavities which they protect, as well as guard against the entrance of other foreign substances; the mustache acts as a respirator; and the beard protects the larynx from the action of cold. (3.) Parts furnished with hair are more sensitive than are those without it. This is because the hair, being planted at an angle to the skin, acts as a lever on being touched, shoves to one side the hair-follicle, and thus causes a slight irritation of the neighboring cutaneous nerves. This sense of touch in man is but little developed. (4.) Of the hair as a promoter of beauty, little need be said, as it will be conceded by all that the hair is an adornment.

Uses of the Sebaceous Glands.—The sebaceous glands furnish an oily secretion to the hair which renders it pliable and soft, and gives it lustre. This secretion is a constant one in health, and is the result of a fatty degeneration of the lining cells of the glands, which slowly fill up with the oily matter, burst, and discharge their contents. As the sebaceous glands empty into the hair-follicles in their upper third, they are admirably located for lubricating the hair just before its exit into the outer air. The oil is upon the outside of the hair and also within it. It reaches the interior of the hair by capillary attraction, as will be readily understood by referring to the anatomy of the hair, and noting its fibrillar construction and the spaces between its fibres. PINCUS believes that the capillary attraction is, to a certain extent, assisted by the pressure exerted upon the hair in the part of the follicle, just above the entrance of the sebaceous gland, its narrowest part or neck, the diameter not being large enough to admit of the escape of the hair and the sebaceous matter at the same time, without considerable squeezing, so that the sebaceous matter is pressed into the hair. It is probable that the hair, besides being supplied with oil by the sebaceous glands, secretes a cholesterine fat. This is shown by the investigations of Liebreich in regard to lanoline.

MUSCLES.—The *arrectores pilorum* muscles pass under and around the sebaceous glands, to be inserted into the hair-follicle. By their contraction they straighten the direction of the hair bulb, and, in consequence, erect the point of the hair, literally causing the hair to stand on end. This function is observed in man chiefly in the occurrence of "goose flesh," from the action of cold. The contraction of these muscles further aids, though to a slight extent, the emptying of the sebaceous glands.

CHAPTER III.

THE HYGIENE OF THE HAIR.

ATTENTION to the care of the hair and the hairy scalp is of special importance to those who belong to families in which premature baldness is hereditary, and it can not be given too early. We should, therefore, instruct the parents as to the importance of giving attention to their children's heads, so that the matter may not be delayed too long, and the hair fall out when it is too late to stop it. Dandruff is regarded by most people as merely an annoyance, and, if not excessive, is neglected. If we could convince the laity that dandruff is the chief cause of baldness, they would eagerly seek relief, the disease could be early checked, and the day of hair-fall very much delayed. The care of the hair is important, not only to those with an inherited tendency to baldness, but to all who wish to preserve their hair in good condition, and, if properly attended to, it will be a prophylactic, not only to diseases of the hair proper, but also to parasitic troubles of all sorts. It is true that this demands the expenditure of a certain amount of time, but it is time well expended, though, I must confess, often greatly begrudged by male patients.

The hygiene of the hair and scalp consists in the proper use of the shampoo; in brushing and combing; in arranging the hair; in the exposure of the hair to air and light; in cutting and shaving it; and in the use of pomades. We should watch over the hair from earliest infancy, and instruct our patients as carefully in regard to its hygiene as we should do in regard to matters of general hygiene.

SHAMPOO.—The first attention that the hair demands is the ridding of the scalp of the newborn child of the *vernix caseosa*. This is the first shampoo, and should be more carefully performed than any subsequent one. Improper management at this time may entail endless worry to the mother and a great deal of suffering to the child, as it is exceedingly likely to set up an inflammation of the scalp. The child is born covered with a fatty matter called the *vernix caseosa*, which is often very thick upon the scalp. Steps are to be taken at once for its removal, which must be effected with the greatest care, and with the avoidance of all force. To this end the scalp is to be saturated with sweet almond oil, which is the most elegant means; or with olive oil or vaseline. It is preferable to use these in their natural state, but if desired, there is no objection to perfuming them with a few drops of the oil of bergamot, wintergreen, or the like. The nurse should do this immediately after she has washed the child's face and eyes. Then after the body has been bathed and the infant dressed, she should wash the head with plenty of warm water and soap, such as pure castile or glycerine soap, either solid or liquid. This should be done very gently, and if the vernix caseosa is not readily removed, she should re-apply the oil and wait until the next day, when it will be easily washed off. Should it still prove obstinate, let her patiently repeat the process until it comes off. In no case should the fine toothed comb be used. For some weeks the infant's scalp should be lightly oiled, as this will prevent any accumulation of sebaceous matter, and protect the tender skin from injury from atmospheric causes until the hair grows, care being taken to wash the head daily to prevent the oil from becoming rancid. When the hair is grown the scalp need not be so often oiled, nor should it be washed more than once or twice a week.

In children and adults, the scalp should be kept clean so as to avoid stopping up the hair follicles with foreign matter, and to prevent any irritation of the scalp, which its presence might cause. This is accomplished by the systematic use of the shampoo followed by careful drying, and the application of some oily substance to the scalp. It may be given as a rule, that a shampoo every second to fourth week is sufficient for the scalp of those who are not exposed to more than the usual amount of dust; while those who are so exposed should shampoo the head every week or two. The practice of daily sousing the head with cold water, as is very commonly done by men, is pernicious, not because the water itself is harmful, but because the scalp is not properly dried afterwards; no oil is applied to take the place of the oil that has been removed by the water, the wet hair cannot be thoroughly brushed, and soon becomes dry and brittle. Women avoid getting their hair wet, and this may be one reason why they are less often bald.

The proper manner of shampooing the head is as follows: Choose some good soap, such as Pears' "Glycerine Soap," Sarg's "Liquid Glycerine Soap," pure Castile soap, the tincture of green soap, or the tincture of prepared olive soap, and with plenty of warm water make a good lather on the head, and rub the head vigorously with the fingers, or with a rather stiff, long bristled brush. Another excellent liquid soap is composed of Castile soap eighty parts, bicarbonate of soda twenty parts, and water one hundred parts (PASCHKISS, 134 ap.). A very little of this makes an abundant lather. If the scalp is very sensitive to irritants, borax and water may be used instead of soap, or a mixture composed of the yolks of three eggs beaten up in a pint of lime water. When the head has been thoroughly shampooed, wash out the lather with a copious supply of warm water, or, where practicable, with alternate douches of warm and cold water, and then dry both scalp and hair with a good bath towel. When all is dry, rub on the scalp, not on the hair, a small quantity of some unctuous substance, such as sweet almond oil or vaseline. Care must be used in drying the hair, specially in women, who should sit before an open fire, or in the sunlight when doing it, and who should not dress the hair until it is perfectly dry. To oil the scalp, the hair should be parted and the oil rubbed in along the part, then another part made, and the operation repeated, and so on till the whole scalp is gone over. In using a fluid oil a medicine-dropper will be found convenient. Should there be an excess of oil upon the hair, a condition which is disagreeable to many, it may be removed readily by pulling the hair between the folds of a towel moistened with ether, chloroform, or cologne water.

BRUSHES AND BRUSHING.—Of far more importance than shampooing is the use of the brush and comb, and much more care should be given to the selection and use of these common toilet articles than is usually bestowed. Too often they are badly made, and generally, specially with men, they are used in a very perfunctory manner. The brush which is to be used upon an infant's head should have long, soft bristles, so as not to scratch or irritate the tender scalp, and should be employed simply in smoothing and polishing the hair. For young children whose hair is well grown, a stiffer brush is necessary, and for adults, two brushes should be used, a stiff one and a soft one. A properly made brush has its bristles placed in little clumps or groups in such a manner that the middle bristles of each group are longer than those of the periphery. The bristles are well set into the back of the brush and the groups are wide apart. Most of the brushes met with in the shops are made with the bristles all of the same length and the groups close together so as to look pretty, but not to perform their proper function. The stiff brush should be used systematically in the morning and with considerable vigor, so as to produce a feeling of warmth in the scalp, and to brush out all particles of dandruff and foreign matter lodged in the hair. Every part of the scalp should be gone over with the stiff brush, and then it should be laid aside for the rest of the day, and the soft one used to assist the comb in parting the hair, and to give smoothness and gloss to it. The stiffness of the brush and the vigor of its employment must vary with the tenderness of the scalp, and in no case should be sufficient to cause a feeling of soreness. Were brushing performed in the manner indicated, the hair would lie properly without the aid of water or pomades, excepting of course in cases of mal-position of the hair, as in the so-called cow-lick, or where the hair is unnaturally stiff.

COMBS AND COMBING.—The comb is next in importance to the brush, its office being to open up the hair so that the brush may reach all parts of the scalp, to part the hair, and to disentangle snarls. A properly made comb has long, thick, wide, perfectly smooth teeth, with well-rounded ends, and set wide apart. In choosing a comb it should be held up to the light, and discarded if any roughness or irregularities are found in the surfaces of its teeth, for such a comb would catch and tear the hair. Combs are usually made with a coarse and a fine half, and there is no objection to this arrangement if the fine part is used only to disentangle the hair. No attempt should be made to pick off crusts from the scalp with the comb. It should be used only as an assistant to the brush, and always with it in the systematic morning brushing. No comb should touch an infant's scalp, and the fine-toothed comb should be rigorously excluded from the toilet case. It is a dangerous instrument, the cause of many a case of eczema, and only of use in removing the ova of lice from the hair. Above all things, the tender scalp of the infant should be spared from its damaging effects.

DRESSING OF THE HAIR OF WOMEN.—Now we come to a place in the discussion of the hygiene of the hair in which fashion often interferes. Examination of old fashion plates and portraits will show how women and men have tortured the hair, twisting it into all sorts of shapes, and smothering it under wigs, false hair and powder. Happily, at present, the hair is worn more simply, but still the crimping or the curling iron is too much used, and the hair is pulled and dragged upon too much in adapting it to the varying demands of the hair-dresser. Sooner or later nature is apt to rebel against fashion, and the hair grows less luxuriantly or falls out. The simplest mode of wearing the hair is the best. It should be combed and brushed smoothly back upon the top of the head, either parted or not as is most becoming, and gathered into a loose braid or coil at the back of the head. Girls should wear a pendant braid; and women whose hair is grown and who gather the hair into a coil, should use large hairpins in fastening it, preferably of rubber or bone, with absolutely smooth surfaces. In doing up the hair care should be taken not to drag upon it; and drawing it into unnatural positions, such as pulling the hair from the back of the head over forwards to near the forehead, should be avoided. If a woman's hair curls naturally, she should be thankful for the favor therein bestowed, but should it not curl of itself, she should not attempt to make it curl by singeing and squeezing it between hot irons, scorching it over a hot pipe-stem, or twisting it up tightly in curl papers.

WIGS, HATS, ETC.—The hair requires for its growth,

and for the maintenance of its health, both air and sunlight, though not necessarily exposure to the direct rays of the sun. It is difficult to prove that wearing of the hat constantly is a cause of baldness, but there are many indications that such is the case, and it is well to avoid keeping the head covered with an unventilated hat. If the occupation compels one to be out of doors most of the time, or exposed to draughts so that a hat or cap must be worn, it should be well ventilated, so that the heat from the head may not become confined, and the hair more or less sweated. The subject of hats as a cause of baldness will be treated of more fully under Chapter VI., upon Alopecia. The wearing of wigs and false hair is bad for whatever hair remains, and should not be practiced. The absurd "water-falls" of a few years ago, and the no less ridiculous powdered wigs of old times are happily things of the past, and should never be revived. If a woman's hair is short and scanty, it is better for her to wear it cut short, and endeavor to stimulate its growth by attention to the scalp, than by wearing false braids to assume a beauty which she has not. Wigs heat the head and sweat the hair. False hair by its weight drags upon the feeble hair it is designed to fortify.

The wearing of night-caps was once a custom, founded upon the need of keeping the head warm in the inadequately heated bedrooms of our ancestors. With the improvement in house building and heating, the custom has passed away, and should not be revived, as it excludes the air from the hair continuously for a good part of the day. Of course where there is no hair, and the bald individual is sensitive to the cold, there can be no objection to keeping the head covered with a wig by day and a cap by night.

HAIR CUTTING.—All men wear the hair short, and employ a barber at varying intervals according to their fancy. As far as the health of the hair is concerned, it is immaterial whether it is cut at longer or shorter intervals, but it is essential that it should be well cut, and a good barber is desirable. It should never be "shingled," as the barbers term an operation which consists in cutting the hair by a to-and-fro motion of the shears, as this tears and roughens the hair. The hair of children, whether they be boys or girls, should be kept cut short until the seventh or eighth year of age, as the growing hair is a drain upon the nutrition of the body, and at this time of life all the nutritive forces should be expended in the growth of muscle and bone. The hair of a girl after she has reached her eighth year should be allowed to grow, as the less the hair is cut the finer it is. But should the girl be so situated that her scalp and hair can not be properly cared for, then she will have a better chance for a good head of hair in later life if it is cut when she is young. The hair of women is seldom worn short, although of late some young women have seen fit to adopt the style of wearing the hair like a man, along with his coat and waist-coat. It is quite common for the long hair of women to be split at the point. This should be looked for, and if found, the hair should be cut above the cleft. All ragged ends should be lopped off, and all weak hairs should be cut off near the head.

SHAVING.—The shaving of the beard is regulated largely by fashion. Physiologically it is best not to shave, for if we do, we rob ourselves of a useful protection to the throat and lungs. As shaving makes the hair grow coarser, it is often resorted to very early by the youth, for the purpose of rendering the down of the lip or cheek more apparent. It would be better to endure the down for a time, as the growth of an elegant soft beard would be the reward. If one must shave, he should do it himself, and see that his razors are kept sharp. He should shave himself, so as to avoid the risk of infection with ringworm of the beard. If the razors are dull, they are apt to set up an inflammation of the hair-follicles or skin. For shaving, a mild soap that forms a thick lather should be used, and after the operation, especially in cold or windy weather, the face should be powdered with simple rice flour or fine corn-starch.

POMADES.—PUNCH'S advice to a man about to marry is equally applicable to the use of pomades. It was: "Don't." Their regular use upon the healthy scalp is uncalled for. They are dirty, soon become rancid, and emit a foul odor, unless this is covered by some strong perfume, and they soil whatever the wearer's head comes in contact with. If the rules already given are followed, the hair will be smooth and have sufficient lustre for beauty without pomades. If the scalp is diseased, the proper remedies should be applied. In the following pages it will be indicated when and what pomades should be used. Most of the greases advertised for the cure or prevention of baldness or grayness are useless, and some harmful. The powers of some have been vaunted upon grounds that are rather absurd, as for instance bear's grease, because the bear is well covered with hair. "Bandoline" and the like sticky substances, as well as hair dyes, should not be used, as the former is bad for the hair, and the latter are not infrequently followed by loss of health from the poisons they contain.

In some cases the hair becomes matted together in a tangled mass, especially that of women during prolonged illness. From whatever cause arising, care and patience will usually enable the mass to be unravelled and the hair saved. To do this it must be attacked a little at a time with oil, soap and water, and the fingers, and picked apart and combed straight. By proper care the condition is avoidable in most cases. It would be very exceptional when a patient could not bear the combing of the hair with a coarse comb once a day, followed by plaiting it into one or two plaits. When this is done gently and quietly it will prove refreshing, and will prevent any trouble with the hair during convalescence. If it cannot be done, then it is best to cut off the hair to one half or one third of its length, so that it will be less liable to tangle.

As the hair sympathizes with the general health of the body, the latter should be maintained in good condition by a wise conformity to the laws of health. By the proper combination of the hygiene of the body with that of the hair, it is possible for even one who is predisposed to premature baldness to ward off the evil day for years; and one who comes of a strong-haired family should, as a rule, not become bald or have any essential disease of the hair.

PART II.

THE ESSENTIAL DISEASES OF THE HAIR.

Canities.— Discolorations.— Alopecia.— Alopecia Areata.— Atrophia Pilorum Propria.— Hypertrichosis.— Trichiasis.— Sycosis.— Folliculitis Decalvans.

CHAPTER IV.

CANITIES, OR GRAYNESS OF THE HAIR.

SYNONYMS.—Trichonosis cana; Trichonosis discolor; Poliothrix poliosis; Trichonosis poliosis; Trichosis poliosis; Spilosis poliosis; Poliotes; Whiteness of the hair; Blanching of the hair; Atrophy of the hair-pigment.

Canities may be congenital or acquired; partial or complete; sudden or slow in its onset. The most common form is the acquired, which usually begins as a graying of a few hairs, and proceeds more or less slowly till it affects all the hair of the head and face.

CONGENITAL CANITIES usually occurs in the form of tufts, sometimes in round patches, the more or less white hair showing conspicuously among the normally colored mass. These cases are rare, but in some families are hereditary, a white tuft of hair occurring in a large number of their members. Among the cases of this sort is that reported by GODLEE (103), who, in 1884, met with a girl, six years of age, who had a patch of white hair growing from a white patch of scalp. The child's mother had a precisely similar patch, and stated that it had occurred in two of her sisters, in one brother, in her father and paternal grandfather. One of her sisters had four children, all of them girls, similarly affected. When the whiteness of the hair is general, it is associated with delicate pink skin and eyes, and is part of that condition known as albinism.

ACQUIRED CANITIES may be premature or senile. Most often grayness does not begin before the thirty-fifth or fortieth year of age. If it occurs before that time, it may be considered, for convenience, as premature; if after that time, as senile. Both forms are exceedingly common, many persons becoming quite gray between the twentieth and twenty-fifth year, while it is exceptional for any one to attain the age of fifty without more or less gray hair on the head. There are so many variations in the manner of turning gray, that it is hard to give a definite rule; but the hair of the temples usually changes its color first, then, after a greater or less length of time, that of the vertex and whole head. Sometimes the beard is the first to whiten, but it more commonly follows that of the head. The hair of the pubes and the axillæ is the last to turn gray, and often it escapes entirely. When the process is due to some passing cause it may cease upon the removal of the cause, and cases of normally colored hair growing in after the fall of the white hair have been noted. A very rare instance of this has been reported to me by Dr. J. W. Warner of this city. The case was well known to him, and was one of relapsing canities affecting a gentleman living in Sharon, Conn. This gentleman's hair and beard changed from black to white, and back again three times in thirty years. The change from black to white was always rapid, while that in the reverse direction was slow, taking some five years for completion. Then would come a pause of some years when the color was normal, and then it would become white again. During all the thirty years his health was good, he was able to attend to an active out-of-door business, and never used any hair dye. Isdell (104) has reported the case of his father, who, in 1861, and when sixty-two years of age, had perfectly gray hair on his head and beard, while in 1882, when he died at the age of eighty-three, his hair was of its natural dark color with the exception of a few gray hairs on the temples. This case is of the same order as those of a new growth of teeth and hair in extreme old age. While instances of white hair becoming normal in color again are rare, it is not uncommon for the grayness to remain partial and apparently stationary for a number of years. Generally, however, grayness is progressive and permanent, whether it is premature or senile.

The hair in canities is usually unchanged excepting in color, but it may be drier, stiffer, and even coarser than normal. As a rule, there is no change in the color of the scalp. Where we find gray tufts upon pale yellow patches of scalp, the disease is rather vitiligo than canities. The change in the color of the hair takes place at its root first. It usually changes to gray on account of the mixture of the essential color of the hair with the whiteness produced by absence of pigment and the presence of air. Exceptionally, hairs are met with whose point and root are white, and the intermediary part normal in color ; or the point and root normal, and the intermediary part white ; or the distal ends are gray, and the proximal ends of normal color. The last variety is most often due to splitting of the end. Gradually, as the pigment becomes more and more deficient, the white color gains the ascendency, the whole hair is blanched, and finally becomes of a yellowish or snowy white. The darker the original color of the hair, the more prone it is to turn gray ; brunettes are more often gray than blondes, and become so earlier in life. Canities may exist for years without alopecia, as there is no direct connection between them. In the senile form, however, alopecia is apt to come on as another senile change, and Landois states that incipient baldness usually follows senile canities in from one to five years. Men are more often affected with canities than are women.

Sudden change of color of the hair from its normal hue to perfect white, has been too well authenticated to allow of a doubt as to its occurrence, though its possibility has been denied by good authorities, who have questioned the correctness of the observations reported. But both medical and lay history record many instances of the phenomenon, and several recently reported cases demonstrate its possibility. Thus Landois (108), in 1866, reported a case observed by himself, occurring in the person of a man thirty-five years of age, who was admitted into the hospital suffering with delirium tremens. His delirium took the form of great terror whenever any one approached him. On admission his hair was of blonde hue, and remained so up to the evening of the third day. On the morning of the fourth day, the hair both of the beard and scalp was noticed to have become gray. Some of the hairs were white from root to point, some only at their points, some only at their roots, while some were blonde and white at different points. Another interesting case of this kind was reported by Raymond (116) in 1882. The patient was a French woman, thirty-eight years old, with black hair. She was of a nervous and impressionable temperament. In July, 1881, she was greatly affected on account of parting with her son, and for about six weeks she could not sleep. In January, 1882, she was utterly prostrated by the loss of a large sum of money on the Bourse, and since then had been exceedingly nervous, finding it impossible to remain quiet in any position for any length of time. She lost her appetite, could not sleep, and had pains in various parts of her body, especially in her head, shoulders and tibiæ. These pains were of two characters, the one constant, and the other neuralgic and lancinating. On the head and face there were various painful points. When she received the news of her loss she was menstruating, and the flow was immediately suppressed. On January 31st she had a terrible attack of neuralgia. By two o'clock in the morning of February 1st her hair was of normal color, at seven o'clock it was almost completely discolored. Upon the sides of the head the color was preserved in part; upon the upper portion most of the hair was fiery red. The remainder of the hair had become completely white. The hair on the rest of the body was unaffected. The pains still continued. On the next day most of the red hair had become white, and was rapidly falling. In fifteen days nearly all the hair had fallen out, only a few hairs remaining on the lateral and occipital regions. Up to the thirtieth of March there was no return of the hair, nor of the color in those remaining.

RINGED HAIR.—Ringed hair is an anomalous variety of blanching of the hair, in which the affected hairs are marked by alternate rings, one ring having the normal color, and the next one being white. This disease is of very rare occurrence, and but few cases have been reported. WILSON (122), in 1867, reported a case of this kind, which, I believe, is the first on record. The patient was a boy between seven and eight years of age; the disease was first noticed when he was between two and three years old, and was increasing. It affected only the head. The hair-cylinder was uniform, the brown or normal segments measured $\frac{1}{30}$ of an inch in length, and the white segments were about one half as long.

Under this name LESSER (110) reported another case in 1885. This occurred also in a child four and a half years old, otherwise healthy. It was born entirely without hair, excepting its eyebrows, which were normal. Soon after birth the scalp presented the appearance of goose flesh, which continued. The hair began to grow by the second year. At time of presentation of the case, there was observed a lichen pilaris appearance along the border of the hair, on the neck and temples. The hair of the whole head was extraordinarily short, measuring even after two years growth, from 1 to 9 cm.; it was dry and brittle and inclined to tangle. The color was brown. The longer hairs were normal. The short hairs were ringed in great part, and microscopically showed alternate swellings and contractions of the shaft, the former being spindle-shaped and forming the light or white ring by direct light. The distance between the contractions averaged 0.6 cm. On a great number of the ringed hairs there were appearances characteristic of monilethrix, to which disease this case more properly belongs.

ETIOLOGY AND PATHOLOGY.—Senile canities and most cases of the premature or presenile form, are due to an obscure change in the nutrition of the hair-papilla, which interferes with the production of pigment. Whatever the nature of the change may be, only this function of the papilla seems to be interfered with, as the hair-forming function is in normal activity, judging from the fact that the hair in many cases is in full vigor. EHRMANN (101), on the other hand, holds that the pigment is formed in the papilla, but fails to reach the hair on account of absence of certain cells in the hair-root, which he believes are the active agents in transferring the pigment from the papilla to the hair. His views are given more fully in the chapter on Physiology of the Hair. According to PINCUS (113), in the beginning of canities the pigment slowly leaves the middle layers of the papilla and remains only in the external layers. With the increase of the canities, only a portion of the external layers of the papilla will produce pigment, which in straight hair will run in streaks parallel to the long axis, and in curly hair will run in a spiral. The blending of the colored and uncolored streaks will produce the gray color, which will gradually change to white as the pigment is less and less produced.

Our study of the physiology of the hair, has taught us that the color of the hair (see page 42) is very much influenced by the amount of air contained in the cortex. In cases of sudden blanching of the hair, the change of color is dependent upon the formation of air-bubbles between the hair-cells of the cortex, its presence rendering the cortical substance opaque and obscuring the color of the pigment. This is proven by placing one of the affected hairs in hot water, ether, or turpentine, when the air-bubbles will be driven out and the hair will resume its normal color. This same infiltration of the hair with air-bubbles will be found also in cases of ordinary canities, though usually merely secondary to some interference with pigmentation.

There are various agencies which act as predisposing or exciting causes of canities. Age or senility is one of the most prominent of these. Heredity exerts a marked influence, most of the members of certain families turning gray at an early period of life. That the nervous system works actively in the production of grayness, is shown by the occurrence of sudden blanching of the hair under the influence of fear or great nervous shock; by the formation of symmetrical white bands or tracts of hair during acute outbreaks of insanity, which disappear during convalescence, as noted by SHAW (117); and by the hair becoming white in parts affected by neuralgia, as of the fifth nerve. Hence canities may be regarded as a tropho-neurosis in some instances.

Local diseases or injuries of the scalp, such as wounds, repeated epilation, prolonged shaving, have been known to have local or general canities follow them. The hair after alopecia areata comes in white and may remain so, but usually the white hairs fall out to be replaced by normal colored ones. WALLENBERG (119) reports a case of entire loss of hair after scarlatina, in which the new hair came in white and remained so, the skin at the same time losing its pigment, and becoming milk white. In this case there was a good deal of air in the cortex. The hair has been known to turn gray in winter, and to become darker in summer. Prolonged residence with much exposure, either in a cold or hot climate, is given as a cause of premature grayness. Albinoes, we know, are most frequent in the negro races, which inhabit the hot countries. On the other hand, HOLDER (124 ap.) says that gray hair is very rare amongst the American Indians. Excessive mental application, or prolonged nervous strain, will sometimes induce canities, which becomes permanent and progressive. In like manner dyspepsia of various forms, excesses of all kinds, chronic debilitating diseases such as syphilis, malaria and phthisis, profuse and frequent hemorrhages, have been given by various writers as causes of canities. It has been noted that in women who early cease to menstruate the hair is apt to become prematurely gray, while those who menstruate late in life often retain the color of their hair. In this as in many other diseases, no one cause or group of causes can be proven to be *the* cause, but a study of those given will show that they have one thing in common, that is, a lowering of vitality.

The curious phenomenon of "ringed hair" is ascribed by WILSON (122) to the development of a gaseous fluid within the hair, and he thinks that either the white, opaque and smaller segments were developed during the night, and the larger and normal segments grew during the day, or the separate segments were the product of alternate days. The gas may have been generated at the time of the formation of the abnormal segment, or the cells which composed that segment may have been originally filled with an aqueous fluid, which evaporated quickly, and was replaced by air penetrating from without. LANDOIS (109) does not agree with WILSON, but believes that we must assume an intermittent activity of the trophic or vaso-motor nerves of the papillæ, through whose influence a hair tissue is formed, in which a periodic development of gas takes place. LESSER (110), in whose case the hair-shaft was affected with *trichorrhexis nodosa*, likewise found air-bubbles in the white or swollen parts, which gained entrance from without through the dry and cracked cuticle of the hair. He offers no explanation for the intermittency of the rings. In his case there was evidently some error in the nutrition of the scalp. BEHREND (3) regards the disease as a stage of *trichorrhexis nodosa*, and says that the process affects already formed hair. As yet no answer can be given to the question, "What gives rise to ringed hair?"

TREATMENT.—As a rule nothing can be done to permanently restore the color to white or gray hair. If the malady is due to neuralgia, the cure of this will sometimes be followed by restoration of color. The administration of iron, phosphorus, and sulphur, has been advised on theoretical grounds, and may be tried if the canities seems to be due to physical debility; but no promise of success should be made to the patient. As cases have been reported in which the hair has grown darker under the long-continued use of jaborandi by the mouth, or pilocarpine subcutaneously, these drugs might be tried. Acetic acid also seems to have a decided tendency to increase the pigmentation of the hair, and might be tried. The rules of the hygiene of the scalp should be at the same time enforced. Plucking the white hairs is worse than useless. All that can be done for canities is to artificially restore the color by means of hair dyes, and their use is to be strongly advised against. Happily the custom of dying the hair is falling out of fashion.

HAIR DYES.—HEBRA and KAPOSI (15) give directions for dyeing the hair black by "henna," which is the Persian name for a small shrub found in the East Indies, Persia, the Levant, and along the African coasts of the Mediterranean, where it is frequently cultivated. The botanical name is *Lawsonia Alba;* in England it is called Egyptian privet, and in the West Indies it is known as Jamaica mignonette. In the East it is used for dyeing red the nails of women, the beards of men, and the manes of horses. The preparation of henna consists in reducing the leaves and young twigs to a fine powder, catechu or lucerne leaves in a pulverized state being sometimes mixed with them. When required for use the powder is made into a pasty mass with hot water, and then spread upon the part to be dyed.* In an hour the hair will be red. A paste of powdered indigo plant is now applied, and then damp heat, and in a few hours the hair will have a fine black color, provided that the process has been regulated by experience and good judgment. LEONARD (64) gives the following formulæ for dyeing the hair black:

No. 1.

Bismuthi citratis	℥j =	50
Alcoholis	ℨv =	33
Aquæ rosae		
Aquæ destillat.	āā ℥ij =	ad 200
Ammoniæ	q.s.	

M. Sig. Apply in the morning

---

* *Encyc. Brit.*, 9th ed., art. *Henna.*

No. 2.

Sodii hyposulphit. . ℥ xii = 60
Aquæ destillat. . . ℥ iv = ad 200.

M. Sig. Apply thoroughly in the evening.

Nitrate of silver in the strength of from five to ten grains to the ounce of water may be used for a black dye, the hair being saturated with it and allowed to dry in the sunlight. If it is desired to hasten the process, the application of a solution of sulphuret of potash, from twenty grains to two drachms to the ounce of distilled water, will set the dye instantly. McCall Anderson recommends the following for a permanent black dye: First, a solution of bichloride of mercury, two grains to the ounce of water, followed by a solution of hyposulphite of soda, one drachm to the ounce of water. Lead may be used in the form of the sugar of lead, ten to twenty grains to the ounce of water, applied to the hair and followed, when nearly dry, by a solution of the sulphide of ammonia, about one-quarter the strength of the British Pharmacopœia. These dyes, by means of nitrate of silver and mercury, are dangerous on account of the metals they contain.

For a brown dye, Pfaff (70) recommends a pomade composed as follows:

Ol. ovorum rec. press.
Med. oss. bovis . . .āā 50.
Ferri lactat. . . . . 2.50.
Ol. cassiæ ether. . . . 1.50.

M.

But the number of dyes is legion, and these must suffice for examples. Before the application of any dye, the hair should be thoroughly cleansed with soap and water.

CHAPTER V.

DISCOLORATION OF THE HAIR.

Synonyms.—Tricolorosi. Trichonosis seu Trichosis decolor.

Under various conditions of health and disease, the hair has been known to undergo changes in color, other than that of turning gray. Some of these changes depend upon causes acting from within, and modifying in some way the pigment formation. Some are due to external agencies, and have the nature of dyes.

Just as serious illness will cause the hair to fall out or to turn gray, so will it induce changes in color. Some instances of this have been reported. Thus Rayer (31) cites two cases of Alibert's, in the first of which a young woman after a long and serious illness lost a fine head of blonde hair, and upon recovery had the loss made good by a growth of very black hair. In the second case, a man during sickness lost his brown hair, which was replaced after recovery by bright red hair. He further cites a curious case in which every time the patient, a young woman, was attacked with fever, her blonde hair became tawny red, to return to its original hue upon recovery. Beigel (123) reports a case in which, after typhus fever, the blonde hair of a woman fell out, and was replaced by coal black hair. Smyly (136) saw the hair of a patient suffering with suppurative disease of the left temporal bone, change from a mouse color to a reddish yellow. The patient was an infant. The left temporal bone was the seat of the suppuration, but the right side of the head was affected with the change in color of the hair. The right eyebrow was likewise affected. There was also a profuse yellow perspiration, so that the pillow and the skin of the right side were stained yellow. This case was one of dyeing. C. REINHARD (135) has reported a case of periodic change in the color of the hair of an idiot boy who suffered with epilepsy and paralysis of the legs. He was subject to violent outbursts of temper, during which his reddish-blonde hair would change in forty-eight to sixty hours to a blonde yellow color. The change of color seemed to begin at the points of the hair, and affected nearly all the hair. After some seven or eight days, the original color would return. There was no disease of the scalp, excepting that during the stage of quiet, there was a scanty secretion of sebaceous matter. Microscopical examination showed that the light hairs contained a good deal of air in the cortex and medulla; that they were dry and inclined to split at their points. Their cuticle cells were displaced; and the medulla cells were very much shrunken. This would be sufficient to explain the change of color, as our study of the physiology of the hair has already taught us.

That it may be possible to influence the color of the hair by internal medication, is inferred from the remarkable case reported by PRENTISS (134). He had a patient with light blonde hair who suffered from pyelonephritis with anuria, for the relief of which condition muriate of pilocarpine was administered hypodermically in doses of from 0.01 to 0.02 gramme. The use of the drug was begun on December 16, 1880, and the hair commenced to grow darker by the twelfth day. On the twelfth of January, 1881, the color had become chestnut brown, and by the first of May it was almost pure black, although the pilocarpine was stopped on the twenty-second of February. The hair became not only darker, but coarser, and grew more vigorously; and the hair of the axillæ was as much changed as that of the head. The hair was in every respect normal, and the change in color seemed to depend upon increase of pigment. With the change of color in the hair, the color of the eye changed from a light to a dark blue. The same author (104 ap.) has reported another case in which a patient with Bright's disease took from twenty to thirty drops of fluid extract of jaborandi several times a day. After a year's treatment his hair turned of much darker color. These cases would tend to show that jaborandi increases the nutrition of the hair and the activity of the pigment-forming cells. It would appear that the cause of the change of color was the jaborandi, because the tendency of exhausting diseases, such as pyelo-nephritis and Bright's disease, is rather towards loss of hair and canities. Jaborandi does increase very markedly the cutaneous circulation, and to this action we must look for the explanation of these cases.

We have no means of explaining with certainty these changes of the hair from internal causes. Of course they are due to some influence upon the pigment-forming cells, but that does not explain the matter. The cases reported are few, and only in those of PRENTISS and SMYLY does an adequate explanation suggest itself. The other changes in color of the hair, the consideration of which falls within the scope of this chapter, are dependent either upon a deposition of colored particles from without, or upon chemical action. From these causes we have green, blue and yellow hair, as well as various anomalous shades.

GREEN HAIR.—Green hair occurs in workers in copper. A number of cases have been reported; the following one by PETRI (132) will serve as an example. An old worker in copper, seventy-eight years of age, presented the curious appearance of having brilliant green hair. He was in good general health. Upon his gums was a well-marked green line. Both his beard and scalp hair were green. The color did not affect the whole length of the hair, but was darkest and most intense at the point, and for a distance of 3 cm. therefrom. From that point it grew gradually less pronounced, till at 10 cm. it had disappeared. Beyond this the hair was gray. The microscopical examination of the green hairs showed a deposit upon the epidermis of the hair, of small, sometimes bluish, sometimes indefinitely colored, sometimes darkly contoured or yellowish, sharply cornered, pyramid-shaped masses of crystals, with transparent edges. Here and there little patches of dirt were found. The crystals could be seen only with very strong light. By the addition of *liquor ammon. caust.*, the color changed to dark or blackish blue, which is the chemical reaction of ammonia with the salts of copper. The color could be washed out of the hair, and was doubtless due to particles of copper oxide floating in the atmosphere of the work-shop, and deposited on the hair. BILLI (124) reports a case of green hair occurring in a patient of his, whom he was treating for trichophytosis capitis. For the ringworm a wash of corrosive sublimate was ordered, and an ointment of the yellow oxide of mercury. The next time he was seen his hair was a brilliant green, the color being disseminated over the whole head, and the hair colored from bulb to tip. The microscope showed that all the elements of the hair were colored. The patient was not a worker in copper, nor had he used any other application to his scalp. Cases of green hair have been reported in which no cause was discoverable. Thus ORSI (131) met with a case in a railroad hand, who was forty-nine years old, and whose hair was gray. Suddenly his hair became green, the green hairs being intermixed with the gray and white. The scalp hair alone was affected. Washing with vinegar, ether, alcohol or a dilute solution of potash, did not affect the color. The microscope showed that the roots of the affected hairs were thick and fibrous; the cortex by natural light was violet, by artificial light greenish in color; while the medulla was yellowish. When the hair was cut off, gray hair grew in. Even in this case, since the hair came in of normal color, there was, doubtless, some external agency at work. In some cases of canities, a greenish yellow tint will be observed, a transition stage from the original color to the gray, and this also might be an explanation of the foregoing case.

BLUE HAIR.—Blue hair is met with in workers in cobalt mines and in indigo works. The color is generally easily removed by washing, and under the microscope BEIGEL (123) found in a case of his own, that the whole hair was not uniformly colored, but that particles of indigo were deposited in an irregular manner upon the cuticle of the hair. In some exceptional specimens a number of hairs were embedded for some distance in a mass of indigo, or stuck with neighboring hairs. The edges of the epithelial scales were more pronounced than in the normal condition, owing to the deposit along them of a fine blue dust. The blue color did not penetrate into the substance of the hair.

YELLOW, BROWN, AND BLACK HAIR.—Yellow hair has been observed in patients suffering from icterus; red-brown hair occurs in handlers of crude aniline; and black, or rather coal-black, hair is often seen on those who work in coal. In all these cases of change in color of the hair from the deposit of foreign matter, the element of time was marked, and the lighter the color of the hair was originally, the easier did it become altered.

CHANGE OF COLOR AFTER DEATH.—The color of the hair may change after death. Thus HAUPTMANN (127) reports a case in which the hair of a corpse, exhumed after twenty years' burial, was found to have changed from a dark brown to a red color.

COLOR ALTERED BY CHEMICALS.—The action of various chemical agents will change the color of the hair, the change generally being transitory. This is illustrated by the action of the various dyes and bleaching fluids used by the hair-dresser. Further, bicarbonate of soda, used for a long time, will change the color of dark hair to a dirty red brown. Chrysarobin colors the hair, as well as the skin, a mahogany red. Naphthol I have seen change white hair to a bright corn-yellow color. Chlorine gas will bleach the hair. Sweat acts upon the color of the hair, and we often find the hair of the axillæ of persons who sweat much of lighter color than that of the head. LEONARD (64) cites a case, in which the brown head hair of a young man turned to a positive red after a few years' residence in the hot climate of Sumatra. It is not uncommon for the color of the hair to grow darker or lighter under the influence of exposure to sunlight.

ANOMALOUS CASES.—In some cases, according to OESTERLEN (130), hair will be met with which is more intensely pigmented towards its root than towards its point, depending upon an irregular deposit of pigment, or alteration in the texture of the hair. One of the most curious anomalies is reported by SQUIRE (137) as occurring in a young man sixteen years old, the left side of whose head was piebald like a tortoise-shell cat. The condition dated from birth, and there was no assignable cause. The light-colored patches were auburn, the dark patches brown, and they were abruptly limited. The opposite or right side of the head was covered with dark-brown hair, of the same tint as the dark patches on the left side. The curious condition known as ringed hair will be found described in the chapter on Canities in this work.

CHAPTER VI.

ALOPECIA.

DERIVATION.—The term is derived from the Greek word ἀλώπηξ, meaning a "fox," on account of its resemblance to the appearance presented in the "fox-mange."

SYNONYMS.—Capillorum defluvium; Athrix depilis; Phalacrotes; Depilatio; Trichosis athrix; Gangræna alopecia; Atrichia; Defluvium, seu Lapsus seu Fluxus pilorum; Lipsotrichia; Vulpis morbus; Pelada; Oligotrichia; Calvities; Psilosis; Trichorrhœa; Ophiasis; Calvitie (Fr.); Kahlheit, Fuchsräude (Ger.); Calvezza (It.); Baldness (Eng.)

DEFINITION.—An abnormal loss of hair, arising from any cause, which usually affects the scalp, but may invade any portion of the body.

VARIETIES.—1. Alopecia adnata, or congenital baldness. The form occurring in infants born either totally or partially without hair. 2. Alopecia senilis. The absolute or relative baldness of old age. 3. Alopecia prematura, seu presenilis. Baldness occurring before the time when the hair usually falls, on account of advancing years. It may be idiopathic or symptomatic. 4. Alopecia areata, in which the hair falls out in circular patches.

The older writers used various terms to denote the different phases of baldness. These, happily, have fallen into disuse. They were the following: Madesis or Maderosis, used to denote a thinning of the hair. Phalacrosis, to denote baldness beginning at the forehead. Ophiasis, to signify baldness occurring in a serpentine line, running from the occiput towards the ears, and sometimes towards the forehead. Opisthophalacrosis, baldness beginning at the back of the head. Hemiphalacrosis, baldness affecting one half of the head. Anaphalantiasis, used to designate loss of the eyebrows, and sometimes general baldness. Oligotrichia, to denote thinness of hair. According to FOURNIER, alopecia should be used to denote the process of hair-fall, and calvities the completed baldness.

1. ALOPECIA ADNATA.

SYMPTOMS.—This is the atrichia of the old writers. It is congenital baldness, as the name indicates, and is either partial or complete. Most infants come into the world with a good equipment of hair. In rare instances the child is born partially bald, and more rarely completely so. Hairless races have been reported. One of these is said to exist in Australia. HILL (157).

ETIOLOGY AND PATHOLOGY.—The disease is due to arrested development of the piliary system; and microscopical examination of sections of the skin in inveterate cases, shows a deficiency of hair-follicles. In a case reported by JONES and ATKINS (160), of a boy who never remembered having had hair, microscopical examination of the scalp revealed only a few aborted hair-follicles, forming shallow pits in the epidermic layer with open extremities looking downwards. In two cases reported by SCHEDE (179) in 1872, one of which was a boy thirteen years old, and the other a girl six months old, brother and sister, there was no hair whatever on the body, not even lanugo hair. Microscopical examination of a section from the boy's scalp showed large, well-formed sebaceous glands, opening directly on the scalp. In the neighborhood of some of these glands, just above their lower extremity, were found a great number of "atheromas" separated from the glands by connective tissue, and apparently developed in a species of short tubes, which in structure corresponded with that of the outer root-sheath, and were no doubt the beginnings of the rudimentary hairs. LUCE (166), in 1879, reported a case in which the hair did not grow until the sixth year, and in which the appearances of keratosis pilaris were present. HUTCHINSON (158), in 1886, reported a case of congenital absence of hair in a boy three and a half years old, the son of a woman who had been almost wholly bald from alopecia areata since she was six years old. In this case there was a withered condition of the whole integument, and an absence of nipples. A somewhat similar case is reported by DE MOLÈNES (115 ap.). The mother had had alopecia areata some years before giving birth to the child whose case is reported. At birth there was a scarcely perceptible down upon the scalp, no eyebrows, and hardly visible eyelashes. At five months the eyelashes fell out and the scalp became white and smooth. There was no keratosis pilaris. Teeth and nails were normal. Under stimulating treatment during three years the hair grew in, excepting on a small place behind the left ear. In some families there is an hereditary predisposition to congenital alopecia.

PROGNOSIS.— The prognosis is usually good. The hair in most cases will grow in course of time, although it may not be as abundant as it should be. According to MICHELSON (171), when congenital alopecia takes the form of circumscribed patches, the prognosis is bad. Delayed dentition and diseased nails will often be noted in children affected with this disorder.

TREATMENT.—As to treatment, the disease usually requires none, as it generally remedies itself. If it is very disfiguring, or the parents require something to be done, stimulating remedies may be used, as in alopecia areata. If there is keratosis pilaris, the free use of soap frictions every day, followed by inunctions with oil, will serve to remove the epidermis which clogs up the hair-follicles, and will thus give the hemmed-in hair a chance to grow.

2. Alopecia Senilis.

SYMPTOMS.—This form of baldness is that which occurs in old age. It is often preceded, accompanied, or soon followed by other signs of advancing years, such as graying of the hair, fall or decay of the teeth, and diminution in keenness of the sight. The hair usually first becomes gray, loses its lustre and suppleness, and then falls out either slowly or rapidly. At first there may be only a general thinning of the part about to become bald, but at last absolute baldness sets in. The fall of the hair usually begins at the vertex, and it is not uncommon to see this region alone affected for some time, giving the appearance of the tonsure. It may begin at the anterior superior part of the skull. Generally the course of the disorder is from behind forwards, or from before backwards, but the whole top of the head may be affected at once. In most cases the lateral and posterior parts of the skull are spared, and the bald patch is bounded on either side and behind, by a semi-circle of hair, running along the temples to the neck. It is always symmetrical. NEUMANN (27) draws attention to the fact that both in senile and premature alopecia, the region commonly affected corresponds quite nearly to that supplied by both supra-orbital nerves, and in very extended cases the regions supplied by the temporal and the occipitalis major and minor are also included. A certain amount of seborrhœa sicca is sometimes present. The scalp appears smooth, stretched, shiny, sometimes oily, and thinned.

ETIOLOGY.—This form of baldness is but one expression of that general lowering of nutrition, incident to advancing years. The age of forty-five is that usually chosen to designate the dividing line between man's fullest development and commencing decay, and it is at about this period that senile baldness begins. Women are far less often bald than men. Why this is so we do not certainly know. In the section on premature baldness, some reasons for their exemption will be given. Neither canities nor seborrhœa are causes of this form of alopecia. KAPOSI (19) says that atrophy of the scalp tissue does not precede, but follows it after a considerable lapse of time.

PATHOLOGY.—There is a lessening of the subcutaneous fat, and an atrophy of the corium, the connective-tissue fibres of which have undergone, in part, fatty, and in part, colloid degeneration. The sebaceous glands are in some places shrunken and in some widened; the hair-follicles are filled with epithelial scales, the remains of the root-sheath, and often contain a thin hair. In many follicles the hair-papilla has disappeared. KAPOSI (19). PINCUS (71) found in senile baldness that there was a rapid relative increase in the shedding of the "spitzen" hairs, (that is, those hairs which have not been cut), and a marked and increasing diminution in the diameter of the long hairs; also an absence of seborrhœa sicca in most cases, and a well-marked atrophy of the scalp. Like in other senile changes, so here the retrograde process begins in the arterial supply to the scalp, and we find a fibrous endarteritis narrowing the lumen of the cutaneous arteries, till finally the capillary circulation about the hair-follicles is obliterated.

PROGNOSIS.—The loss of hair is permanent. In some rare cases the hair has grown again in old age, and this has been attended by the development of new teeth.

TREATMENT.—When the scalp is atrophied and bound

down to the skull, the hair-follicles are too far gone for any stimulation to affect them, and our best endeavors to restore the lost hair will be in vain.

3. Alopecia Prematura.

There are two main varieties of premature baldness; Alopecia Prematura Idiopathica, and Alopecia Prematura Symptomatica.

Alopecia Prematura Idiopathica.

Symptoms.—Alopecia prematura idiopathica is that form of baldness which begins at any time before middle age, and arises uninfluenced by any antecedent or concomitant, local or general disease. Usually the hair does not begin to fall before the age of twenty-five years, and it is apt to progress slowly. The hairs which first fall out, are replaced by those of less vigorous growth, and these in their turn are shed to make way for still weaker ones, and so the process is repeated until complete baldness results. In its general course it is similar to the senile form. Like it, it is symmetrical, it begins most often upon the vertex, forming the tonsure, it progresses slowly, has the same boundaries which it does not go beyond except in extreme cases, and leaves a smooth, shiny, bound-down scalp. It often begins anteriorly and recedes at the sides, giving that wide and high forehead thought to be indicative of wisdom. It may leave for a long time a little islet over the middle and anterior part of the skull. Unlike senile baldness, it is not preceded by canities, or any other senile change. In some cases the fall of the hair will be rapid, then cease for a time, to begin again and progress steadily. The beard in all forms of baldness is affected rarely, a luxuriant beard being very often associated with a more or less bald head.

Etiology.—The disease is in many cases hereditary,

it being not uncommon to meet with families in which the fathers and sons for many generations lose their hair early in life. PINCUS (173) says this is due to a markedly stretched condition of the aponeurosis of the occipito-frontalis muscle, which becomes hereditary in certain families. According to him, "there are but two predisposing causes of alopecia prematura. They are: 1. Inheritance. 2. A chronic eczema or impetiginous eruption on the scalp, in the years preceding puberty. The latter is the most frequent, and is often connected with symptoms of relative or absolute debility." It is very prevalent in those leading sedentary lives, especially so in brain-workers. JAMIESON (16 ap.) would explain this on the theory that the nerves supplying the scalp are in direct connection with those supplying the pia mater and dura mater, and that an irritable condition of the brain due to cerebral congestion would reflexly interfere with hair-growth. The continuous wearing of caps or of close-fitting unventilated hats, a practice very common in this country, is assigned as a cause. Some hatters claim that according as the head is long or wide, the baldness commences on the forehead or crown. F. A. KING (161) says "baldness of the vertex is due to compression, by stiff hats, of the anterior temporal arteries in their course over the frontal protuberances; of the posterior temporals at or near the parietal ridges; and of the occipital behind. The reason why baldness occurs in different places in different individuals is probably due to differences in the shape of the head. The little tuft of hair often observed on top of the forehead, is nourished by the two supra-orbital arteries, which escape pressure, by passing over the forehead in the slight concavities between the frontal eminences." The existence of the islet of hair in front is probably dependent upon the fact that it lies over the frontalis muscle, and is not upon so tense a substratum as the other hair of the vertex. Lack of care of the hair is an active cause. ELLINGER (150) believes that the daily use of water on the head is a frequent etiological factor, and has found this the case in eighty-five per cent. of his cases. He says "the form due to water begins at the forehead, and precedes upward and to the sides; it occurs in people who wear the hair long, and is due to the fact that the water at the point of exit of the hair makes an emulsion with the sebum and the scales of epidermis, which, hardening, forms a plug to the hair-follicle, causes a damming up of the sebum in the follicle, and subsequent atrophy of the hair." In an article upon alopecia by POHL PINCUS (177) it is held that all forms of the disease are due either to primary or secondary induration of the scalp. Some causes of primary induration are said by him to be the action of cold, as by an ice-bag, gleet, leucorrhœa, great depression of spirits, or anxiety of mind, which the subject struggles against; while in those cases in which the subject succumbs entirely to his depression, these do not cause baldness.

It has been noted that eunuchs generally have hair neither on the face nor pubes, if they are castrated before puberty; and if castrated after puberty, they lose whatever hair they had on those regions. HOLDER (124 ap.) says that the American Indians never grow bald. They have no hair on their bodies, and but little on the pubes, and on the face of the males. Women less often become bald than men. The reasons given for this exemption, are that they do not wear their hats as much as men, neither are their hats so close fitting, nor made of so impermeable materials as are men's hats; that they give more attention to the care of their hair than men do; that they carefully avoid wetting their hair; that they are not so abundantly covered with hair as are men, and therefore do not suffer so great a drain upon the hair-forming elements; that the hair is not so often cut; and that there is a greater amount of subcutaneous fat in women than in men, and this is preserved longer in them. PINCUS says "the reason why they suffer less, is due to the fact that in them the spaces between the connective-tissue fibres, in the deeper and middle layers of the scalp, are much larger than in men, the skin of women during their life preserving more of the characteristics of the skin of children." In an analysis of one hundred cases of baldness appended to this chapter, the remarkable fact is brought out that there is a tendency for baldness to be hereditary in the line of sex. Thus in all the cases in women there was a distinct history of the affection occurring on the maternal side.

That the beard is not affected is because the underlying tissues are not so stretched, as is the case with the part of the scalp most often affected, that is, the region over the occipito-frontalis aponeurosis. This reason would also hold good in regard to the exemption from baldness of the other hairy regions of the body, and of the parts of the scalp hair which usually do not fall out.

PATHOLOGY.—From birth up to the age of twenty years the scalp undergoes a continual change in structure, which consists in increasing thickening and tension of the middle layer of the connective tissue, this being most marked in the regions most commonly affected by alopecia, namely, over the aponeurosis of the occipito-frontalis muscle, and least marked on the temples. In both alopecia prematura simplex and alopecia senilis the connective-tissue which binds the scalp to the underlying parts, suffers still further changes. It undergoes a narrowing of its meshes with a thickening of its fibres, till at last in many places the meshes disappear. At first the hair loses its typical length, but not thickness, its lustre is somewhat diminished, and both the quantity and quality of daily fall is slightly increased, more short hairs being shed in proportion to long hairs than normally. At the same time seborrhœa sicca generally becomes pronounced, though in about ten per cent. of the cases it is wanting. The thickening of the subcutaneous tissue progresses slowly, but at last in from six months to five years, the hair begins to lose in thickness, and to fall more rapidly, and baldness sets in. The diameter of the hair is in direct proportion to the diameter of the papilla; and as the papilla is more and more pressed upon and reduced in size by the increase of the connective-tissue, the diameter of the hair is steadily lessened. At first the papillæ are unaltered in constituence, later they are obliterated, and then no more hair can be produced, and the part is bald. If the process of induration greatly increases, it goes beyond the aponeurosis, and the whole head becomes quite bald. PINCUS (173).

PROGNOSIS.—If the disease has not progressed too far, we may sometimes stay its progress. If the scalp is so tightly adherent to the subcutaneous tissues that it does not readily slide upon them, we can not hope to better the condition. A well-marked history of heredity renders the prognosis unfavorable. We can best watch the progress of the case by having the patient save all the hair that falls from his head for three days, putting each day's fall by itself, and then counting the pointed and cut hairs, if the case occur in a man, or the hairs over and under six inches, if the case occur in a woman. If the uncut hairs of a man, or the hairs under six inches of a woman, exceed one quarter of the mass, the disease is progressive, and the prognosis is unfavorable; if under one quarter the prognosis is better.

TREATMENT.—In persons who have reason to expect early hereditary baldness, prophylaxis is of the greatest importance. This consists in close attention to the hygiene of the scalp, and of the general health. When the disease has once set in, it must be combated by means of stimulating applications, as in symptomatic premature baldness. SHOEMAKER (182 and 183) recommends that to persons in apparent good health, who are unconsciously losing their hair, either mercury, tincture of ignatia, or sulphurous acid should be administered. Either the bichloride of mercury or calomel is given by him for a short period, and then the tincture of ignatia in ten-drop doses three times a day with a bitter tonic, and thus the treatment is varied from time to time, according to the patient's condition. He also advises the use of the oleate of iron locally in these cases. The use of jaborandi or pilocarpine would seem to hold out some hope for the cure of this affection. In 1879 two remarkable cases of hair growing on perfectly bald heads were reported by GEO. SCHMITZ (180), after using two or three hypodermics of pilocarpine for some eye disease. In one case the man was sixty years old and quite bald, and yet a vigorous growth of gray and black hair took place and covered his scalp. In the other case the man was thirty-four years old, with a small bald spot upon which the hair grew so as to cover it. COTTLE (52) speaks highly of the power of either of the following lotions to stop the fall of the hair, namely:

Ac. Acetici,	℥ ss., say	6
Pulv. boracis,	ℨ j., "	1.50
Glycerine,	ℨ iij., "	4.50
Spts. vini,	℥ ss., "	6
Aq. rosæ,	ad ℥ viij.,"	100

M

Or, Liq. ammon. acetat., . ℥ ij., say 25
     Ammon. carbonat., . ʒ ss., "   .75
     Glycerine, . . . ʒ iij., "   4.50
     Aq. sambuci, . ad ℥ viij.," 100
  M

PINCUS (175) advises in acute idiopathic loss of hair the avoidance of all stimulation of the scalp, and temperance in eating, drinking and mental excitement. Where the fall of the hair is progressive and chronic, and while it is still merely a thinning of the hair, he directs the application to the scalp, for two to five minutes on two or four successive days of each week, of a lotion composed of:

    Bicarbonate of soda . . 2
    Distilled water . . . 100
  M

rubbing in one to two tablespoonfuls with a soft hair-brush or a sponge. On the first or second day of the interval between the applications, some oil is to be rubbed into the scalp. This treatment is to be continued for a year, and then if the disease still progresses, more powerful remedies are to be used, such as will be spoken of under the treatment of alopecia furfuracea.

MAYERHAUSEN (132 ap.) reports good results from the use of static electricity. MAPOTHER (131 ap.) believes that as the hair contains sulphur, silicon, iron, and manganese, a dietary composed largely of oatmeal and rye bread will promote its growth.

ALOPECIA PREMATURA SYMPTOMATICA.

SYMPTOMS.—As its name indicates this form of baldness occurs as a consequence of some other local or general disease. It may occur at any time before old age and be either circumscribed or diffused, depending upon its cause. It may be very rapid, as after fevers, the hair falling out by handfuls, in which case it is named "defluvium capillorum;" or it may be very slow, as in seborrhœa sicca, taking years to produce complete baldness. Other parts of the piliary system may suffer besides the scalp hair, as in trichophytosis, spyhilis, etc. In some cases all the hair of the body has fallen out at once, as after sudden nervous shock. When dependent upon non-symmetrical diseases, as variola, pustular diseases in general, and the parasitic diseases, the baldness will be asymmetrical.

Its main varieties are: alopecia furfuracea or alopecia pityrodes; alopecia syphilitica; defluvium capillorum; and a class comprising the non-symmetrical cases arising from local lesions, which has been named by T. Robinson (75) "alopecia follicularis." The most frequent cause of alopecia prematura symptomatica, is seborrhoea sicca or pityriasis simplex, and on this account our first attention must be given to

Alopecia Furfuracea.—In this there is always some scaling of the scalp, it may be in so slight a degree as to constitute the commonly called "dandruff," or pityriasis simplex; or the process may be so intense as to produce the thick, easily friable, grayish-white, greasy scales of seborrhœa sicca. This kind of baldness is met with in all ages, but its most serious form occurs usually between the twentieth and thirtieth year of life. It has two stages. In the first stage, simple seborrhœa sicca or pityriasis is present, and the hair is dry and falls out slightly. The subject notices that his clothing, especially about the shoulders, is covered more or less thickly with grayish epithelial scales mingled with sebaceous matter, and that these fill his brush; and that a few hairs fall out of themselves or are pulled out by combing. This stage lasts from two to seven years, as a rule. Now the second stage begins, when to the seborrhœa is added a rapid fall of the hair. The location usually affected is the same as in alopecia senilis, the top of the head, from vertex to forehead, and sometimes over the temples, and the tuft of hair just over the forehead is preserved longest. The patient does not become bald at once, as there is at first only a thinning of the hair; then a diminution in the length and diameter of the hair; but the atrophy continues till at last only a few lanugo hairs are left, which in their turn fall out and complete baldness results. The bald scalp appears white or of rosy hue; is often stretched over the sutures, though frequently easily moved over the aponeurosis of the occipito-frontalis muscle; and it seems thinned. The seborrhœa keeps pace with the intensity of the disease, until the hair falls out markedly, when it lessens, and when baldness is fully established it is no longer present.

Such is the usual course of the disease. In infants seborrhœa generally gives rise to thick crusts, which, being removed, bring away with them the first growth of the hair. In them it does not produce permanent baldness. In women the region affected is generally that occupied by the part. It is affirmed that they are more prone to seborrhœa sicca and fall of the hair than men are, but that the process in them is only passing, and the fallen hairs are soon replaced. Seborrhœa sicca will also produce a general thinning of all the hairy scalp, but this is usually an acute and temporary trouble.

PINCUS (71) says: "In alopecia pityrodes the definite proportion between the hair growth and fall is disturbed, the latter becoming excessive. The longer the seborrhœa lasts the more the hair growth will fall behind. When the proportion of short hairs to the total fall is as 1:8, the average length of the hair being five inches; or as 1:10, the average length of the hairs being from two to three inches, the loss is abnormal." If under treatment the disease is checked, and the process has not lasted long enough to cause destruction of the hair bulbs, improvement will be shown by a lessening of the fall of the hair, and by and by a growth of lanugo hairs and then of good strong colored hairs.

By UNNA (186) a method is proposed for proving the improvement, which is as follows: The patient is directed to gather the fallen hairs into a little bundle with the roots all looking one way, so that the condition and number of them can be seen at a glance. So long as the fall is rapid, many hairs will be found which have just passed the papilla stage, whose knob-like roots are still long, and often have a drawn-out epithelial projection. The less the fall is the less this form of root will be found, till only the rounded full roots are met with, which shows that the hairs have been a long time in the "bed-hair" stage.*

ALOPECIA SYPHILITICA next claims our attention. It occurs most often early in the disease with the early specific lesions, but may occur later, with the tubercular and gummatous lesions. Syphilitic exanthems may be present on the scalp, or the alopecia may be the only symptom of the disease. It has no definite time of appearing, coming as early as the third month, or as late as the end of the second year. When it is the result of the syphilitic cachexia, seborrhœa is often present, and there is a general thinning of the hair all over the head, with the formation of irregular patches of bald-

---

*By "bed-hair" is meant a hair which has been cast off from its papilla, but not shed from the follicle. It is one of the regular stages in the life of every hair, and shows that it has attained its full development.

ness which do not tend to form circles. In the patches we often find tufts of long hair, and the head has a peculiar ragged look, as if it had been badly cut with dull shears, which is quite characteristic. In some cases the middle region of the scalp is alone affected. When it is due to a pustular or ulcerating lesion the baldness is localized at the seat of the lesion, and cicatricial tissue not infrequently takes the place of the normal scalp tissue.

Other regions besides the scalp may be affected, but always at the same time with the scalp. The broken arch of the eyebrows is characteristic and is most often seen in women. Alopecia of the eyelashes is less frequent. The hair of the pubes is quite frequently attacked; specially, according to FOURNIER (152), is this the case in women.

Usually there are no subjective symptoms, and the fall is more rapid than in other forms of alopecia. It comes in both benign and malignant cases, more often in the latter. Some authorities say that it is not so common as it used to be in former times when it was the custom to salivate the patients, and hence it is inferred that it is due to the administration of mercury. While we know that alopecia is one manifestation of chronic mercurial poisoning, yet as alopecia occurs in syphilitic cases in which no mercury has been used, we can feel assured that syphilis of itself is a sufficient cause of the baldness.

DEFLUVIUM CAPILLORUM is the form of baldness which follows acute diseases, especially fevers, or occurs in the course of some cachexia such as mercurialism. Usually the hair does not fall out till after convalescence has set in, when the fall will be very rapid. It is more apt to be a general thinning, attacking all parts of the scalp, than a localized baldness, and in most cases is associated with seborrhoea. As a rule it is not very intense, and rarely produces absolute baldness. At times, however, the hair of the whole body will fall with great rapidity so that the disease will have a strong resemblance to alopecia areata maligna.

ALOPECIA FOLLICULARIS.—The appearance presented by this form will vary with the cause. When due to pustular diseases, as impetigo, the patches are not larger than from the size of a dollar to that of the palm and we may have cicatrices. When due to some diffuse inflammatory disease, as erysipelas, the patches are quite large and irregular, and the scalp is hyperæmic. When due to favus or ringworm the hairs are altered ; in the former case they are lustreless, dry, brittle and sometimes split longitudinally ; in the latter they appear as if gnawed off near the roots. The scalp, too, is altered ; in favus in old cases it is more or less atrophied ; in ringworm it is usually covered with thick scales forming a crust.

ETIOLOGY.—Alopecia prematura symptomatica has many causes. We have already mentioned seborrhœa sicca, syphilis, fevers, impetigo, erysipelas, variola, and parasitic diseases. Besides these may be mentioned, violent shocks to the nervous system and mental distress; parturition, lupus erythematosus, psoriasis, lichen ruber, lichen scrophulosorum, lepra, and other cachexiæ. Sweating of the head will cause the hair to fall. Removal to the seashore is often attended by increased fall of the hair. Scarlatina has been followed by permanent baldness.

The baldness following fevers, and with syphilitic and other cachexiæ, is due in most cases to seborrhœa, but may be purely a nutritive trouble, the bulbs being badly nourished, the hair becoming loose and falling out. This view is supported by the fact that the hair does not fall out till some time has passed since the onset of the fever, and grows in again when convalescence is fully established. GIOVANNINI (121 ap.) says that the baldness of syphilis is due to a deep folliculitis pilaris, and not to anæmia or atrophy. The baldness of the pustular diseases, such as variola and the syphilides, and of the destructive diseases, such as lupus erythematosus, is due to the destruction of the hair-follicles. The baldness which is found to follow the use of mercury, excess in venery and intemperance, is due to their damaging effect upon the constitution of the patient. Anything, in fact, which will tend to impair the full vigor of a man, may secondarily contribute to the production of baldness, especially if he has a predisposition thereto. This it may do by its effect upon the nutrition of the hair, or more commonly through a pityriasis simplex or seborrhœa sicca.

BROCQ (111 ap.) believes that many cases of so-called alopecia furfuracea are really due to keratosis pilaris, which is capable of destroying the hair-follicle when it is attended by an inflammation. He reports the case of a boy with small patches of baldness on the scalp which appeared sieved and showed no sign of lanugo hairs. He had keratosis pilaris both of the scalp and body.

LASSAR and BISHOP (163) maintain that alopecia furfuracea is contagious, and is frequently transmitted by the agency of barber's brushes and combs. Hence, they state, women are less often affected than men, because they are not so much exposed to infection at the hands of the barber. They base their opinion upon the following experiments: They took a twenty-five years old student, of sound health and without nervous tendency, belonging to a family in which baldness was uncommon. This student had been growing bald for five or six years. The baldness was absolute only over the forehead, but the hair was very thin from there to the vertex, the sides being spared. In the thin part, the hair was short and dry. The hair of the neighboring parts was somewhat lustreless and brittle, coming out easily on slight traction. There was only a slight, dusty pityriasis. There was a little itching, and the scalp showed some excoriated places made by the patient's nails in scratching. The itching had been present for some years, especially when he was working or reading. The hair loss had been gradual till the preceding summer, when during a foot tour in hot weather, it had increased more rapidly. From this student's shed hair and scales, a pomade was made by chopping them up fine and mixing them with vaseline. This pomade was spread over the back of a guinea pig and of a rabbit, and the animals watched while kept in the best hygienic surroundings. In the course of three weeks there formed upon the backs of these animals patches of absolute baldness as large as the palm of the hand; other places showed marked thinning of the hair, which came out with the slightest traction; and there was present a mealy desquamation similar to that on the student's head. To control this experiment, it was repeated on another guinea pig and rabbit, and upon yet a third rabbit; in these cases the hair being taken from No. 1 and No. 2 respectively. The results were in all cases similar, only attained more rapidly. In all the cases the baldness preceded from the infected into the sound parts.

It must not be thought that every case of seborrhœa of the scalp will be followed by loss of hair. In those who inherit naturally vigorous hair this may not occur. It is when there is an inherited tendency to loss of hair that seborrhœa will quite surely cause alopecia.

In 1874, MALASSEZ (169) and CHINCHOLLE (142) described a vegetable parasite as present in pityriasis simplex, oval in shape, which disappeared with the disappearance of the pityriasis. MALASSEZ described it as "constituted of spores without tubes of mycelium, generally oval in shape, seldom spherical, and very small. It inhabits the corneous layer of the skin, and penetrates the follicles, but does not reach the orifice of the sebaceous gland. It is generally very abundant, and is the cause of the disease." Recent investigations by BIZZOZERO tend to show that these spores are found generally upon the normal human skin (603). UNNA (141 ap.) has found cocci in seborrhœal eczema of the scalp from which he has made cultivations. He has produced in a rabbit progressive falling of the hair with pityriasis by inoculations with these cultivations.

Much that has been given in the etiology of idiopathic alopecia prematura, especially in regard to the use of water on the head, the wearing of hats, etc., could be repeated here, since they tend to produce pityriasis and in that way alopecia furfuracea.

PATHOLOGY.—As we have learned, the majority of the cases of premature symptomatic baldness that are not due to destructive diseases are due to seborrhœa or pityriasis simplex. We will, therefore, give our attention first to the pathology of alopecia furfuracea *seu* pityrodes. In seborrhœa sicca there is a too hasty casting off of the lining cells of the sebaceous glands, and as these are continuous with those lining the hair-follicle, it is probable that a like increase in the shedding of the cells of the hair-follicle takes place. This must be followed by interference with the nutrition, and loosening and falling of the hair. If the process can be stopped before the papillæ become atrophied the hair will grow again. If the process has continued unchecked for from six to ten years, the papillæ will probably be destroyed, and the baldness will be permanent. The normal life of a hair is one year or more; it may be three months or less. The shorter time the hair lives, the shorter and thinner it will be. In this form of alopecia, the quantitative proportion of the short ("spitzen") hairs to the total shed hairs is increased markedly without the absolute daily hair loss being as markedly increased. The characteristic of the first stage is that hairs suffer an increasing diminution in their normal length, the later growths often having a markedly shorter length of life than those that have gone before. The lessening of the length of growth is due to a shortening of the typical life. The characteristic of the second stage is a reduction in the diameter of the hair. The hairs occur in bundles of from three to five together. The life, length, and diameter of the hairs in the same group vary considerably, and they do not all die at the same time. In fever processes, however, a whole group often falls at once. PINCUS, (173).

MALASSEZ (170), a supporter of the parasitic theory of alopecia pityrodes, explains its occurrence as follows: The parasite causes an irritation of the follicular walls, followed by their hypertrophy. This hypertrophy closes the cavity of the follicle, causes a fibrous transformation of the follicle, and final fall of the hair.

In defluvium capillorum, when it does not depend upon a pityriasis or a seborrhœa, we may find an explanation for the sudden fall of the hair in the arrangement of the blood supply as suggested by UNNA, (82 a) and given on page 38 of this book.

PROGNOSIS. — This will vary with the cause. The prognosis of alopecia furfuracea is good if proper treatment is begun while the hair is only thinned, and the scalp is not actually bald. Then we have fair ground for hope if there is no predisposition to the disease and lanugo hairs are present. When the scalp is atrophied and bound down to the skull, the prospects are bad. PINCUS says that the nearer to puberty the disease begins, the more rapid will be its course.

The prognosis of defluvium capillorum is usually good. This is especially the case after fevers and parturition. It may be generally stated that the prognosis of alopecia, due to local disease of the scalp, is the more favorable the more superficial the disease is which causes it; thus, it is good after eczema and erysipelas, bad after lupus and the ulcerative diseases, in which there is destruction of tissue and production of cicatrices.

Before beginning the treatment of any case we must explain to the patient that rapid results cannot be expected, and success will depend largely upon the care with which he carries out directions, and upon his perseverance. At least one year of treatment must be stipulated for. I have found that women will give much more attention to the physician's directions than men, and this is probably one reason why results are more satisfactory with them.

DIAGNOSIS.—Alopecia prematura symptomatica is diagnosed from the other forms of alopecia by, usually, the presence of a seborrhœa or some local disease, by its running a more irregular course, and by a less marked atrophy of the scalp. From senile alopecia the absence of other senile changes will distinguish it; and the lack of smooth, circular, oval, or serpiginous patches serves to differentiate it from alopecia areata.

Alopecia syphilitica is diagnosed by its sudden invasion; by its non-inflammatory and non-pruritic character; by the ragged appearance it gives to the hair with its irregular patches of baldness; sometimes by the presence of syphilides on the scalp or elsewhere on the body; and by the history of infection.

TREATMENT.—*Prophylaxis.*—It should be especially urged upon those predisposed to baldness that, by proper care of the hair and scalp, much may be done to prevent the early fall of the hair. The prophylactic treatment consists in brushing and combing the hair, washing the scalp, the avoidance of the abuse of local applications, and the attention to certain hygienic laws; matters which have already been fully considered in the chapter on hygiene.

Curative Treatment.—When due to seborrhœa or pityriasis, the first thing to be done is to remove all crusts and scales. This is accomplished by the use of soap and water when there is only scaliness, and for our soap, the tincture of green soap, composed of equal parts of sapo viridis and alcohol or cologne water, is the best. This acts both as a cleanser and a stimulant. The soap should always be washed out with a copious stream of water, using it hot and cold alternately, when convenient. If there is a tendency to dryness of the hair, as there generally is in these cases, after drying the scalp and hair carefully, a little oil or vaseline should be rubbed into the scalp. Or instead of the soap, if the scalp is very irritable, we may use a shampoo of eggs—the yolks of three eggs being beaten up in one pint of lime water, to which half an ounce of cologne water may be added. Rub this thoroughly into the head after carefully brushing it, and wash out the same as after using soap. Where there are thick crusts this will not suffice, and we must use oil. Let the patient saturate the head with sweet almond oil, put on an oiled silk cap, and keep it on all night. The next morning wash the head with borax and water. This is to be repeated every night until the head is clean, and afterwards an occasional wash with the tincture of green soap or borax and water will suffice for cleanliness. Sopping on alcohol, with or without two to five per cent of tincture of benzoin, will hasten the drying of the hair after washing.

Cases of alopecia furfuracea as a rule need stimulation, and to this end a multitude of hair tonics have

ALOPECIA.

been used, such as carbolic acid; tincture of cantharides; tincture of cinchona; tincture of nux vomica; tincture of capsicum; ammonia; chloral; corrosive sublimate; and the like. The good these substances do is by their stimulating properties, and no one of them can boast of any specific action. They must be made strong enough to cause the scalp to glow, but not to irritate it to the point of inflammation.

Carbolic acid may be used as strong as two per cent. in alcohol: tincture of capsicum, and tincture of cantharides of the strength of ʒj—iij (4.0 to 12.0) to ʒj (30.0); chloral, up to ʒj (4.0) to ʒj (30.0); tincture of nux vomica say ʒj (4.0) to ʒj (30.0); corrosive sublimate gr. i—iij to ʒj; aq. ammon. fort. may be used in some cases pure, but better diluted to the point of toleration.

Or ointments containing these or other substances may be used. A good one is

Hydrarg. ammon.,	gr. lx.,	say 9
Hydrarg. chlor. mitis,	gr. lxxx.,	" 18
Petrolati,	ad ʒj.,	" 100

M.

Sulphur also acts well, in the strength of ʒss—ij, to ʒj. UNNA (186) recommends:

Sulph. precip.,	10
Adeps,	100

M.

rubbed in every evening, at first from before backward, and then around and around. Every three or four days the head is to be washed. So soon as desquamation lessens, rub in every second evening, and so gradually decrease. If the scalp is irritated, he substitutes ungt. zinci oxid. for the lard in the above prescription. Sulphur has proved itself of value in my hands, and it is the remedy in which I have most confidence. The disagreeableness of using ointments upon the scalp is much lessened by using a very little of the ointment, and rubbing it, or having it rubbed, thoroughly into the scalp, and not smeared on the hair. Thanks to the kindness of Messrs. Daggett & Ramsdell, of this city, I have been able to use for the past two years a most elegant *Sulphur Cream* composed of :

℞ Ceræ albæ,	ʒ vij.
Ol. petrolati,	℥ v.
Aq. rosæ,	℥ ijss.
Sodæ biborat.,	gr. 36
Sulphur,	ʒ vij.

This is not very greasy, and it acts efficiently.

PINCUS (175) advises in the first stage, when the scaling is pronounced and the hair begins to fall, a solution of bicarbonate of soda strong enough to redden the patient's forehead after rubbing it a few minutes. This is to be rubbed thoroughly into the scalp, a compress to be applied over it, and an oil silk cap to be worn all night. One objection to this is that it stains the hair a dirty reddish brown. In the second stage, when the hair fall is pronounced, ℞ Tanin gr.lxxx (6.00); ungt. rosæ ℥ j (30.0) is to be rubbed in every night, and the head cleansed two or three times a week. A lotion of Ol. sabinæ gr.v—xxx(0.30—2.0) to alcohol ℥ j (30.0) M. applied every night is better, as it may be interrupted for two or three weeks at a time, while the tannin cannot be stopped more than six days. A hood is to be worn during the night with either of these. The ol. sabinæ often causes headache, nausea, vertigo, and sleeplessness, which is an objection to its use.

POHL PINCUS (177) advises the use of a lotion composed as follows:

Ac. lactic,	0.5 to	1.0
Ac. boracic,	2.0 to	5.0
Aq. destilat,		220.
Spts. vini rect.	30 to	40.

M.

Two to three teaspoonfuls of this are to be rubbed into the head once or twice daily for three or four minutes. Or an ointment may be substituted composed as follows:

Ac. lactic,	1 to 3
Ac. boracic,	8 " 12
Adeps, vel	
Vaseline,	100
Ol. bergamii,	q.s.

M.

Two to four pea-sized masses of this are to be rubbed into the bald places once or twice daily for three minutes. After using either of these for two or three weeks, he makes a pause of a few days, and then for one week uses:

Sodii carbonat.,	3 to 8
Adeps, vel	
Petrolati,	100
Ol. bergamii,	q.s.

M.

in the same manner as the previous ointment, and so alternates his remedies for a year.

LASSAR (163) managed the case cited under etiology upon the antiparasitic plan, with the result of causing a growth of strong new hair by the end of the eighth week. The patient's head was daily washed with a strong tar soap which gave a good lather, which was rubbed in for fifteen minutes. The soap was then washed out with water, at first warm, then gradually cooler, at last cold. Then a wash composed of equal parts of a solution of corrosive sublimate (1. to 300.0), spts.cologne and glycerine, was thoroughly applied. The head was then dried and a napthol solution (napthol. 0.5; spts. dilut. 70.0; aq. destil. 30.0) was rubbed in. Finally a one and a half per cent. solution of carbolized oil was slowly poured over the head. In 1888 (130 ap.) he gives as a substitute for this troublesome treatment the following:

℞ Carbolic acid,	1.5
Sublimed sulphur,	6
Horse-neck fat,	100

M.

He also commends oil of turpentine with equal parts of oil or alcohol.

PASCHKISS (134 ap.) believes that where there is a good deal of seborrhœa oily applications should be avoided, and the scalp should be washed daily with tincture of green soap; or a soap composed of castile soap 40 to 50 grains, potash or soda 10 grains, and water half a pint. The suds should be left on from ten minutes to several hours. In women this is impracticable, and we must substitute for the soap a two to five per cent. solution of soda. If this fades the hair the color will be restored by the use of oil. We may use benzol with equal parts of alcohol. If these means do not cure, resort may be had to naphthol, resorcin, or ichthyol. After the seborrhœa is somewhat lessened, we should apply to the scalp every day either

℞ Quininæ sulphat.,	1.50
Spts. vini gallici,	65
Aquæ cologn.,	ad 100

M. or

℞ Tannin,	1 to 5
Alcohol, q. s. ad solut.,	
Ol. amygdalæ dulc.,	40

M.

SHOEMAKER (31 ap.) recommends equal parts of oleate of iron and oil of ergot or any other oil; also sopping on the tincture or fluid extract of soap bark.

HEITZMANN (156) recommends in these cases the use of crude oleum rusci in the proportion of 10 per cent.

to 20 per cent. in an ointment of vaseline and paraffin, with enough fragrant oil to cover the smell of the tar. This is to be alternated with sulphur and white precipitate ointments. He claims that twenty-four per cent. of two hundred cases were temporarily benefited, and in a small percentage the improvement was lasting.

IHLE (159) recommends the use of resorcin as follows:

Resorcin pur,	5
Ol. ricini,	45
Spts. vini.	150
Bals. peruv.	0.5

M.

This is to be rubbed into the scalp daily with a piece of flannel. It forms an agreeable mixture with a slight odor of alcohol. This drug, as well as *icthyol*, has also been highly recommended by UNNA.

So much for local measures. As seborrhœa is an indication of lowered vitality, the general health must be cared for, and tonics given when indicated.

In *syphilitic alopecia* our main dependence is upon the internal treatment—mercury or iodide of potassium being used according to the stage of the disease. Locally, if any lesions are present, we may use a lotion of bichloride of mercury, or an ointment of the ammoniate of mercury. Stimulating remedies, as in alopecia furfuracea, may be used with advantage. *Defluvium capillorum* takes care of itself in most cases. Its treatment, when needed, is that of alopecia furfuracea. *Alopecia* arising *from local diseases* needs the treatment applicable to the special disease present, which will be given in the appropriate chapters of this work. In pustular diseases affecting the scalp, if the hairs are early extracted before the follicles are destroyed, much will be accomplished to prevent alopecia.

Loss of Hair: A Clinical Study of its Causes, Founded on One Hundred Cases.

I have chosen the term "loss of hair," rather than baldness, because in my tables I have placed a number of cases in which absolute baldness was not present, but a general thinning of the hair, which if not checked would lead to baldness. Cases of defluvium capillorum coming on after acute illnesses are also included.

Of the 100 cases, 65 occurred in men and 35 in women.

Of these the nationality was: Canada, 1; Ireland, 1; Germany, 3; United States, 95.

Condition:

Married..............................12 men, 16 women.
Single ....45 " 16 "
Widowed.... .. ..3 "
Not recorded........................ 8

Occupation. Of the men:

Architect, 1
Bookbinder, 1
Brokers, 5
Clergymen, 2
Clerks, 8
Dentist, 1
Electrician, 1
Farmers, 2
Grocer, 1
Lecturer, 1
Librarian, 1
Lithographer, 1
Lawyers, 5
Manufacturers, 2
Mechanics, 2
Merchants, 2
Physicians, 15
Presser, 1
Waiter, 1
None, 5

Of the women:

Housewives, 16
Gloves, 1
Singer, 1
None, 16

Age at beginning:

From 10 to 20 years................ 7 men and 7 women
"   20 to 30 "   ....44 "   " 11 "
"   30 to 40 "   ....12 "   " 11 "
"   40 to 50 "   .... 1 "   " 3 "
Over 50 years..... . ...   2 "

The greatest number of cases began in the twenty-fifth and twenty-sixth years, viz., 9 in each year. The next most frequent age was twenty-two, with 8.

The baldness or loss of hair took the form of a general thinning in................................... 6 men, 19 women.
It affected the crown and temples in.............. 13 "
" " alone in ....32 " 10 "
" temples in....................... . 4 " 2 "
" " and tonsure in .... 5 "
" tonsure in .... 4 " 2 "
" occiput in.... ... . 1 "
" . parietal region in .... 1 " 1 "

The following diseases were noted as complicating the loss of hair:

Anæmia, 4 cases.
Chancroid, 1 case.
Chorea, 1 case.
Constipation, 1 case.
Dyspepsia, 11 cases.
Dyspepsia and constipation, 5 cases.
Endometritis chronica, 1 case.
Gonorrhœa, 1 case.
Gout, 1 case.
Headache, 4 cases.
Influenza, 1 case.
Malaria, 7 cases.
Measles, 1 case.
Menopause, 1 case.
Metrorrhagia, 1 case.
Overtraining, 1 case.
Parturition, 3 cases.
Peritonitis, 1 case.
Pneumonia, 2 cases.
Spermatorrhœa, 1 case.
Sunstroke, 1 case.
Urethritis chronica, 1 case.
Uterine fibroids, 1 case.

The cases of loss of hair following influenza, measles, parturition, pneumonia, peritonitis, and sunstroke were what are called defluvium capillorum, and came on from one to five months after convalescence.

The scalp and hair were found to be diseased in 88 per cent. of the cases, as follows:

Atrophied and bound down, 2 cases.
Canities, 4 cases.
Eczema capitis eight years before hair-fall, 1 case.
Seborrhœa congestiva, 2 cases.
Sweating head, 5 cases.
Heat of head, 4 cases.
Pityriasis, 10 cases.
Seborrhœa oleosa, 3 cases.
Seborrhœa sicca, 55 cases.
Fragilitas crinium, 2 cases.

In 46 per cent. of the cases there was a history of baldness in the family, as follows:

	Men.	Women.
Father only	7 cases.	
Father and paternal uncle	2 "	
" " " cousins	1 "	
" " " grandfather and brothers	1 "	
" " brothers	4 "	
" brother, and maternal aunt		1 case.
" and maternal uncles		1 "
" mother, and brother	2 "	3 "
Paternal uncle and brothers	1 "	
" " "	3 "	
" " maternal grandfather and uncles	1 "	
Mother	3 "	2 "
" and maternal grandfather	1 "	1 "
" " sisters		1 "
" " brothers		1 "
Maternal grandfather	1 "	
" uncles	1 "	
" " and brother	1 "	
Brothers	2 "	
In family on both sides	4 "	1 "

As possible contributory factors it was noted that

1 patient constantly wore a close-fitting cap.
2 patients became worse after a short residence at the seashore.
2 smoked to excess.
21 soused their heads daily in cold water.

What deductions can we draw from the preceding figures?

1. As to sex. We find that 65 per cent of the cases occurred in men. This is in accord with the well-known fact that men are more frequently bald than women. Therefore masculinity is a predisposing cause of loss of hair.

2. As to nationality, no deduction can be drawn, as a preponderance of Americans was to be expected.

3. As to condition. Although we find 61 per cent. of the patients were unmarried and but 28 per cent. were married, this does not allow us to draw any inference, because experience teaches that most men do not mind becoming bald half as much after they are mar-

ried as before they marry. Men tell me again and again that they would not mind growing bald if they were married. It is a common experience that men are not so particular about their personal appearance after marriage as before, as they have other and more important things to think of. It is also to be observed that our tables show that the number of married and unmarried women is the same. With women the case is different from what it is in men, as the condition of the hair is for women a most important consideration from the standpoint of personal appearance, a fine head of hair being to them a matter of pride.

4. As to occupation. Here, too, I should hesitate to draw an inference. If we took the figures alone we would be compelled to think that doctors were especially liable to become bald. It is certainly suggestive that 26 per cent. of the cases occurred in professional men—that is, in architects, clergymen, dentists, lecturers, librarians, lawyers, and physicians. If we add to these the five brokers, who certainly live under a constant nervous strain, we have 31 per cent. of the cases occurring in brain workers. This is not intended to throw any slur upon the rest of the occupations mentioned in the tables. It, however, tends to suggest that active brain work does predispose to loss of hair.

5. As to age. It is evident from my tables that the majority of the cases of loss of hair begin before the thirtieth year of life—namely, 69 per cent. We also see that in 67 per cent. of the men and 31 per cent. of the women loss of hair began between the twentieth and thirtieth years. There is also a very marked difference between the sexes in the next decade—18¾ per cent. in men and 31 per cent. in women. We also learn that the most critical years are from the twenty-second to the twenty-sixth. Though the present paper

is not intended to touch upon treatment, I would say that it is between the ages of twenty and thirty-five that we are most justified in expecting good results from it.

6. As to the location of the loss of hair. We find that 42 per cent. of the cases affected the crown alone, and 13 per cent. affected both the crown and temples. In 25 per cent. of the cases there was a general thinning of the hair, and this was three times more frequent in women than in men. The tonsure alone was seen only six times, but it occurred with the receding temple eleven times. These figures bear out the well-observed fact that the top of the head is the location of baldness.

7. As to complicating diseases. In only 52 per cent. of the cases did this factor enter into the field of our study. Of these, anæmia, or diseases inducing anæmia, constituted nearly 79 per cent.—namely, anæmia, chorea, constipation, dyspepsia, endometritis, gout, headaches, malaria, metrorrhagia, overtraining, spermatorrhœa, menopause, chronic urethritis, and uterine fibroids. Acute and general constitutional diseases were met with in but 9 per cent. of all the cases. Sexual disorders were met with only four times. A fair deduction from these figures is that anæmia is a cause, or at least a predisposer, to loss of hair, and that the popular idea that loss of hair is due to sexual excesses is wrong.

8. As to diseases of the scalp and hair, other than its fall. The scalp was diseased in 82 per cent. of the cases. Seborrhœa or pityriasis was present in 70 per cent. of the cases, and sweating or heat of the head was seen in 9 per cent. A history of an antecedent disease of the scalp was seen in but one case. From this the deductions are that seborrhœa is a most active cause of loss of hair; that sweating or heat of head is

a far more important symptom of danger to the hair than writers on baldness have commented on; and that an antecedent disease of the scalp, other than seborrhœa, is by no means so common as Pincus would have us believe.

9. As to heredity. In 46 per cent. of the cases there was a history of baldness in the family. It is certainly a most surprising fact, brought out by my tables, that in all the cases of loss of hair in women there is a distinct history of the affection occurring on the maternal side; while in the thirty-five cases in which there is no history of maternal heredity the men only are affected. It will be interesting to note if further statistical studies show the same results. From the tables now presented it is fair to deduce that loss of hair is markedly hereditary, and that it tends to descend in the same sex.

10. As to contributing factors. In 21 per cent. of the cases we find a history of daily sousing the head. This is by no means as great a percentage as is given by some other observers, but still great enough to suggest that the habit is detrimental to the hair. The other factors noted are so few as to be useless for deductions. They would tend to show, however, that the popular idea that wearing close-fitting head gear is a frequent cause of loss of hair is not true.

Summary.—From the foregoing study of loss of hair we can summarize as follows:

1. Men are far more prone to baldness than are women, the proportion being as 65 to 35.
2. Neither the married nor the unmarried state exercises any influence in the production of baldness.
3. It is probable that active brain work and nervous mental strain predispose to baldness.
4. The majority of the cases of baldness occurring before middle life do so between the twentieth and

thirtieth years; and this is more marked in men than in women.

5. Anæmia, or diseases that predispose thereto, are active causes of baldness.

6. Seborrhœa in all its forms is an active cause of baldness.

7. Sweating and heat of head may be regarded as danger signals, foreshadowing loss of hair.

8. Heredity is a pronounced predisposing factor of baldness, and it shows a tendency to descend in the same sex.

9. The daily sousing of the head is pernicious to the preservation of the hair.

The study of etiology is helpful chiefly as it teaches us how better to treat diseases. From the analysis of the causes of loss of hair, as here given, we may, I think, learn that it is essential for us to put our patients who are losing their hair in the best possible physical condition; to cure, or at least alleviate, any disease of the scalp that may be present; and to forbid their sousing the head in water.

CHAPTER VII.

ALOPECIA AREATA.

SYNONYMS:—Area Celsi, (v. Bärensprung); Area occidentalis diffluens, serpens, seu tyria; Alopecia circumscripta, (Fuchs); Alopecia occidentalis, (Wilson); Porrigo seu tinea decalvans, (Bateman); Vitiligo capitis, (Cazenave); Ophiasis; Phytoalopecia, (Gruby); Teigne

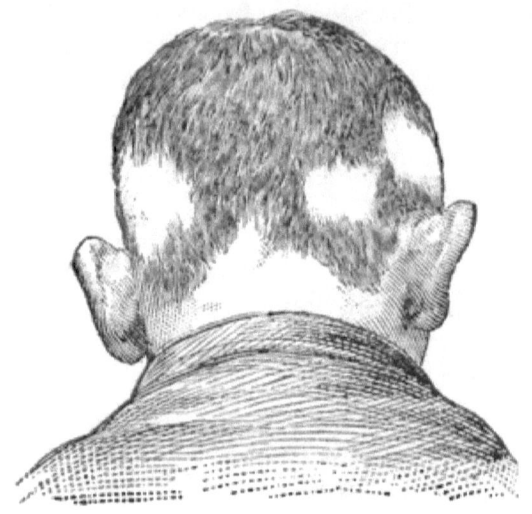

Alopecia Areata.

pelade, (Bazin); Pelade acromatosa; Pelade decalvante, or, ofiasica ; Pelade (Fr.); Die kreisfleckige Kahlheit, (Ger.); Circumscribed baldness, (Eng.).

DEFINITION.—A disease of the hair characterized by its sudden fall and the production of perfectly bald, smooth, usually circumscribed and circular patches ; which generally affects the scalp, but may invade other hairy parts; runs a chronic course without concomitant symptoms, and tends towards spontaneous recovery.

SYMPTOMS.—The disease usually begins suddenly, the patient finding to his surprise a bald circular spot upon the head without knowing when it formed. In a number of cases there will be a history of severe, often periodic, and localized headache preceding the hair-fall for weeks or months. In other cases some slight pruritus, burning, or pain may ante-date the alopecia. In any event the patch forms suddenly, the hair falling out at once. The size of the patch varies; it may be very small, even pea-sized, or it may be as large as the palm of the hand. When it exceeds the latter size, it is usually formed by the coalescence of smaller patches. It tends to gain its full size at once and to remain circumscribed, but in some cases it pushes out into the surrounding parts, either preserving its circular shape, or becoming irregular or serpiginous. There may be only one patch, but more commonly there are two or three patches, which appear simultaneously or successively. The shape of the patch, while usually oval or round, may be in stripes or bands. This is seen most often after injuries to the scalp, as in SCHUTZ's case (219 ap.) resulting from injury to the scalp by barbers' shears. This irregular form of the patches is sometimes called alopecia neurotica by those who believe that the usual form of the disease is parasitic. The disease most frequently affects the scalp and may be symmetrical. Its next most common seat is the beard. It may, however, affect all the hairy parts of the body, and in some cases cause universal baldness.

The appearance of the disease is striking, and not readily forgotten when once seen. The bare rounded spot stands out conspicuously from amongst the surrounding hair. It is perfectly bald and smooth, without any scaliness. In some cases there are a few black points in the patch, which have been taken for broken-off hairs, but are really the remains of hair-roots which have not fallen out. In some cases there will be a few straggling hairs remaining. It seems as if the patch was depressed and atrophied, but this appearance is probably due to the fact that the hair and roots have fallen out of the part. In color, the affected area in its earlier stages may be slightly reddened from hyperæmia, but in most of the cases that we meet with, it is white and polished. To the touch the skin seems soft and pliable. Sensation may be normal in the patch, but often somewhat blunted, as shown by the amount of stimulation the diseased surface will tolerate.

The hairs at the margin of the patch at first, and if the disease is spreading, will be loose, dry, brittle, atrophied at the root and come out readily on slight traction. Later, and when the patch has attained its full development, they will be firmly seated in their follicles, and normal in appearance.

Recovery will be indicated by the appearance of fine lanugo hairs in the patch. These may fall out again to be followed by a new crop of white hair, which will remain and, growing stronger, develop into strong colored hairs. Kaposi says that, if the disease has been severe and general, restitution of the hair will not take place; and that even in milder cases it may go no further than the production of lanugo hairs. The disease is comparatively rare. Neumann describes a form of this disease in which there is marked anæsthesia of the scalp.

Some of the French authors, judging from their descriptions, would seem to have an entirely different idea of alopecia areata than we have, and to have confounded it with other diseases. Thus, Cazenave (48) describes cases in which a change in the color of the hair took place, and designates the disease vitiligo capitis. This is probably the pseudo-pelade of some French authors. GRUBY (214) says that the patches are covered with a whitish dust formed entirely of cryptogamia. HARDY considers the disease as having two stages: First, an early parasitic contagious stage; and second, a neuropathic stage in which the parasite has disappeared. COURREGES (53) regards the disease as having three stages: First, one in which there is discoloration of the skin and fall of the hair, itching of the scalp and pityriasis. This stage is short. Second, the appearance on the bald spots of delicate downy hairs. This period may last for years. Third, the period of recovery and growth of strong hair. Further, he makes two varieties of alopecia areata, namely: la pelade achromateuse, and la pelade decalvante. The first is the porrigo decalvans of BATEMAN, and the vitiligo de cuir chevelu of CAZENAVE. It corresponds to our usual and typical form of alopecia areata. The second form is one in which a rapid fall of the hair of the whole head or body takes place, and corresponds to what we should regard as an unusually marked form of the disease. COURREGES' views as to these divisions are not peculiar to himself, but are quite commonly held by French authors. BAZIN (190) affirms that true pelade (alopecia areata) is very rare, but that there is a false form which is often the consequence of tinea tonsurans badly treated or left to itself.

The disease has a tendency to recur, and sometimes is hereditary. THIN (257) reports a case in which a father and three of his children had the disease; and HARDAWAY (216) has met with two cases in which relapses took place in the Spring of successive years and attacked new areas each time.

ETIOLOGY.—The etiology of this disease has been a

field of battle for many years, and dermatologists are divided into two great camps in regard to the question of its being parasitic or non-parasitic. The large majority of the older authorities are against the parasitic theory; thus of forty-two authors consulted as to this point, and they were taken without selection, fifteen were for the parasitic theory, and twenty-seven against it. In the first group were such men as Anderson, T. Fox, Thin, Gruby, Bazin, Hardy, and Eichhorst; in the second group, Duhring, Kaposi, Neumann, Schwimmer, Michelson, Veiel, Vidal, Horand, Duckworth, Pye-Smith, Liveing and Wilson were found. GRUBY (214) was the first to allege a parasitic origin for the disease, and in 1843 described the parasite, which he named "*Microsporon Audouini.*" BAZIN, in 1862, brought forward the claim of "*Microsporon Decalvans*" to the honor of being the cause of the disease. THIN (529), in 1881, described yet another fungus in this disease which he called "*Bacterium Decalvans.*" VON SEHLEN (267) in 1885 entered the field as the discoverer of *the* parasite in alopecia areata. ROBINSON (215 ap.) in 1887 also described a coccus as the cause of the disease.

It is true that a single positive result from the study of a disease is of vastly more value than many negative ones, and we should have the etiology of alopecia areata settled upon a firm foundation if the positive results of those who believe that they have found a parasite were in accord. But they are not, as will be seen further on when discussing the pathology of the disease. Such disparity in the findings of the investigators of this disease, and the fact that many other well-trained and expert workers in pathological histology have sought for the parasite and have not found it, are in striking contrast to what obtains in the other parasitic diseases of the skin. At present we are forced to acknowledge that there is not sufficient evidence offered to warrant us in regarding this disease as a dermato-mycosis. Besides the absence of proof of the presence of a parasite, the suddenness of onset of the disease, its occasional general appearance, and the want of any trace of fungous growth or *débris* upon the affected areas, are all against the parasitic theory. Dermato-mycoses are slow in development, usually appear primarily in a number of small foci, and do not involve the whole surface at once, and generally scales or crusts are present upon the diseased areas.

The most probable cause of the disease is a trophoneurosis. It is true that no nerve lesion has yet been demonstrated in connection with the disease. But the sudden onset of the malady, all the hairs in a given area being at one and the same time loosened, seems to indicate that some profound disturbance in the nutrition of the affected part has taken place, probably dependent upon some trophic nerve disturbance. The neurotic theory of the disease finds further support in the occurrence of periodic headaches and disturbances of sensation before the outbreak; in the presence of anæsthesia in the patches to a greater or less degree; in its frequent occurrence in children who have unstable nervous systems; in its following upon severe nervous shock and injuries to the scalp. UCHERMANN (263) has reported a case of alopecia areata in a boy which followed a blow on the head with a stone, and involved the whole head. MICHELSON (231) has met with a case following a fall. TYSON (262) has recently reported three cases of complete fall of the hair following rapidly upon nervous shock. MICHELSON (231) also suggests that the disease may be due to a closure of the cutaneous vessels of the affected area, and consequent interference with nutrition, the closure being due to a connective-tissue increase of the intima.

It is possible that the neurosis may express itself in a constricting action upon the blood-vessels, either those going to the middle follicle region or to the papilla, and a consequent preponderance in the productivity of one or the other region, and a fall of the hair. JOSEPH (223) has recently produced in cats lesions exactly resembling alopecia areata in man, by cutting out the second cervical ganglia. His experiments have been repeated by others with similar results. MIBELLI (198 ap.) found that cutting the second cervical nerve without extirpating the ganglion was sufficient, and that the fall of hair was not confined to the district supplied by the nerve, but was seen in regions supplied by other nerves. PONTOPPIDAN (209 ap.) saw typical patches of alopecia areata follow the operation for removal of an enlarged cervical gland in the left carotid region. In further support of the neurotic theory ASKANAZY (147 ap.) cites a case following removal of a tumor from the right submaxillary region and attended by facial paralysis; and another of a hypochondriac who suffered from severe headache and burning of the top of the head.

The disease is comparatively rare. BULKLEY met with it one hundred and nineteen times in twelve thousand cases occurring in New York city. It would seem to be rather more common in Glasgow, ANDERSON having had one hundred and fifty-three cases in ten thousand. In London RADCLIFFE-CROCKER met with it two hundred and fifty-three times in ten thousand public cases, and eighty-two times in two thousand private cases. His youngest patient was three years old. It befalls both sexes, though it is more common in males. Children are often affected with the disease. They formed about one-sixth of BULKLEY's patients. According to his tables, the disease in the upper classes is most frequent between the ages of twenty and forty, no fewer than twenty-five out of thirty-five cases occurring between those ages. He met with it much less frequently in public than in private practice. I have met with the disease twenty-nine times in four thousand consecutive cases in dispensary practice up to 1887. Of the patients, seventeen were males and twelve females. The majority of the cases occurred between the twentieth and fortieth year of age, namely, eighteen. There were ten patients under twenty years of age, and but one was over forty. The youngest was a girl of five years, and the oldest a man of forty-four years.

Up to comparatively recent times the disease was not considered contagious. The most earnest advocates of the contagiousness of the disease are the French. BUCHIN (46 ap.) cites a number of cases of contagion. VAILLARD and VINCENT (225 ap.) report an outbreak of the disease in which forty-four soldiers were attacked. CLEMENCEAU DE LA LOQUERIE (53 ap.) gives many like instances, as in a school in Amiens in which fifty out of four hundred children had the disease, and an epidemic of eighty cases in a regiment at Montpellier. He regards the modes of conveyance to be by the barbers' shears, wearing of infected caps, cushions of public conveyances, and the like. FEULARD (170 ap.) reports the occurrence of forty-four cases in one company of a regiment stationed in Paris, and states that in 1891 and 1892 it occurred in the proportion of about 3.30 per 1,000 of the army. He thinks that the hair clipper is responsible for its prevalence. It is most frequent in the cities. In Germany EICHHOFF (168 ap.) reports several instances of apparent contagion. CROCKER and HILLIER in England also report cases of contagion. In this country the only epidemic of a disease simulating, if not identical with, alopecia areata is reported by PUTNAM (210 ap.) as occurring in an asylum for girls near Boston. The cases were examined by Drs. White and Bowen of Boston. In this epidemic sixty out of sixty-five girls were attacked by the disease. MORROW has also met with cases apparently contagious. In the light of these observations, it seems impossible to deny that the disease is contagious at times. It is equally impossible to assume that errors in observation have been made in all these cases.

Contagiousness of the disease is, however, contrary to the experience of most observers, and at best is a rare exception. The disease has been seen in the same patient coincidently with ringworm of the head, and probably such occurrences have led to mistakes in diagnosis. The "tinea decalvans" of TILBURY FOX, in which perfectly bald circumscribed spots occur with parasites in the neighboring hairs, may have been of this kind. LIVEING (226) thinks that it is just possible that trichophytosis may in some cases so interfere with the nutrition of the hair as to favor the subsequent development of alopecia areata. ALDER SMITH (79) draws attention to the fact that a patch of trichophytosis capitis may be changed into a perfectly bald, smooth place by the application of an ethereal solution of boracic acid. There seems to be some relation between alopecia areata and ringworm, as in a country where one is common the other abounds. Until further light is thrown on the subject, it is probably best to hold that there are some cases that are contagious and may be parasitic, and some that are non-contagious and neurotic.

The following are considered to be *predisposing causes, viz.:* The nervous diathesis; disturbances of the general nutrition of the body, as from recent constitutional syphilis (SQUIRE, 34 a); menstrual disorders (NAYLER, 52); arsenic (WYSS, 272); parturition and pregnancy (GRAHAM, 213). DE TULLIO (167 ap.) reports a case of progressive spread of the disease in a suspected case of syphilis in which the iodide of mercury had been administered during three years.

PATHOLOGY.—Hairs extracted from the margin of an advancing patch of alopecia areata show marked atrophic change, and are seen to terminate abruptly in a pear or club-shaped extremity. Sometimes a portion of the root sheath is attached to the plucked hair, sometimes not. As we approach the free end of the hair an oval swelling will sometimes be found tapering again towards the extremity of the hair, which is often split. DUHRING (202) draws attention to the fact that these changes differ from those found in senile alopecia only in the suddenness of their occurrence. Sometimes ampullary swellings are found near the root, composed of a concentration of granular pigment matter; and sometimes there are adherent follicular and sebaceous matters round the shaft. Below the swelling a stricture is apparent, which again passes into the deeper bulb elements above the papilla and at the base of the sac. Sometimes bright refracting granules are seen investing the hairs, which are not easily removable by ether, and no doubt suggested the fungus theory, but are in reality fatty particles. The swelling of the cuticular scales of the hair simulates a fungus. DUHRING (202) says that what is described as the *microsporon Audouini* is an accumulation of an appreciable amount of sebaceous matter, broken up epidermic scales, and *débris* about the roots of the hair; further, that sebum when subjected to a reagent has a tendency to split and break up into fine particles, which adhere so closely and with such tenacity to the hair as to accurately resemble spores.

According to GIOVANNINI (175 ap.) the disease begins by a perivascular infiltration by leucocytes, especially about the lower part of the hair-follicle, where they invade the circular connective-tissue layer. From there they often enter between the cells of the matrix, internal root-sheath, lower part of the neck of the hair, or both root-sheaths. Either in the matrix or internal root-sheath of the infiltrated hairs the cells undergo karyokinesis, diminishing in number until they disappear completely. At the same time the cells of the matrix, neck of the hair, and internal root-sheath degenerate. Degeneration is accompanied by disappearance of pigment. Destruction of the hair-bulb follows, and then that of the neck of the follicle and the internal root-sheath. The follicle itself atrophies, but usually not to the degree of destruction. A new hair generally forms in the follicle, unless infiltration continues, in which case it will fall. If the process becomes chronic the follicle and the sebaceous glands will eventually be destroyed.

All those who describe the parasite in this disease give explicit directions as to the preparation of the hair for examination, and lay great stress upon the difficulty of finding the fungus. GRUBY (214) described the parasite as follows: The cryptogamia are arranged so as to form a tube or vegetable sheath about the hair. They consist of trunk, branches, and sporules. The trunks have an undulated form following the direction of the hair fibres, are transparent, and have a diameter of .002 to .003 mm. They bifurcate at times, giving off branches at an angle of thirty to fifty degrees. The branches are distinguished from the trunk by the sporules which accompany and cover them. The sporules are oval or round, the diameter of the former being 0.002 to 0.008 mm., and of the latter 0.001 to 0.005 mm. They are transparent, and do not contain molecules in their interior, and swell in water. These are the *microsporon Audouini*. They commence to develop at the surface of the hair, 1 to 2 mm. from the epidermis. They are first seen parallel to the axis of the hair, and spread by immediate contact from hair to hair. BAZIN (190) gives a similar description. T. Fox (210) says the fungus occurs in the form of very delicate waxy mycelial threads. THIN (259) found, after careful preparation, minute round or elongated rounded bodies in the hairs which resembled in shape and refractive power his "*bacterium foetidum.*" These he believes are the cause of the disease, and names "*bacterium decalvans.*" They were in position and arranged so as to show that they were distinct from the rows and aggregations of minute granules which are found in healthy hairs. They were found frequently in pairs, the long axis of each member of a pair forming a continuous line. Sometimes three were found end to end, with the appearance of one continuous sheath for the three. MALASSEZ (228) describes the spores as occurring in the epidermic scales, in the superficial layers of the epidermis, and occasionally upon the hair. According to him they are spherical or ovoid highly refractive bodies, not larger than 4 to 5 mm. They are double contoured and many small buds project from their circumference. Some smaller spores (2 mm.) were without the double contour, and some still smaller were simply spherical. They were found singly or in groups or chaplets. VON SEHLEN's (267) micrococci were found in the root-sheaths of the hair, but his description of the cases from which the hairs examined were obtained is so strikingly like that of trichophytosis capitis that we are led to think that he made an error in diagnosis; and BORDONI UFFREDUZZI (193) has found micrococci identical with these upon the roots of the hair of the normal skin.

The findings of the different observers vary amongst themselves so much that it is best to await further developments before deciding that the disease is parasitic. It is noticeable that the different parasites described are all superficial to the skin and in or upon the hair. This is not the way in which perfectly smooth bald patches could be produced; for that it is necessary that the hair papillæ should be affected. In this connection it is interesting to note that NYSTROM (237) found spores identical with those described by MALASSEZ upon a napkin hung in a moist corner of a room, and therefore regards them as derived from the atmosphere. MICHELSON (234) has found on normal hairs by cultures the same cocci as described by VON SEHLEN. He also found them on hairs taken from the edge of alopecia areata patches. Thus far his attempts at inoculation of these have been failures, and he regards them as being a normal condition.

ROBINSON, in 1887 (215 ap.), is the first to describe cocci deep down in the follicle as the cause of the disease. Like Giovannini, he found perivascular infiltration of the upper part of the corium by small cells, affecting the papillary portion. With this there was proliferation of connective-tissue corpuscles, fall of the hair, and finally coagulation in some of the blood- and lymph-vessels. The hair-roots showed atrophic alterations. Lanugo hairs were apparently formed in the upper part of the follicle. In chronic cases the follicles were sometimes hairless, and the sebaceous glands were either normal, degenerated, or destroyed. The areas of baldness corresponded in size to the areas supplied by the plugged-up blood-vessels. In the lymphatic vessels and in the walls of a few blood-vessels he found small, round, dark bodies of equal size and grouped in zoöglœa masses. These were not acted on by acids or alkalies. They are cocci, 8 μ in diameter, and in size similar to staphylococcus pyogenes aureus. There were also some diplococci. These were less numerous in chronic cases.

DYCE DUCKWORTH (201) had the opportunity of examining a case of alopecia areata occurring in the person of a boy aged thirteen years who was drowned. The disease first came on when he was four years of age, and had gone and come several times since. After death portions of the scalp were carefully examined, and the results reported, as follows: There was found to be: 1. A distinct atrophy of the hair-follicles and the sebaceous glands in connection with them. 2. Infiltration of the hair-follicles, specially their outer root-sheath, with a new round cell growth. This growth appeared to be perivascular, and tracts of it were found in the middle layer of the corium leading up to the papillary layer. 3. The hair follicles in the affected part were mostly quite atrophied, their nourishment having been cut off by the new growth. In some instances remains of the papillae were seen, but the capillary loops were infiltrated with the cell growth. In other instances the follicles appeared to be making efforts at repair by throwing out numerous digitations. 4. The vitreous membrane of the follicles was in some cases hypertrophied. 5. The sweat glands were practically unaffected, though parts of their ducts were implicated in the new growth. 6. No parasitic elements were found.

WAGNER and SCHULTZE (248) have examined pieces cut from the living scalp with negative result.

DIAGNOSIS.—The diagnosis is easy, as a rule, the disease presenting so striking an appearance as hardly to be mistaken for anything else. As in psoriasis, the symptoms are so well pronounced that students, once having seen a case, do not easily forget it. Sometimes, however, it may be difficult to distinguish it from ringworm of the head, and it must also be diagnosticated from favus, from other forms of alopecia, and possibly from vitiligo.

Differential Diagnosis from Trichophytosis Capitis.

Alopecia Areata.	Trichophytosis Capitis.
1. Occurs suddenly without antecedent lesion, and the patches often attain their full size at once.	1. Begins usually at one point by a small erythematous papule or patch, and spreads from it more or less slowly.
2. Patch usually perfectly circular and does not contain "gnawed-off" hairs, scales, or crusts, but is perfectly smooth and shiny.	2. Patch more or less circular, with broken and gnawed-off hairs in it, and floor covered with thick, grayish crusts or abundant scales.
3. Hairs about patch unaltered, though at times they may be easily extracted.	3. Hairs in and about patch are brittle, break easily when pulled on, and bend at an angle.
4. Occurs only on hairy parts of the body.	4. Occurs both on hairy and non-hairy parts of the body; and patch will sometimes spread from non-hairy to hairy parts, or vice versa.
5. No parasite found, or at least none readily detected.	5. Fungus found abundantly in hair and scales.

The two conditions are less easily distinguished when ringworm of the head has lasted some time, involves a large portion of the scalp, and is in an inactive condition. But even here there will be more or less scaling, and an occasional broken-off stub of hair; and careful search will discover the fungus in the hair or scales.

Favus lacks the circular development of alopecia areata, is more disseminated, has the sulphur-yellow cup-like crusts characteristic of it, or else the scalp is covered with powdery scales. It developes slowly, is inflammatory in character, and leaves cicatricial patches where it has run its course. The microscope shows the fungus abundantly in scales, crusts, and hair.

Alopecia Senilis and *Praematura* develop slowly, beginning either at the vertex and spreading forward, or at the forehead and spreading backward, involve only the upper middle region of the head, and take months or years to produce complete baldness. There is often a history of preceding disease of the scalp, such as seborrhœa. In alopecia areata the baldness occurs suddenly, occupies the lateral parts of the head quite as often as other regions, and generally there is no antecedent disease of the scalp. If it involves the greater part of the head, there will yet be the history of distinct patches at the begining.

Sometimes *syphilitic alopecia* will present appearances very much like alopecia areata. In syphilis we meet with two forms of baldness, one occurring as the result of the cachexia of the disease, and the other from the absorption or breaking down and ulceration of a syphilide. The first form sometimes resembles a severe case of alopecia areata, affecting more or less generally the whole head, and causing great loss of hair. It differs from it in giving a characteristically ragged look to the head, and in showing no tendency to the formation of circles. The history of the case will aid in making the diagnosis. Should there be any cutaneous manifestations of syphilis present, the decision will be easily reached. The second form of alopecia syphilitica resulting from ulcerating lesions is differentiated from alopecia areata by its history and by the cicatricial tissue present.

Bald spots arising from burns and other injuries to the scalp should offer no difficulty in diagnosis; their history and cicatricial appearance show their origin.

Vitiligo should not be confounded with alopecia areata. It is a disease affecting only the color of the hair.

PROGNOSIS.—The disease tends to recover spontaneously, especially in young people. In older people the prognosis is not so good. NEUMANN (27) says the outlook in the anæsthetic form is bad; and the serpiginous form is considered by SQUIRE (34a) to be of less favorable prognosis than the circular form. Cases of general alopecia, especially those coming on suddenly, are more grave than the circumscribed cases. The duration of the disease is variable. Recovery has taken place in six weeks, but this is exceptional. From six months to two years may be given as a reasonable time in which to look for recovery. The patient should be apprised of the tendency the disease has to relapse. Cases of universal alopecia are of bad prognosis and often never recover.

TREATMENT.—Many of the sufferers from this malady show some indication for the exhibition of tonics. When occurring in children they will often be found to be anæmic. They should be taken out of school and allowed plenty of air and exercise until they attain to a better degree of health. Cod-liver oil, iron, phosphorous, arsenic, and quinia are the drugs most indicated as tonics. The use of phosphate-bearing food is advised by some, as oatmeal, cracked wheat and the like. We should endeavor to place our patients under the best possible conditions, to relieve them as far as may be from all sources of anxiety, and to alter anything that may be wrong on the side of the general economy.

The local treatment consists in stimulation of the scalp. In the beginning of treatment it is well to remove by epilation all the loose hairs about the margins of the patches. The best method of effecting this is by pulling the hair between the thumb and an ordinary spatula or stout card held in the hand. This procedure sometimes seems to check the further growth of the patch.

For stimulants, carbolic acid; tincture of cantharides; cantharidal collodion; tincture of nux vomica, veratrine, capsicum, phosphorus, or aconite; sulphate of quinine; strychnine; liquor ammoniæ fortior; sulphur; bichloride; yellow sulphate, and oleate of mercury; croton-oil and castor-oil, each and all have their advocates, and are used either separately or two or more of them combined. As the diseased scalp will bear, as a rule, a good deal more stimulation than the healthy scalp, we must regulate the strength of our chosen stimulant solely by the amount of reaction it causes. Thus, liquor ammoniæ fortior, in full strength, may be freely applied to the scalp, and its use persisted in for weeks without apparent over-irritation of the scalp. As our object is simply stimulation, I can see no reason for combining any of the above-mentioned stimulants, excepting that the castor-oil may be used as an eligible excipient for some other remedy. We may choose as a vehicle for the stimulant either an oil, an ointment, lard, vaseline, or water. It is well to change our stimulant from time to time.

Good results have been reported from the use of *electricity*, the galvanic current being used with one pole at the nape of the neck, and the other brushed over the affected parts; and G. H. Fox has reported improvement in some cases treated with static electricity. Hypodermic injections of $\frac{1}{8}$ to $\frac{1}{16}$ grain of the muriate of *pilocarpin* every few days are well spoken of by some observers.

Sulphur ointment is advocated by THIN (258), and he has published a series of excellent results from this treatment, maintaining that these results are due to the action of the ointment as a parasiticide.

The treatment by *blistering* has in some cases produced rapid results. For this purpose either croton-oil or cantharides may be used. Thus, HORAND (218) advises painting the bald spots with croton-oil, covering with cotton, and wearing a hood over all. If ulceration is caused, the part is to be dressed with olive-oil. When healed, the application of the croton-oil is to be repeated. This method of treatment is to be continued till the hair grows. VIDAL (266) recommends the following: If the subject is a child, the whole head is to be shaved. If the patient is older, the scalp is to be shaved for half an inch around the bald spot. Now apply a blister, which should never be larger than a twenty-five cent piece, to each bald area, putting it on in the morning and taking it off when the epidermis begins to rise. If a large bleb forms, it should be opened. Powder the blisters with starch and cover with linen. If there are numerous bald spots, apply the blisters to them successively. If the hair does not grow after thorough blistering, repeat the operation as soon as the effects of the first application have disappeared. If blisters are not well borne, use sinapisms. When the hair begins to grow, shave, and rub in, every morning and evening, a lotion of

> Liq. ammon. . . . . . 3.
> Alcohol, ... 12.
> Decoct. foliæ juglandis, . . . 100.
> M.

For alopecia areata of the face he uses blisters in some cases. Generally he has the part shaved every day and uses frictions of tincture of cantharides, either pure or with one-fifth to one-sixth of the tincture of rosemary.

COURRÈGES (53), a believer in the parasitic origin of this disease, directs that the head be washed twice a day with a lotion of

> Chloride of ammonia
> Bichloride of mercury, . . āā 1.
> Water, ... 500.
> M.

or that the following ointment be applied twice a day;

> Yellow sulphate of mercury, . 1.
> Lard, ... 30.
> M.

This he does to destroy the parasite. To stimulate the growth of the hair he advises shaving once or twice a week. If there is only one patch or but two or three, these and their immediate neighborhood are to be shaved. Should there be many patches, the whole scalp is to be gone over with the razor.

COTTLE (52) directs that the affected area be painted with acetic acid until the scalp whitens, and then sponged off with cold water. This is to be repeated every three or four days; and in the intervals an ointment composed of three drachms of balsam of Peru in one ounce of benzoated lard is to be applied.

GUIBOUT (14) recommends shaving, frictions with camphorated alcohol, and bathing very often during the day with equal parts of camphor and sulphurous ether.

A. R. ROBINSON (33) speaks favorably of inunctions with a five to ten per-centum ointment of *chrysarobin* preceded by vigorous friction with soap and water and a rough towel. He later, 1887, recommends equal parts of croton oil and any other oil, reducing the strength if too much reaction.

LASSAR reports satisfactory effects in two cases from the use of *corrosive sublimate* lotions, followed by disinfectant oils such as salicylated or carbolised oil. In some cases of alopecia areata I have used a solution of corrosive sublimate, not on account of its parasiticide qualities, but solely for its stimulating effect. The strength of the solution used was three parts of the bichloride of mercury in one thousand parts of water, or say one grain and a half to the ounce. This was applied once or twice a day, and has given satisfactory results. One patient, a man aged thirty years, had upon his head a number of perfectly bald patches of alopecia areata, some of which had lasted for more than a year. He was given the bichloride lotion just indicated. In fourteen days lanugo hairs had begun to grow in all the patches, and in three months they were completely covered with vigorous though white hairs. He had used the lotion alone and shaved the patches once in two weeks after the hair had begun to grow. Unfortunately, this is the only case in which I have been able to watch the effect of this treatment.

In a few cases I think that I have produced benefit by using a pomade of *jaborandi*, made by boiling down the fluid extract to one half its volume and adding this to lard in the proportion of one of the jaborandi to four of the lard. This is to be thoroughly rubbed in twice a day. In the first case, of the recurrent type, the hair in two relapses returned under this treatment in the course of ten weeks. But it had no effect in preventing the formation of new patches. In the second case, occurring in a child with chorea, after the use of various other remedies for three months, the spots continually growing larger, the jaborandi was used, and in ten weeks the disease had ceased spreading, and the hair was growing on every patch. In a third case, affecting the mustache of a medical student, its use was followed in seven weeks by the appearance of new hairs in the patch, shaving being practiced at the same time. In these three cases, the return of the hair took place some six or seven months after the beginning of the disease. Pilocarpine may be substituted for the jaborandi in the strength of two or three grains to the ounce. Lanolin is a good excipient for it, as it penetrates the skin more readily than lard. It should be diluted with one or two drachms of castor-oil to the ounce to make it more fluid, and the whole may be perfumed with one or two drops of oil of roses.

MICHELSON (40), in *Ziemssen's Handbuch* (vol. II. pg. 139), speaks highly of the use of warm salt-water baths five per cent. salt) three times a week and of twenty-five minutes duration. In conjunction with this, twice a week the faradic current is to be applied to the scalp with a brush electrode for ten minutes. Upon the days when the baths are not used, the scalp is to be rubbed with a stronger aqueous solution of salt.

CHATELAIN (162 ap.) advises painting the patches, as well as for a short distance beyond their borders, with iodine one part, collodion thirty parts. This is to be renewed when it falls off. After a few applications the pellicle is to be pulled off, and with it the fine new hairs. If it causes a good deal of dermatitis, it may have to be stopped for a while. The rest of the hair is to be treated with an antiseptic solution.

MOTY (201 ap.) expects rapid results from injecting five or six drops of a solution composed of

℞ Hydrarg. bichlor., ... 4
  Cocain. hydrochlor., ... 2
  Aquæ destillat., ... 100
M.

In a small patch only one injection is made; in larger patches four or five injections are made about the periphery. Intervals of four or five days should be made between the injections. He finds that the hair will begin to grow in about three weeks.

QUINQUAUD (212 ap.) recommends washing the patches in the morning with soap and water, and applying

℞ Hydrarg. biniod., ...20
  "   bichlorid., ... 1.
  Alcohol, ... 40.
  Aquæ, ... 250.
M.

At night the parts are to be washed again, and then a solution of six per cent. of liq. ammon. in equal parts of balsam of Fioraventi and spirits of camphor. After eight days substitute tincture of pyrethrum for the ammonia, and so alternate every eight days. Every sixth day the following ointment is to be used:

℞ Chrysarobin,
  Ac. salicyl.,
  " boric., ... āā 2
  Vaselini, ... 100
M.

RAYMOND (213 ap.) saw recovery take place in four months by Moty's treatment, the hair coming in of normal color, and recovery beginning often at the periphery. Sometimes the method is very painful, but tolerance seems to be established after a time. He has seen one case recover after seven injections of pure water. He thinks the application of Bidet's vesicating liquid gives as good results as Moty's method. He has seen recovery take place in thirty to forty days after using Quinquaud's method. He himself recommends shaving the periphery of the patch and washing its surface with carbolic soap twice a week. Every morning the following wash:

℞ Hydrarg. bichlor., ...50
  Tinct. cantharidis, ... 25.
  Balsam of Fioraventi, . . . 50.
  Aq. Cologniensis, ... 150.
M.

is to be applied to the whole head and rubbed into the patches with a paint brush for one or two minutes. In the evening the spots are to be rubbed with a mixture composed of

℞ Ac. salicyl., ... 2
   β naphthol, ... 10
   Ac. acetic. crystal, ... 15
   Ol. ricini, ... 100
M.

This treatment should cause only redness and should cure in fifty days, the hair often coming in normal.

BUSQUET (161 ap.) speaks enthusiastically of the treatment by 33⅓ per cent. of essence of cinnamon or spikenard in ether. Others have tried it with no benefit.

MOREL-LAVELLÉE (193 ap.) recommends superficial scarifications of the patches before the application of an antiseptic ointment, and expects return of the hair after the third scarification.

MORROW (200 ap.) advises applying to recent patches either chrysarobin eight to ten per cent., or salicylic acid two to five per cent., every three or four days. In more severe cases equal parts of acetic acid and chloroform are to be used two or three times a week at first, and later at longer intervals. Between the applications of any of the foregoing this mixture:

℞ Ol. eucalypti,
   Ol. terebinthinæ, . āā ℥ ss., say 15
   Ol. petrolei crud.,
   Alcohol, ... āā ℥ j., say 35
M.

is to be used daily with massage for about five minutes.

BULKLEY (160 ap.) advises the application of pure carbolic acid to small portions of large patches at a sitting. CUTLER (166 ap.) recommends equal parts of carbolic acid, chloral, and iodine painted on the parts every few days. This has given good results in my hands.

As alopecia areata has a tendency to get well of itself, in its own good time, it is hard to determine how far our remedies are active in hastening a cure. So good an authority as Kaposi (19) has said that remedies have little if any effect on the disease.

Even after the damage to the hair has apparently been made good, it is advisable to counsel our patients to use some mildly stimulating lotion to the scalp, and to pay particular attention to the hygiene of the hair for some months.

CHAPTER VIII.

ATROPHIA PILORUM PROPRIA.

ATROPHY of the hair exists under three forms, namely: Fragilitas crinium, Trichorrhexis nodosa, and Aplasia pilorum propria. In all the hair-shaft is easily friable and splits, or breaks off of itself or upon the slightest traction. The three forms differ, in that in the first there is only a simple or compound cleavage of the hair; in the second, there is also the formation of nodular swellings along the hair-shaft, and a brush like breaking up of the elements of the hair; while in the third, the fracture occurs through the internodular portion.

FRAGILITAS CRINIUM.

SYNONYMS:—Scissura pilorum.—Trichoptilosis.—Trichoxerosis.

DEFINITION.—That condition of the hair in which it is more or less dry, and its shaft is split either at its end or in its continuity. It may be symptomatic or idiopathic.

1. SYMPTOMATIC FRAGILITAS CRINIUM.—This form is by far the most common variety of the disease, and needs little comment here. In the parasitic diseases of the hair—trichophytosis capitis et barbæ, and favus—the hair becomes dry, brittle, and broken off. This condition of the hair is always met with in these diseases and is a diagnostic symptom. In any disease of the scalp if of long continuance we meet with dryness and brittleness of the hair, and this is notably the case in seborrhœa sicca and eczema. In any general constitutional disease, as in fevers, phthisis, scrofula, and the various cachexiæ, in which there is a lowering of

the nutrition of the body, the hair sympathizes, loses its lustre and suppleness, and takes on the condition of fragilitas crinium.

ETIOLOGY.—The causes of this condition are easily discoverable. In the parasitic diseases, the fungus grows in and about the hair and its root, and by its presence causes a degeneration of the hair. In favus, a complete destruction of the hair-follicle and papilla takes place. In seborrhœa and eczema the hair becomes dry, because it is deprived of its proper lubricant, on account of alterations in the sebaceous matter, and because its nutrition is interfered with. In the general constitutional diseases, mal-nutrition is the cause of the fragility of the hair.

TREATMENT. — This will depend upon the disease which the condition accompanies, and will be given in the proper chapters.

2. IDIOPATHIC FRAGILITAS CRINIUM.—In this form, without any apparent disease of the scalp or underlying skin, and often without any general constitutional disease, the hair becomes dry, brittle and split. The cleft in the hair may be either at its free end, in the continuity of the shaft, or even within the bulb. When beginning at the free extremity, it may run for some distance up the shaft. When it begins at the exit from the follicle, the cleft extends for a variable distance, it may be for the whole length of the shaft. In this case and in the case in which the cleft occurs in the middle of the shaft, the filaments will either separate widely or hold together. When the split occurs at the end, the filaments will either separate from each other more or less, or will curl up upon themselves. The disease occurs most often upon the scalp, the beard being the part next most frequently affected. It is by far most commonly met with in the long hair of women. The affected hairs are

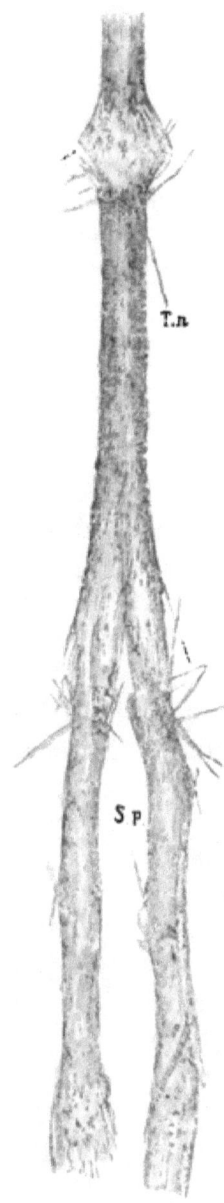

Splitting of the end of a hair. T.n, Trichorrhexis nodosa. S.p, Scissura pilorum. (Michelson.)

scattered here and there throughout the general mass of hair, which may appear normal or be somewhat drier than it should be. In some cases all the hair of a certain portion of the scalp is found broken up into filaments. Besides the splitting, the affected hairs may show no other abnormality, although they are often irregular and uneven in their contour.

In a case reported by DUHRING (280), which is unique, the beard was the region affected. In it the hair began to split within the bulb, and the process was attended by so great an irritation of the follicles as to cause follicular papules and pustules. "The hairs," he says, "were very bristly; sometimes they were of good length, sometimes short, sometimes thicker than normal, sometimes abnormally slender, sometimes straight, sometimes crooked. Sometimes they came out easily when pulled upon, or fell out of themselves, and sometimes they were quite fast in the follicle and broke off, leaving the root behind, when epilation was attempted."

PATHOLOGY. — In most cases the microscopical examination shows nothing of importance, apart from a little irregularity in the contour of the shaft, and the split at the end with its diverging filaments. The bulb of the hair may be normal or atrophied. In DUHRING'S (280) case, there were marked atrophy of the hair-bulbs and fissure of the hair-shaft beginning within the bulb, separation of the segments taking place at the bulb or at varying distances beyond it. The cortical substance presented a dry and brittle look in the narrowed portions, and a spongy luxuriant appearance in the thickened parts of the hair. The medulla was nowhere normal, but was present here and there in broken tracts.

ETIOLOGY.—The cause of this disease is yet undetermined. KAPOSI (19) would explain the splitting of the long hair of women upon the theory that it is due to the distance of the distal extremity of the hair from its source of nourishment—the hair-root or papilla. This is not an altogether satisfactory explanation, as it is by no means always the longest hairs which present the cleft end. Nor would this theory account for the disease as met with in the short hairs of the beard. GAMBERINI (59) would find its cause in want of due care of the hair and in allowing it to grow to excess. But as in the symptomatic form, there is some evident disease affecting the nutrition of the hair, and as in some of the idiopathic cases, there is an atrophy of the bulb; we can feel sure that the idiopathic form also is dependent upon some interference with the nutrition of the hair.

TREATMENT.—In all cases the scalp or underlying skin should be kept in good condition, as is directed in Chapter III. When the disease occurs in the long hair of women, besides attention to the scalp and the brushing and combing of the hair, the cleft ends should be carefully cut off just above the split. If the disease occur in the beard it should likewise be regularly clipped, and all rough rubbing and handling of it should be avoided. In shaving, we have a last resort by which the deformity may be removed, and possibly after a time the hair may grow normally: but this should be ordered only in very aggravated cases in women, as it is an heroic plan of treatment for them, and milder measures will generally be sufficient. In all cases we should strive to correct any thing that we find wrong in the health and well-being of the patient.

Trichorrhexis Nodosa.

Synonyms:—Trichoclasia ; Trichoptylose ; Clastothrix; Trichosyphilis; Nodositas crinium.

Definition.—Trichorrhexis nodosa is that condition of the hair in which nodular swellings occur along its shaft and the hair breaks easily, usually through one of the nodes, and exhibits a peculiar brush-like spreading out of the fibres of the broken-off hair, while the underlying tissues are normal.

This disease was first described by Beigel (276) in 1855, under the title "Auftreibung und Bersten der Haare." Wilks (309) recognized the disease in 1852, but did not publish an account of it until 1857. Wilson (311) asserts that he described it in 1849 in his book, "Healthy Skin," under the title of "fragilitas crinium," and proposed the name of "clastothrix" for it. In 1867 he exhibited to the Royal Medical and Chirurgical Society of London, specimens of what he called trichosyphilis of the beard and whiskers, which he considered as a disease distinct from fragilitas crinium. This was probably a case of trichorrhexis nodosa. Eichhorst (281) reported a case in 1858; since then various authorities have met with the disease and published cases of it. Kaposi was the first to propose for it the name of "trichorexis nodosa," which has been accepted by the profession as the proper name for the disease. The spelling, however, has been slightly changed, and it is now spelt either as "trichorhexis" or "trichorrhexis." We have adopted the latter style, as it is the one used by the best writers of the German school. The disease is rare. BULKLEY met with it but four times in eight thousand cases.

SYMPTOMS.—The disease comes on without any antecedent symptoms, the patient first noticing upon handling his beard that the hairs when pulled through his fingers, feel irregular and knotty, or else that the hair breaks easily and the beard looks ragged. If we examine the beard we find, along the hair-shaft, one or more whitish or grayish, shiny, transparent, nodular swellings, looking not unlike the nits of pediculi, but more oval, and evenly involving the whole circumference of the hair. Should the hair be red, the color of the nodes may be black. The number of nodes upon a single hair varies from one to five, and their diameter varies with that of the hair, being greater in the coarser than the finer hair. The nodes occur nearer the distal than the proximal end of the hair, and usually in the upper third of its length. The hair is exceedingly brittle and fractures upon slight traction, as on combing, or spontaneously, the fracture taking place almost invariably through a node, and the hair fibres of the broken ends separating like the hairs of a paint brush. Very rarely the fracture is internodular and the ends pointed. Sometimes the hair does not entirely break off, but the fibres splinter about the node, and the appearance presented is like that seen when two small paint brushes are pushed together, end to end. The fracture is usually transverse through the node, but sometimes, if there is an excessive amount of medulla present, it is longitudinal.

Sometimes the hair has an irregular contour and is frayed along its entire length. When many hairs are affected, there will always be many frayed-out ends, and the beard will look as if it had been singed. The hairs are usually firmly fixed in the follicles, and the disease exists for many years with no tendency to the production of alopecia.

The disease, in nearly all the reported cases, affected the beard, and therefore occurred in men; but it does occur at times in the scalp-hair and also in women. Thus W. G. SMITH (303) reported a case which occurred in the scalp-hair and was peculiar in that the fracture took place in the internodular portions, and the ends of the broken hairs were not frayed out. There was also one nodose hair found amongst the pubic hair. MICHELSON (40) believes that the condition is far more frequent in the scalp-hair than is supposed, and that the reason it is most frequently observed in the beard is because the hairs of that part are of larger diameter, and allow the condition to be more readily seen. On thin hairs the little nodes can be more easily felt than seen, indeed they can be hardly distinguished by the naked eye. He has met with the disease many times in cases of various forms of alopecia. DEVERGIE (279) has also reported a case that occurred in the scalp-hair of a woman. ABRAMOWITSCH (230 ap.) reports a case affecting the scalp-hair of a man. I have met with two cases in women, both of whom came to me on account of their hair being short. They were both debilitated. The disease has been seen on the axillary hairs and those of the eyebrows.

ETIOLOGY.—The cause of the disease is obscure. BEIGEL (276) believed that it was due to an accumulation of gas within the substance of the hair, which, exercising pressure from within, first gave rise to the bulbous-formed swelling, and ultimately burst the hair. WILKS (310) and WILSON (311) look to nutritive debility for an explanation of the malady. EICHHORST's (281) theory is, that the swellings are caused by fatty infiltration of the medulla, taking place at certain points, and that the splitting is purely accidental, the result of pulling, in brushing and combing the hair. He thinks it probable that a swelling of the medulla first occurs that causes a bulging outwards of the cortical substance till it finally bursts and breaks asunder.

So much for the earlier theories. Of more recent date are those of Schwimmer, Startin, Pye-Smith, Kohn, Pincus, Wolfberg and Michelson. SCHWIMMER (301) says that the disease is a nutritive disturbance, probably a tropho-neurosis. STARTIN (305) holds that it is due to a degeneration of the medulla, a consequent rapid accumulation of cells at one point, and eventually a bursting open of the hair.

PYE-SMITH (289) regards a gradual drying of the cortical substance, and a consequent loss of coherence of its constituent fibre cells, as the cause; this being followed or attended by the breaking up into a granular material and swelling of the cells of the medulla, and finally a rupture of the cortex, there being nothing left to hold it together. S. KOHN (286) believes that the process is analogous to the splitting of the long hair in women, and is to be considered as an atrophy of the medulla taking place at different points, or as a partial disappearance of the same. PINCUS (296) regards the disease as in part due to an interference with the nutrition of the hair and in part to a deficient action of the sebaceous glands. WOLFBERG (312) puts forth the view that repeated mishandling of the thick medullated hairs of the beard, as in violent rubbing in drying after washing, is sufficient to cause the disease in many people, and in this way he has designedly produced appearances microscopically identical with those of trichorrhexis nodosa. He believes this to be the only cause of the disease, and gives the following four reasons: 1. The anatomical appearances. The hair outside of the nodes is perfectly normal. The same appearances can be produced mechanically, and the disease occurs often in diseases of the beard which are itchy and hence cause it to be much rubbed. 2. Location. It occurs most always in the strong medullated hair of the beard, the node formation being favored by the central canal. These appearances are difficult of production in fine non-medullated hairs. 3. Because the explanations heretofore offered are not sufficient. 4. The obstinacy of the disease to treatment. This is easily explained, because it has consisted in rubbing the beard with irritating substances. SHERWELL (302), in his case, suspected a somewhat similar cause on account of his patient's habit of rubbing cologne water into his beard; but subsequently he abandoned the theory. MICHELSON (40) looks upon abnormal dryness of the hair-shaft as the cause.

We have thus ten different views expressed in regard to the cause of this singular affection. To choose any one as the true cause, would be unwise at present. The disease is not parasitic. This I believe is the unanimous testimony. CHEADLE and MORRIS (278) have, it is true, reported a case to which they gave the name, amongst others, of trichorrhexis nodosa, but it was so different in its appearances from the disease in question, that we should rather call it "tinea nodosa." It does not seem to depend upon any diathesis, nervous or otherwise. ANDERSON (274) has reported one case or otherwise. NEWTON (244 ap.), who studied the disease in two cases, is convinced that it is a pure neurosis. ANDERSON (274) has reported one case which shows an hereditary influence. The disease in his case was congenital, or nearly so, and could be traced back as far as the great-grandmother. Those members of the family who had what they called "weak hair," were almost invariably of dark complexion. He gives the following "family tree," those marked with an asterisk (*) having "weak hair." (By this term is apparently meant hair that splits easily.)

Mr. L.	Mrs. L.*

Mrs. B.*

Robert, James, John,* David,	William,	Thomas,* Samuel.

Marion, James,* Jane,* John, Mary,* Margaret,* Thomas.*	John, William, James,*	Thomas, William.*

Thomas, James,* Walter, Margaret,* Mabel,* Maud.

I have met with but two cases of the disease. Both occurred in men and in the beard; and both patients were in the habit of handling the beard. As far as these cases are concerned I would be inclined to adopt Wolfberg's theory. The fracture of the hair is probably always due to mechanical causes.

PATHOLOGY.—BEIGEL (276), who first described the disease, found minute globular bodies in the shaft of the hair which he took to be air-globules. WILSON (311), in his case of syphilitic degeneration of the hair, which was probably a case of the disease we are now discussing, observed "the diseased portions of the hair looking as if composed of a dark cylinder enclosed in a transparent envelope. Closer examination showed that the fibrous structure of the hair was lost, and its place occupied by a dark gummous-looking substance. The essential peculiarity of structure of the diseased hair was arrest of development of the fibrous portion at its cellular stage, the dark cylinder which formed the bulk of the diseased part being composed of large and small nucleated cells commingled with pigment matter, irregularly formed air-cells and fragments of crystaline substance." EICHHORST (281) saw "in hairs which were examined dry or in glycerine, that the individual fibres of the cleft hair were in places covered with fine glistening kernels. As these were not seen in preparations in Canada balsam and in turpentine, and as they were colored very black by hyper-osmic acid, they were doubtless fat drops. The further the swelling had proceeded, the less were they seen. The nodes were exclusively on the medulla, the cells of which could be traced into the spindle-formed broken-up nodes, and appeared unaltered. But when liquor potassæ and hyper-osmic acid were used, the medulla cells were seen to be overladened with drops and masses of fat." WILKS found, interspersed in the broken fibres, a few dark granules. Such were the findings of the earlier investigators. Now let us turn to the more recent investigations and see what they teach us. PYE-SMITH (298), in 1879, reported upon certain specimens of the disease as follows: "First stage; hair swollen in fusiform nodules at intervals of a few lines. Later, the cortical layer has begun to split up and to separate transversely. In the more advanced cases a fine, granular material oozes out from the medullary portion. Lastly the hair breaks across leaving the brush-like broken end formed by the frayed-out cortical fibres. Under a high power the exuded material appears as a uniform, finely granular substance. This has some resemblance to fish roe and might be mistaken for sporules. There is no vegetable substance present. It is not probable that a development of gas causes the hair to split. Apparently the cortex becomes more and more dry, its constituent fibre-cells become less coherent, the medulla cells break up into this granular material and swell, till the rupture of the cortex is complete, and nothing remains to hold the hair together. There is no fatty degeneration present."

S. KOHN (286), in 1881, writes: "The little clear bodies on the outer filaments of the split nodes, which are not altered by different methods of handling, shut out the parasitic nature of the disease, because they are only occasionally met with, never in the nodes nor other parts of the hair, and because they are out of all proportion to the disease. They are merely deposits from the outer air. The theory of atrophy has support in the following microscopic appearances: In treating the hair-shaft with concentrated caustic potash or acetic acid, a swelling of the medulla is caused. This swelling we have never found in a node, but only in the beginning stage of the disease when no node could be seen. In those places where there are nodes, there is no medulla. The nodes under the microscope appear as brush-like masses of split-up fibres. The fibres are partly directed outwards, and in part, the opposite fibres are shoved into each other. If the hair is broken off, we will find the remaining portion looking like a brush with its fibres spreading outwards. Between the stage of swelling of the medulla and the stage in which the broken node is met with, there is a stage in which holes are formed in the cortex. One sees plainly that the single cells of the cortex are loosened and beginning to fall apart. Transverse sections of the nodes show a part of the medulla replaced by air." MICHELSON (40) says, in Ziemssen's *Handbuch*, Vol. XIV., "In some places on the hair there is the appearance as if two brushes were pushed end to end; at a little distance from them the shaft is split into several fasciculi, whose ends are frayed and present a broom-like appearance. As a rule, the partial longitudinal splits are very nearly in the middle of the shaft, and if any medulla is present, it will be destroyed. Sometimes, however, the hair splits into a larger fasciculus embracing the medulla, and into one or more smaller fasciculi. The affected part of the hair, already deprived of its cuticle, bursts, and perhaps, with its fibres spread out in all directions, forms an admirable net to catch atmospheric dust. All sorts of particles are deposited, and sometimes caught so fast that they can not be dislodged by any mode of handling. Specially striking are the dark, sharp-cornered coal-particles, and the blue ultramarine grains. A good deal of air finds lodgment also in the splits and holes. In the microscopical examination, this air will be more or less rapidly expelled from the hair by the fluid medium, and the appearance thus caused was the foundation for Beigel's theory of the development of gas within the hair, as the cause of the disease. We have never found any parasite present. Many hairs in the neighborhood of the affected hairs present an atrophic appearance." HANS HEBRA (16) describes the hair roots as "presenting a shrunken appearance, and approximating to the size of the shaft." GIOVANNINI (237 ap.) says that the hairs have a triangular contour.

From consideration of these various microscopical appearances, we learn that there is first a swelling of the shaft, forming the node; then a transverse fracture, taking place through the node, combined with a splitting up of the hair-fibres; and lastly a lesion of continuity occurs, and a brush-like end is presented by the broken hair. Further, in the node there is no medulla present. Michelson, Pye-Smith, and Wilson seem to find the beginning of the disease in the cortical portion of the hair; while Kohn, and Eichhorst, regard degeneration of the medulla as the first step in the process. The hair-roots are either unchanged or else slightly atrophied.

LESSER (288) has reported a case of "ringed hair" which presented the appearance of trichorrhexis nodosa, excepting that, in his case, the fracture always took place through the constricted portion, which he believes to be the diseased part. The beginning of the fracture in his case was marked by a transverse breaking of the cuticle, which often ran in a circular manner quite around the hair, and the raised epidermis formed a sort of collar. In a case of W. G. SMITH (303), the fracture likewise occurred in the internodular portions. These last two cases were doubtless instances of aplasia pilorum propria. In parasitic diseases of the scalp, the appearances of trichorrhexis nodosa are not infrequently observed.

TREATMENT.—Thus far all modes of treatment have proved very unsatisfactory. All sorts of applications have been made to the affected parts, generally of a stimulating character, and particularly the various forms of mercurials, but without curative effect. GAMBERINI (59) recommends either bathing the part with a lotion, composed of 15 grammes of subcarbonate of potassium to 150 grammes of dilute alcohol (say 3 drachms to 4 ounces); or using inunctions of tannic acid or oil of cade. SCHWIMMER (301) advises that an ointment composed of a half gramme (gr. vii) of oxide of zinc, 1 gramme (gr. xv) of washed sulphur, and 10 grammes (3 ijss) of simple ointment, be rubbed in morning and evening.

But the chief reliance is placed upon shaving, with the hope that it will stimulate the nutrition of the hair, and that after a time the hair will grow in a proper manner. WOLFBERG (312) founds his treatment upon his theory, and advises that the hair be left entirely alone. As other treatments have failed of doing any good, it would seem worth while trying this expectant plan in cases in which handling the beard is a possible etiological factor.

APLASIA PILORUM PROPRIA.

SYNONYMS:—Aplasia pilorum intermittens; Monilethrix; Moniliform hairs; Cheveux moniliformes.

This is the third variety of atrophy of the hair, that has often been mistaken for trichorrhexis nodosa, as it too presents nodes along the hair-shaft. It differs, however, from that condition principally in two particulars, viz.: the nodes are here the normal parts of the hair, while the internodular portions are the abnormal parts; and the fracture takes place through the internodular parts.

SYMPTOMS.—The disease is congenital in nearly all cases, and appears in infancy. A few cases have come on later in life. Usually the child is born with apparently normal hair, but in the course of a few weeks the hair breaks off either over the whole head or in patches, and the scalp assumes the appearances of keratosis pilaris or of trichophytosis capitis, being covered by small, scaly, elevated cones, and it is somewhat reddened. Pustules may form on it. Sometimes complete baldness results, and many cases of congenital alopecia are doubtless due to this disease, the scalp having been destroyed, as evidenced by the presence of many small cicatricial points scattered over it. From the little scaly hair-cones the short, stubby hairs protrude. They look as if scorched, and some are bent. They are very brittle, and easily break on slight traction. They are seldom longer than a quarter of an inch; they may present simply as black points. If examined with care many of the hairs will show fusiform swellings with contractures between, through which fracture has taken place. After a time partial recovery may take place, but it is never complete. All the hairy regions of the body may be affected, and there may be a general keratosis pilaris.

Apart from the loss of hair, the subjects of the disease may be in good health; sometimes they may be in poor physical condition.

ETIOLOGY.—In many cases the disease is inherited, and seems to be a deformity of the skin rather than a disease, just as is icthyosis. It is also prone to descend in the same sex, and to affect more than one member of a family, a peculiarity shared by several other dermatoses. Thus SABOURAUD (243 ap.) traced the disease back to a great-grandfather and found evidence of seventeen cases in the family. LESSER (241 ap.) reports a case of a man and two sons, and a history of eight cases in the same family in two generations, the descent being direct and through the males. HUDELO (239 ap.) reports the case of a girl whose mother had the disease, that was hereditary on the mother's side back to her great-grandfather. A sister of the patient had weak hair. TENNESON (251 ap.) met with three cases, all girls, in a family of five, the father of whom had the disease, as well as his brother and sister; and PAYNE (246 ap.) reports two brothers with the disease. Isolated cases do arise, as well as those in which there is no history of inheritance. We do not know the cause of the disease.

PATHOLOGY.—Under the microscope the nodes on the hair show better than when viewed by the naked eye. Indeed, in some fine hairs the nodes are only seen by the microscope. The hairs will be found to have on them at regular intervals alternate strictures, or narrow places, and swellings. The latter are about 1 mm. long, fusiform in shape, of darker color than the narrow portions, and about three times longer and three times wider than they are. The difference in color is due to the fact that the constricted parts contain neither medulla nor pigment, and may consist of the cuticle layer of the hair alone. The hair-bulbs are atrophied. It has been noted by BEATTY and SCOTT (234 ap.) and by SABOURAUD (249 ap.) that the hair examined in sections of the skin shows a deformity of the Huxley's layer of the root-sheath at the part corresponding to a node. The nodes are all along the hair from root to point. Fracture takes place through the internodes, and frayed-out or brush-like ends may or may not be found. SABOURAUD found that the constrictions formed at two days' intervals, and JAMIESON'S (240 ap.) experiments led to the same result.

TREATMENT thus far has been of none effect. Stimulation might be tried, but the prognosis is bad.

Under the name of *end atrophy* CROCKER describes a case, reported by MCMURRAY (242 ap.), in which the distal ends of the hairs were bulbous and of lighter shade than the rest of the hair.

Besides the foregoing atrophic conditions, there are certain other somewhat allied deformities that are here noted.

There is also a condition of the hair called "*Phagmesis,*" in which feathers adorn the body instead of hair. T. ROBINSON (299) cites a case of this nature, which occurred in a boy who was exhibited in Bremen, and was reported upon in *Bauerle's Magazine for* 1831. The boy's head is said to have been covered with feathers in place of hair.

FERBER (282) reports two peculiar cases of change in the texture of the hair. Both patients were nervous individuals, and their hair in a few hours would change from being soft and curly, to become straight and bristly. The change followed, in one case, nightly pollutions, and in the other any special deterioration of the always feeble health. After a time the hair would return to its usual condition.

NODULI LAQUEATI is that condition of the hair in which it seems to tie itself into knots. The hair is usually dry and curly. According to MICHELSON (40), the condition is common in trichorrhexis nodosa. The loop of the knot forms an excellent net to catch dust and flying particles, and hence under the microscope, all sorts of dust elements are found with the hair. A case of this sort is reported by L. D. BULKLEY (277), which occurred in the pubic hair of a man who was troubled with itching and sweating of the genitals. The hair looked as if invested with the nits of pediculi, but the microscope showed that the appearance was due to the presence of a double knot on each hair, composed of several turns.

I have recently met with a case of undoubted trichorrhexis nodosa of the beard, in which this knotting of the hair occurred. The patient was of unsound mind, and kept constantly pulling at his beard, and to this habit was ascribed the knotting. MICHELSON ascribes the disease to improper combing, and pulling of the beard through the fingers.

CHAPTER IX.

HYPERTROPHIA PILORUM.

SYNONYMS:—Hypertrichosis; Hirsuties; **Trichauxis; Polytrichia; Dasyma; Dasytes;** Trichosis hirsuties; Poils accidentels (Fr.); **Superfluous hair;** Hairiness. (Eng).

DEFINITION:—A growth of hair which is either abnormal in amount or occurs in places where, normally, only lanugo hairs are present. While it is normal for a man to have a beard from four to six inches in length, it would be abnormal for him to have one reaching to his feet. Again, while normally the hair grows thick and strong upon the cheeks, chin, and upper lip of a man after puberty, should such a growth occur upon the face of a woman it would be abnormal.

SYMPTOMS.—Hypertrichosis may be general or partial, congenital or acquired. Of these the general form, *hypertrichosis universalis*, is very rare, while the acquired form, or *hypertrichosis partialis*, is the most frequent, and is familiar to every one in the cases of the unfortunate bearded women.

Hypertrichosis universalis is generally congenital, and, in spite of its name, does not affect the whole body. Hair never develops, even in this disease, in places in which normally no hair is found. It is not met with, therefore, upon the palms of the hands, the soles of the feet, the backs of the last phalanges of the fingers and toes, the inside of the labia majora, the prepuce, or on the glans penis. Subjects of this malady are usually born covered more or less thickly with hair, which may be light or dark in color. This continues growing longer, coarser, and darker till it reaches its full development. As a rule the long hair covering the body is fine, resembling more the hair of the head than of the beard, as is also the case with the hair on the face of these persons. It follows a definite direction in growing, and this is away from certain well-defined centres. Thus on the back it grows on each side downward and outward from the spinal column; on the forehead away from the middle line, following the lines of the eyebrows; on the face, also, from a line running down the middle. With this excessive growth of hair there is usually combined a deficiency of teeth, specially marked in the upper jaw. MICHELSON (353) has seen a family which was very hairy, in many members of which there was a defect of all five back teeth, the alveolar processes for the same being wanting.

These homines pilosi are met with in all parts of the world. Thus we have records of the Kostroma family from Russia, a father and son. They first attracted notice some twelve years ago, and were described in a number of European medical journals. The father died about four years ago in Paris. The son was on exhibition in this country in 1886. Nothing is known of the parentage of the father, nor have I found any record of the boy's mother. An excellent likeness and description of the father is given in the last volume of Ziemssen's *Handbuch der speciellen Pathologie und Therapie*. The son was in 1886 sixteen years of age, and looked like a well-developed boy, though somewhat under the average height. He appeared to be muscular, and was active and energetic in his actions. He seemed intelligent, and certainly showed acuteness in the rapidity and correctness with which he gave me change for a dollar bill when I bought his photograph.

He spoke his native language with great rapidity, and has picked up some German and English words. His agent said that he was docile, and his health was good. His head was covered with a luxuriant growth of fine, glossy hair of blonde color, some six inches long. This extended further down on the neck than is usual. The scalp was normal, white and soft. Coming forward,

The Dog-faced Boy.

the hair grew well down on the forehead, and then continued over the whole face, though on the face it was finer and lighter in color than on the head. The facial hair did not resemble in texture the usual hair of the beard, even where it grew in the places ordinarily occupied by that growth in men, but was much softer. Upon the upper lip there was a space, occupying about the middle third, where the growth was very scanty. Under the eyes the hair also was absent for a space of about half an inch. Otherwise the whole face was covered with long hair, growing from the inside and outside of the nose, and continuing all down the neck. There was also a luxuriant growth of hair from the inside and outside of the ears. Under each eye there was a group of three or four black hairs. The hair of the face was some four inches long. Upon the body the most remarkable growth was down the spinal column, where the hair stood out not unlike a horse's mane. The rest of the trunk and the extremities were completely covered with hair, but not very much more so than in not a few other hairy people. The backs of the hands presented nothing remarkable. On the body the hair was exceedingly fine and delicate, and more fluffy than that on the head and face. There was a cast in his left eye, and he was near-sighted. He had only five teeth, two upper canine, and two lateral, and one middle lower incisor. The alveolar ridges show no sign of there ever having been any more teeth. The teeth he has are badly shaped and discolored. His father is said to have had no teeth till he was seventeen years old, and then only four in the lower and one in the upper jaw. The London *Lancet* in 1873, reported the boy as having four incisor teeth in the lower jaw, so he must have lost one.

Other instances of universal hypertrichosis have been reported. Thus the case of Barbara Ursler is cited by STRICKER (370) as occurring in the seventeenth century. This woman's whole body was covered with blonde, soft, curly hair, and she had a thick beard reaching to her girdle. In a book published in 1642 and entitled *Aldrovandi Monstrorum Historia* there is an account of a hairy family consisting of the father aged forty, a son aged twenty, and two daughters aged eight and twelve. They came from the Canary Islands, and were covered with hair, excepting that the daughter's lips, nose, neck, breast, and hands were smooth. In 1851, CHOWNE (328) reported a case of universal hirsuties occurring in a Swiss woman twenty years of age. Her body was covered with hair excepting on the breasts and chest, which were free of the growth. In BEIGEL's (44) book on *The Human Hair*, accounts of several cases of this deformity are given, *viz.:* that of Julia Pastrana, a Spanish dancer; and of Shewe Maon and his daughter Maphoon, in India. The latter's second child was hairy like its mother. In the father and daughter there was an absence of the canine and molar teeth. But it is useless to multiply examples, as the foregoing cases are sufficient.

Of *partial congenital hypertrichosis* we have an immense number of examples. This condition is apt to be of the nature of nævus. It must be held in mind that the distinction between a localized hypertrichosis and a nævus is made mostly upon the color of the underlying skin. In the former case the skin is perfectly normal, while in the latter it is pigmented and may be otherwise altered. Thus we have, in the *Lancet* of 1869, an account of a Mexican woman who had a nævus pilosus extending, like a pair of bathing trowsers, from the umbilicus anteriorly and the sixth dorsal vertebra posteriorly, to about half-way down the thighs, covering the buttocks. CUMMIN (329) mentions the case of a lady who was noted for the beauty of her face, whose body from breast to knee was covered with a profusion of black, thick, bristly hair. WALDEYER (83) reports the case of a girl nine years of age, who had a lock of hair running from the first to the fourth lumbar vertebra, and a smaller one from the third to the fourth cervical vertebra. These localized and partial cases of hypertrichosis are most frequently met with in the sacral or lumbar region, and not infrequently are associated with spina bifida. ORNSTEIN (281 and 282 ap.) says that sacral hypertrichosis is common in Greeks, and reports two cases of tails in Greek soldiers, one a quarter of an inch long and cone shaped, the other not quite so long and stumpy.

Partial acquired hypertrichosis is more common than the congenital variety, and takes the form either of an excessive growth of hair in regions where it is usually found, or of the development of hair in regions usually hairless or only provided with downy or lanugo hairs.

The following cases are instances of excessive growth and precocious development. CHOWNE (328) speaks of a boy, eight years of age, who had the whiskers of a man. BEIGEL (321) has seen a six-year-old girl with pudenda like a twenty-year-old woman, both in shape and hair. This form is called "Hetero-chronie of Hair" by BARTELS (318). As cases of excessive growth may be cited the following: LEONARD (64) mentions the case of a man in his neighborhood whose beard measured seven feet six and a half inches in length. Other instances of excessive length of beard are met with in medical literature, such as that of the carpenter at Eidam, whose beard was nine feet long, and who was accustomed to carry it in a pocket devoted to the purpose; and that of the Bürgermeister of Braunau whose beard reached to the ground. WILSON (378) met with a lady who was five feet five inches in height, whose hair, when she walked, trailed three or four inches on the floor. Many men have an excess of hair upon the chest and shoulders. Hair is generally more developed upon the forearm than upon the upper arm, and upon the leg than upon the thigh.

As men grow old they are apt to have long hair grow from the nostrils and the ears. These are instances of the growth of strong hair where normally lanugo hairs alone are present. But these cases are interesting only as curiosities and as subjects of study.

BEARDED WOMEN.—The growth of the beard in women is the form of hypertrichosis which concerns

Bearded Woman.

us most, as it is the deformity which we will be called upon to cure. It has been called heterogenic. As women grow old, especially after they have passed through the climacteric period of middle life, a slight mustache or a few straggling dark hairs on other parts of the face often appear. These will seldom annoy them much, as they are accepted as evidences of advancing years. The case is very different when a young woman is afflicted with a beard, and many of the patients who apply for relief from their facial hair are between twenty and thirty-five years old. The hair generally begins to grow so as to be noticeable at about the eighteenth year of age. To get rid of the trouble the tweezers are first resorted to, and this only makes matters worse. Then depilatories are tried which have but a passing effect, and some of them leave bad scars. Sometimes burning is attempted, and as a final refuge the razor is used. All the time the hair grows coarser and more abundant. Were this all, though it would be bad enough, these cases would not so greatly need our aid. The deformity is only the beginning of evils. These women shun company, keep themselves shut up all day, their health deteriorates, and, constantly brooding over their misfortune, they are prone to become hypochondriacal and melancholic. Anyone devoted to dermatological practice must have seen these nervous, sensitive women, whose health is broken and spirits depressed on account of, it may be, no very formidable facial hirsuties. The amount of hair presented by these cases varies. Perhaps the commonest growth is a mustache alone. In most of my cases the hair has grown thickest and coarsest under the chin and upon the front of the throat. It is rare, even in the best developed cases, to have much hair under the lower lip. Sometimes the growth is as complete, as heavy, and as coarse as is met with in men. An excellent account of such a case is described by DUHRING (330). The skin of many of my cases has been coarse, muddy, greasy, and studded with acne.

From time to time cases of *transitory hypertrichosis* have been reported. This has been noticed during the treatment of a fractured limb, the hair being much more prominent upon the part that has been kept quiet and warm. In some of these cases the increase is probably more apparent than real, the hair not having been rubbed off by friction. Likewise, after injury to nerves the hair sometimes becomes hypertrophied, only to fall off after recovery. Continued irritation of a part, as by blisters, may stimulate hair-growth which may or may not be transitory. The most interesting of this group of cases are those instances of hirsuties occurring during pregnancy or amenorrhœa, and disappearing again some months after parturition. WILSON reported a case of delayed appearance of menstruation in which hair grew upon the face. After the menstrual function was established, the hair ceased to grow and gradually fell off. THIN (291 ap.) and GOTTHEIL (263 ap.) have reported cases of hirsuties that disappeared after conception.

ETIOLOGY.—The cause of hypertrichosis is very obscure in some of its forms, while in other varieties we can more readily discover it. In universal hirsuties heredity plays an important part. Such instances as those of the Kostroma father and son; of Shewe Maon, and his daughter and grandchild, and others like them, attest this fact. But hereditary tendencies will not explain the first appearance of these congenital cases. VIRCHOW (375) endeavored to account for them upon the theory of nervous influence, founded upon the fact that in the Kostroma people the lack of development of the teeth and jaws was in the same zone of nervous influence as was the over-development of the hair on the forehead, nose, cheeks, and ears; these regions all being supplied by branches of the trigeminus or fifth cranial nerve. PARREYDT, quoted by GEYL (56 ap.), thought that those who had large teeth were apt to have a profuse growth. This theory is not borne out by facts. Atavism is another theory to account for these cases. When we remember that the fœtus is completely covered with hair of some length and coarseness, though not coarse, there is good ground for believing that UNNA's* theory of congenital hypertrichosis is right, namely: "That it is due to a persistence of the fœtal or primitive hair; the change of type between the primitive and permanent hair not taking place." While normally the change in type does take place and the primitive hair of most of the body is replaced with lanugo hair, in some individuals for some unknown cause the primitive hair remains, grows stronger, and we have the homines pilosi.

At the present time it is hardly necessary for us too gravely to discuss the theories of maternal impressions, fecundation of the human female by a hairy animal, and the like.

The cause of acquired hirsuties is, in some cases, not far to seek. Heat and moisture will apparently increase the growth of hair, just as they favor the growth of vegetable life. Thus the hair has grown luxuriantly under the stimulation of poultices, and on the limbs when confined in a fracture box. To these factors must be added an increase of the flow of blood to the part, which will stimulate hair-growth independently of heat and moisture. At least PRENTISS' (134) case of hair growing more luxuriantly and coarser under the use of pilocarpin, which causes hyperæmia of the skin, would seem to indicate this. Hypertrichosis following injury to nerves is probably dependent upon vaso-motor disturbances. The growth of hair upon exposed parts, as upon the arms and chests of laboring men, sailors, and the like, is due to the local irritation of the sun and wind.

Now we come to the more obscure cause of facial

---

*Ziemssen's *Handbuch*, Vol. XIV., p 56.

hirsuties in women. In the majority of cases it will be found that the deformity is hereditary on the female side, the mother, maternal aunt, or maternal grandmother having hirsuties. To account for this, numerous hypotheses have been formed. Probably the one most generally accepted is that it is in some way connected with derangement of the uterus and appendages. Because in some bearded women there has been some evident derangement of the sexual organs, it has been affirmed that some similar derangement is present in all, just as many of the laity believe that the too free use of alcohol is *the* cause of rosacea. In the cases I have met with, the majority were as free from uterine trouble as the rest of their sex. While it is true that some of these women are of masculine build, and have a masculine voice, most of them do not exhibit these characteristics. The heaviest bearded of my female patients was the mother of three children, and this experience is not unique. In some cases, however, there does seem to be some relation between the reproductive organs and the growth of the beard. Several instances illustrative of this have been given above. Recently, I have met with a case of a woman with a dark but not very heavy beard, which began to grow about one month before the birth of her fourth child. The appearance of hair on the face of women who have ceased to menstruate would suggest such relationship. It is a common idea with women who have a good deal of hair on the face, that they have brought it on themselves by their endeavors at removing a very slight, hardly perceptible growth. This is an error. Though undoubtedly shaving or cutting the beard may somewhat increase its coarseness, and to a certain extent stimulate its growth, still it cannot make new hairs grow. It is exceedingly likely that even if these women left the hair entirely alone it would in time become of itself coarse and dark, though it would be more silky.

The question of the inheritance by the daughter of the physical character of the father or male ancestors is worth investigating. By this I mean, whether she resembles the father in her general build more than the mother.

We may sum up the evidence on the etiology of facial hirsuties in this way: While at times there appears to be a relation between the uterine, or, more properly, the menstrual function, and the growth of hair on the face, shown by a decrease or deficiency of the first, and an increase of the second, still in the majority of cases no such relation is discoverable, and it must be viewed as a deformity or freak of nature.

An interesting study of the relation between hirsuties in women and insanity was made by HAMILTON (339). He regards hair-growth on the face in women as the inevitable result of the overactive and continuous exercise of the uterine and ovarian functions. He believes it to be of neuropathic origin, connected with disorders of the fifth cranial nerve; and when it occurs upon the face of an insane person it is indicative of an unfavorable form of insanity, especially if the subject has not reached middle life. SHAW (364) speaks of the hair of chronic lunatics changing from fine to coarse and increasing in quantity, specially during or after violent outbreaks of insanity, and ascribes it to the effect of long-continued increase of temperature of the scalp, leading to a greater activity of the hair-bulbs.

TREATMENT.—The treatment of hypertrichosis is simple and efficacious, though laborious and tedious. For general hypertrichosis we can practically do nothing. This, not because we cannot destroy the hair so that it will not grow again, but because of the great amount of time it would take to destroy it. Happily, a superfluity of hair on the body does not incommode the bearer of it, nor, in most cases, do him any damage. One case, however, is upon record where the consequences of too much hair resulted in serious damage. It is that of a woman in old times whose face was very beautiful, and who made an advantageous marriage. But the husband was disgusted to find her body covered from breast to thighs with a profuse growth of stiff coarse hair, and upon the strength of this he was divorced from her on the next day.

The only form of hirsuties which urgently calls for relief is that occurring upon the face of women. Until recently there was no cure for this, but in 1875 MICHEL (380), of St. Louis, devised a method for removing the hairs in trichiasis by means of electrolysis, which was taken up by HARDAWAY (341), of the same city, for the removal of superfluous hair. PIFFARD (29) in 1876 spoke of the removal of superfluous hair from hairy nævi by this method; and LEON LE FORT, in his edition of Malgaigne's "Manuel de Médecine opératoire" in 1877, published his method of destroying hair by electrolysis, which he had used since 1875. As MICHEL published his method in 1875, to him belongs the priority. The question is often asked: "Is the removal, by this method, permanent?" This question may be answered, "It is, without a shadow of a doubt." The object being to destroy the papilla, and that being very small and often placed at an unexpected angle to the surface of the skin, it is not possible always to accomplish this at the first attempt. The amount of success on going over the face the first time will vary with the operator, and, according to his skill, there will be a return of from twenty to fifty per cent. of the hairs removed. It will be necessary, therefore, to go over the face a second or a third time, but then there will be no return. At times, after the dark coarse hairs have been removed there will be found a number of finer and lighter hairs. This appearance is due partly to the uncovering of these hairs, and, partly it may be, to lanugo hairs becoming stronger under the stimulation of the operation. Cases occurring in young women are much more tedious in their treatment than those in women past the climacteric, because while in the latter the growth is limited in amount, in the former new hairs form from new follicles, just as in a young man's beard, and it is impossible to tell when the process will stop. It is well always to tell these patients that they must be prepared for a long course of treatment, as only the hairs that have appeared can be destroyed, and we have no means of preventing the formation of new hairs; that these new hairs do not form in the old follicles, but are simply developed in time from those already in the skin. They will have to be treated from time to time during a number of years until all the hair papillæ have produced their hairs and these have been destroyed. In most cases, with proper care and the use of a fine needle, the amount of scarring will be very slight, amounting to nothing more than fine punctate cicatricial spots. In some peculiarly irritable skins it is very difficult to prevent the formation of plainly visible scars. If the proper conditions are not observed, the operator must expect to produce a good deal of disfigurement.

The amount of pain experienced by the patient will vary greatly. Some women will complain bitterly of a current of half the strength that other women will bear with ease. Certain parts of the face are far more sensitive than others. The most sensitive points, according to my experience, are over the ridge of the lower jaw on each side of the chin, and upon the upper lip. On the whole, the pain does not amount to much. After a time the skin seems to become tolerant of the action of the current and the patient no longer

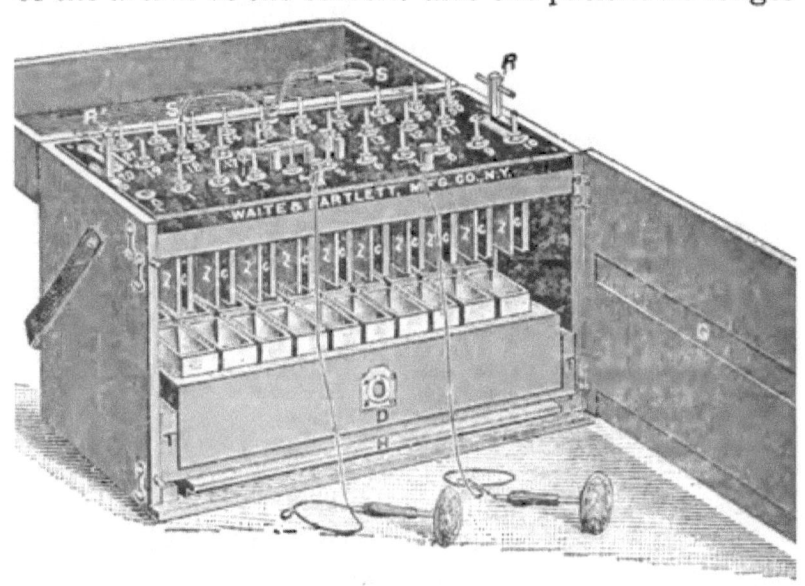

Twenty Cell Galvanic Battery.

complains. Hyperpigmentation may be produced by

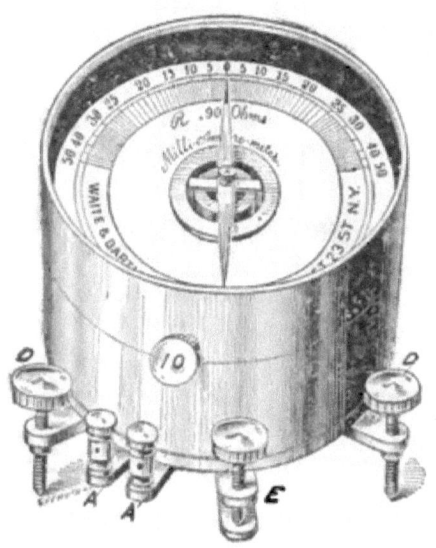

the operation. This is a very rare complication, and is only mentioned by way of warning. I have seen it in two cases. Instead of the redness which is always produced about the point of puncture fading away in a few days, it gives place to a dark-brown stain which may remain for a number of months.

The instruments necessary for the operation are a good twenty cell zinc-carbon (galvanic) battery, a sponge electrode, a proper needle-holder, a fine needle, a pair of epilating forceps, and, if the operator's eyes

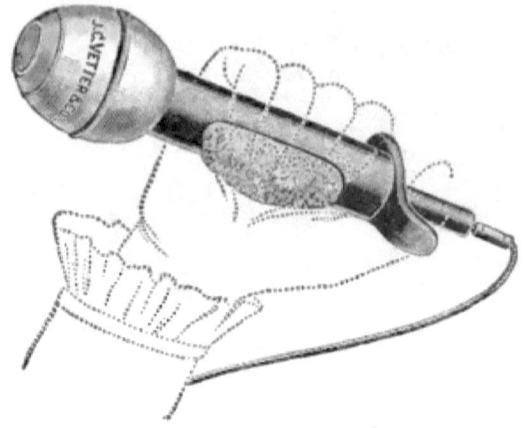

Leviseur's Sponge Electrode.

are not good, a lens of low power. A galvanometer is not essential, but is an aid to exactness in working.

Needle-Holder.

Any sponge electrode will answer. There are various patterns of needle-holders, any one of which may be

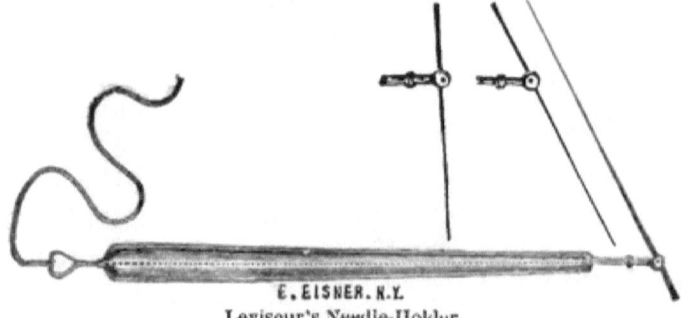

Leviseur's Needle-Holder.

used. I prefer one which is not longer than three or three and a fourth inches, with a diameter of about three-sixteenths of an inch. It should be long enough to be held with ease, and not too long to be readily manipulated. If the woman has a large bust, a high chest, and a short neck, a short holder will be found a great convenience when working under the chin. There is a form of needle-holder in the market furnished with an attachment for cutting off the current and letting it flow again after the needle is in place in the follicle. This I cannot recommend, as it causes a very uncomfortable shock to the patient, much more than that experienced in using the method detailed below. Another needle-holder, invented by LEVISEUR (272 ap.), is ingeniously arranged so that the needle may be placed at varying angles. The most essential instrument is the needle, and for the best success this must be carefully chosen. At first the finest cambric needle was used, and it did fairly well. HARDAWAY (342) recommends a needle made of iridium and platinum, claiming that it will follow the direction of the hair-follicle and hit the papilla more surely than the steel needle will. I have used these needles, but have not found them so satisfactory as the steel "broaches," which are fine-pointed steel needles used by jeweller's and obtained at any jeweller's supply store. They come in many grades of fineness. It is advisable to have two grades, No. 5 for the coarser hair, No. 7 for the finer, and for the lip. Care should be exercised in selecting them, as they are so fine that their points are sometimes broken while in the packages. A lens is generally not needed. PIFFARD (359) has invented a needle-holder with a lens attachment, which he has found useful. Good eyesight and a steady hand are the essentials to be supplied by the operator. If he has not these, he had best not attempt the operation. A good light is necessary for the operation, that is, one that is steady and not glaring. A cloudy day with changing light is try-

-ing to the eyes. I have been able to work with much comfort on dull days, even under the chin, by spreading a white napkin over the patient's throat and upper chest. An operating or reclining chair is a comfort, and the patient should be so placed that the part to be operated on is on a level with the operator's eye.

The operation is done in the following manner: The patient, being in position, is to be given the sponge electrode attached to the positive pole of the battery, and told to hold it in one hand. The hair to be extracted is to be seized with the forceps, and put slightly on the stretch in the direction in which it naturally grows. The needle-holder is attached to the *negative pole*, and held like a pen-holder. The needle is then inserted parallel with the hair and into the follicle. One soon learns to know whether the follicle is entered or not by the sense of touch. When the follicle is entered the needle glides along smoothly; when it is not entered a sense of resistance is communicated to the fingers as the skin is punctured. The depth to which the needle is to be thrust will vary with the case. Roughly speaking, it is from one-sixteenth to three-sixteenths of an inch. The needle being inserted, the patient is told to place the palm of the disengaged hand over the sponge electrode. Watching the effect on the skin, a slight hyperæmia will be noticed about the point of insertion of the needle, which almost instantly gives place to a blanching of the tissues. In a few seconds there will be frothing about the needle, and in from half a minute to a minute, the hair will come away upon the very slightest traction.

Such is the operation. Care must be taken to use the negative pole, for otherwise, instead of an electrolytic action, a charring of the tissues will result, and permanent blackish spots will be left in the skin. The hair must not be pulled on with any force, for the ease with which it leaves the follicle is a guarantee of the completeness of the operation. The hairs must not be extracted in close proximity, because the inflammatory action thus set up will lead to more or less ulceration and subsequent prominent scars. At first, at least some three-sixteenths of an inch should be left between the hairs; later, when the susceptibility of the skin is known, they may be taken out nearer to each other. It is best only to extract the coarser hair and to leave the lanugo hairs alone. The strength of the current to be used will depend upon the quality of the patient's skin and the recentness of the filling of the battery. Eight cells are the fewest I have used, and fifteen the greatest number; that is, a current of from one half to one and a half milliampères.

The immediate effect of the operation is the production of a number of wheals which are evanescent. On the next day only red points are seen, and in a few days no traces of the operation are visible, or else minute cicatricial points, when the skin is examined with the greatest care. In some skins, however, the reaction is much greater. The finer the needle is, the less the reaction will be. The patient should be directed to bathe the face in *hot* water after the operation, and to anoint it with cold cream. If these directions are carefully carried out, and the requisite skill in the introduction of the needle is acquired by practice, an absolute cure will be the result.

While the just detailed operation is the only one which is at all certain of success, various other means have been tried to destroy superfluous hair, and, for completeness, must be given in this place. Depilatories are of very ancient date, and there are many vaunted ones now in the market. Many of my patients have used many kinds. That they have come to be operated on by electrolysis tells the tale of the failure of depilatories. Still, they are useful in some conditions. When the patient's hair-growth is very fine they may be advised, as fine hair is not so favorable for operation as is coarse hair. When the hair-growth is partly fine and partly coarse it is a good plan to remove the whole by a depilatory, and thus lighten our work, as the coarse hairs will naturally appear first in growing, and then can be removed before the fine hair obscures them. Depilatories act only on the hair above the surface and not on the roots. ANDERSON recommends one composed as follows:

> Sulphuret of barium, . . ℨiss=6.
> Oxide of zinc, . . . ℨvj=24.
> Carmine, ... gr.j=06.
> M.

Some of this powder is to be mixed with enough water to make a paste, applied to the part and washed off in three minutes. DUHRING (10) gives the following formula:

> Sodii sulphid, ... ℨij=8.
> Cretæ preparat, ... ℨvj=24.
> M.

This is to be made into a paste with water and applied as a thin coating, and left on for ten or fifteen minutes. As soon as it causes heat of the skin, it is to be washed off, and the part is to be anointed with a bland ointment.

LEONARD (64) tells us that in Eastern harems, where it is the fashion to destroy the axillary and pubic hair, they use a composition called "rusma" made of

> Arsenici tersulphuret, . . ℨss=2.
> Calcis, ... ℥ss=16.
> Farinæ tritici, ... Ɉij=2.5.
> Aquæ ferv. qs. ut. ft. pasta.
> M.

This is applied with a wooden spatula, to the thickness of a knife blade, and left on for from five to ten minutes, or until it begins to sting, when it is scraped off with a blunt-edged knife, and the skin washed with warm water and dusted with rice or starch powder.

NEUMANN (27) gives an extended list of depilatories in his work on the skin. All of them require careful watching, as it is possible that their action may go too far.

Nothing need be said about cutting the hair, shaving it, and pulling it out by tweezers. The most ancient method of getting rid of hair is by fire—burning it off.

C. HEITZMAN (344), in 1881, reported the results of a number of experiments with hypodermic injections, the hair being extracted and then the needle introduced and the injection made. He first tried the tincture of iodine, one part in ten, and extracted fifty hairs. More than one half returned in three months. He then used equal parts of carbonate of potash and water. Nearly all returned. Then caustic potash, one to six or four, was injected. Twenty-four hairs were extracted from his own arm and in six months only six returned. He thinks that in the strength of one to four with a fresh solution it is capable of destroying hair under symptoms of suppuration. The needle should be run obliquely under the skin and given a rotatory motion. Many other attempts to destroy hair by means of introducing needles dipped in various substances into the follicle have been made with indifferent success. It would seem that the medicament would be rubbed off against the walls of the upper part of the follicle before it could reach the papilla, and that any good effected would be due to exciting a suppurative action at or near the surface of the skin.

BULKLEY (324), in 1878, reported several successful

cases treated by means of a three-cornered surgical or glover's needle, which was passed into the follicle and rotated while in the region of the papilla so as to break up its substance. The hair is to be seized with the epilating forceps, put slightly on the stretch, and the point of the needle placed at the mouth of the follicle. The needle is to be slowly pressed in and the hair pulled out when the needle will pass into the follicle. Generally no bleeding occurs, and the operation is not very painful and no scarring results. This is a good method of treatment where there are only a few hairs to be removed and a galvanic battery is not at hand.

CHAPTER X.

TRICHIASIS AND DISTICHIASIS.

These diseases belong to the domain of the opthalmic surgeon, rather than to that of the dermatologist, but they are usually included in systematic treatises upon the skin, under the section upon Hypertrichosis. They occur not infrequently: the last report, (1886,) of the Manhattan Eye and Ear Hospital showing 20 cases in 458 cases of diseases of the eyelids.

Trichiasis is a congenital or acquired misplacement of the cilia, so that they are directed backward toward, and scratch upon, the globe of the eye; combined with a growth of lanugo hairs with the same inclination backwards. This condition is usually acquired, though there may be some original irregularity of the cilia. These are not only misplaced so as to point backward, but are often twisted or distorted. As a rule both the upper and lower lid are affected; and both eyes may be involved in the disease. The lanugo hairs that are met with develop after the distortion of the true cilia has existed for a greater or less length of time. They grow from every portion of the tarsal margin, and even from the mucous membrane of the lids.

This disease is dependent upon some long-continued inflammatory disease of the eyelids, which causes a distortion of the lid and interference with the direction and nutrition of the hair-follicles. It therefore very commonly follows upon granular and purulent opthalmia.

Distichiasis is a congenital or acquired condition in which the cilia grow in two distinct rows, the inner row being directed inwards so as to impinge upon the cornea. It differs from trichiasis in an entire absence of lanugo hairs, the supernumerary hairs being normal cilia. Sometimes the inner row cannot be seen until the lid is everted. The condition may be partial or complete, usually the former. According to MICHEL (380) generally the outer third of the upper lid is affected alone, the deformity is symmetrical and bilateral, and of embryonic origin. The second row of cilia springs from the flattened tarsal margin near its posterior edge. Where this arrangement of the cilia is not congenital, it is dependent upon the same causes as trichiasis. Both diseases cause irritation of the cornea, opacity of the same, panus, and ulceration, matters which do not concern us here.

The *treatment* is palliative, as by epilation; or radical, by operation. Until recent times the opthalmic surgeon has made a radical cure by cutting out either a piece of the lid so as to shorten it and somewhat evert it, or by excising a part of the tarsal cartilage with the offending hairs. For these matters, books on the diseases of the eye are to be consulted. To MICHEL (380), of St. Louis, is due the credit of first destroying the supernumerary and distorted hairs by means of electrolysis. The operation is the same as is given in the preceding chapter. MICHEL states his preference for a No. 8 cambric needle, because its action is not so limited as is that of a finer one.

CHAPTER XI.

SYCOSIS.

Derivation.—Gr. σύκωσις, a Fig.

Synonyms.—Sycosis non parasitica; Sycosis menti; Sycosis barbæ (Celsus); Mentagra (Plenck); Acne mentagra; Folliculitis barbæ (Köbner); Folliculitis pilorum; Herpes pustulosus mentagra (Alibert); Lichen menti; Acne sycosis (Morris); Coccogenous sycosis (Unna); Fr.: Sycosis non parasitaire; Dartre pustuleuse mentagra; Adénotrichie (Hardy); Ger.: Bartfinne, Bartflechte, Fikosis; Eng: Barber's itch.

Definition.—A chronic follicular and peri-follicular inflammation of the long hairs, chiefly affecting the bearded portions of the face; characterized by an eruption of papules, pustules, and nodules perforated by hairs; by the formation of infiltrated patches; and by a greater or less amount of crusting.

Symptoms.—It is only of comparatively recent years that this disease has been recognized as a separate entity, and it is still regarded by some authorities as merely a form of eczema. But the experience of the majority of dermatologists has taught them, that the disease is quite independent of either eczema or trichophytosis barbæ. The disease begins by the formation of a number of red inflammatory papules and nodules, which are more or less conical, usually raised above the surface of the skin and always perforated by hairs. Their appearance is preceded and accompanied by disagreeable local sensations, such as pricking, burning, and smarting, and at times by a feeling of tension in the part, on account of swelling of the skin. In acute

SYCOSIS. 183

cases there is considerable redness of the skin between the papules, and the inflammation may be so intense as to give rise to enlargement of the neighboring lymphatic glands. The papules and nodules vary in size, from that of a millet seed to that of a pea, and are isolated or grouped, not every hair-follicle in a diseased part being affected by the peri-follicular inflammation.

Sycosis.

Only in very severe outbreaks or in acute exacerbations do the papules and tubercles tend to run together and form infiltrated patches.

The papules and nodules soon change into pustules, which preserve the same characteristics of grouping and are likewise always pierced by hairs. These pustules, conical in shape and perforated by hairs, are pathognomonic of the disease. In old cases they are met with in the infiltrated patches, arising apparently without the preceding appearance of papules and nodules. The pustules show no tendency to rupture, but the pus accumulates below, wells up alongside of the hair, appears upon the surface of the skin and dries into thin crusts. The amount of crusting is never very great, far less than in eczema of the beard, and is appreciable mainly when the beard is growing. If the crusts are removed from a well-developed patch, the skin looks as if pitted, with the hairs standing in the pits. If the inflammation is very intense, we may meet with small cutaneous abscesses here and there, instead of pustules. According to A. R. ROBINSON (33) the amount of pus production varies with the individual attacked, being more rapid and abundant in the robust than in the scrofulous; in acute than in chronic cases.

The hair, if of any length, is early affected in appearance, becoming lustreless. It is at first firmly seated in its follicles, and when pulled upon gives rise to pain, and if extracted its root-sheath will appear as a clear glassy cylinder. Later, as pus forms more abundantly in the peri-follicular tissues, and the follicles themselves are involved in the process, the hair becomes loosened and easily extracted, when its root-sheath will be found swollen with pus. If the pus production is excessive, the hairs will fall of themselves or upon the slightest traction. When this occurs the hair papilla may be so damaged that no new hairs will form. In chronic cases the beard is markedly thinned, though permanent loss of hair is the exception.

The disease may attack any part of the bearded face, and may be met with in other hairy regions, as the neck, the eyebrows, scalp, axilla, and pubes. But the beard is by far most often the site of the disease, the other situations being affected in the order in which they are named. Occurring in the beard, it may be limited to a single region and show no tendency to spread. Thus, it is met with very frequently upon the upper lip alone, or at times only upon the chin. It may attack the whole bearded face in an acute outbreak, or it may involve it by extension from a limited area during a number of successive outbreaks. In chronic cases it is usually symmetrical. The course of the disease is chronic and made up of a number of acute exacerbations. If left to itself it may produce a good deal of deformity, the tubercles and pustules breaking down, ulcerating, and leaving cicatrical tissue and more or less baldness.

A typical case of sycosis presents the following appearances. Upon a single region, two or more regions, or upon the whole bearded portion of the face, there will appear a number of isolated or grouped papules, nodules or pustules pierced by hairs. The skin about the lesions is reddened and swollen, it may be indurated, and there is a slight amount of crusting. There is no tendency for the disease to spread to non-hairy parts, but very commonly the eyebrows will be similarly affected, and a blepharitis will be present. In some chronic cases there will be much loss of hair and some scarring, and perhaps few pustules, but a red, thickened, slightly scaly skin.

When the case is watched for a time, marked exacerbations will arise, often without apparent cause, last for a few days, and then the disease will sink into a subacute condition.

When the disease affects the vibrissæ of the nose, by extension from the upper lip, the Schneiderian membrane becomes swollen and exquisitely sensitive.

ETIOLOGY.—The etiology of the disease is not settled. Statistical tables show that it occurs about six times in every thousand cases of skin diseases. It probably occurs more often than this, as some cases recover rapidly under domestic treatment or even when left alone. It is non-contagious. UNNA (312 ap.) and some others believe that the disease is due to the entrance of pus cocci into the hair-follicles. It is seen in men almost exclusively, as we might expect, and attacks them most frequently between the ages of twenty-five and fifty. Both the well nourished and the poorly nourished, the rich and the poor, suffer from it. I have seen it very frequently in tailors, and in them it has proved very obstinate. Eczema is often a forerunner of sycosis, the one process passing over into the other. A nasal catarrh is the cause of the majority of cases occurring on the upper lip. Shaving with a dull razor against a stiff beard is sometimes an exciting cause, though those who do not shave are by no means exempt from the disease. An irritant applied to the skin may excite it, such as exposure to intense heat, the dust of a workshop, cosmetics and the like. Exposure to inclement weather is regarded by WILSON (38) as the principal cause. One of the worst cases I have ever met with was directly traceable to a poultice applied to the face for the relief of a neuralgia.

Given a hyperæmic or irritable condition of the skin of the face arising from any internal or external cause, the hairs, especially if they are coarse, may excite the disease, acting as irritants when touched or moved. HEBRA (394) thinks that some cases may be due to an abnormality in the growth of new hairs. WERTHEIM ascribed the inflammation to irritation of the hair-follicles by hairs whose diameter was relatively too large for their follicles.

PATHOLOGY.—A. R. ROBINSON (400) who has made an exhaustive study of the pathological anatomy of this disease, teaches us that it is " primarily a peri-follicular inflammation of the skin. The first changes which take place occur around the follicle in the peri-follicular region, and are those which are usually observed in vascular connective-tissue inflammations. The transuded serum penetrates the hair-follicle, and as the inflammation proceeds and the pus and serum increase in quantity, the follicle becomes more and more affected. Its sheaths become softened and more or less destroyed, and a portion of the pus may enter the follicle through the ruptured sheaths. The cells of the external root-sheath become swollen and soon begin to break down; similar changes occur in the cells of the hair-root: they swell, the protoplasm becomes more granular in appearance, and there is evidence of commencing destruction. After the rupture of the follicle-sheaths, or even before, the cells of the hair-root and of the root-sheaths rapidly become broken up and changed by the transuded serum entering the follicle. If pus corpuscles have also entered the follicle, the hair-root is infiltrated with a sero-purulent matter; it does not in every case enter it in large amount. In the pustular stage the principal changes take place within the follicle; the hair-root and its sheaths are broken down and separated from the follicle sheaths, so that the hair lies loosely within the follicle.

"As the inflammation progresses, the connective-tissue around the follicle becomes crowded with pus cells, as far as the surface of the skin. If the hair is allowed to remain within the follicle until expelled by the accumulating pus, the root-sheaths and soft parts of the hair are destroyed, and only the hard part remains. The follicle-sheath, and the peri-follicular tissue are more or less destroyed, and the Malpighian layer becomes ruptured at the neck of the follicle. The pus reaches the surface by breaking through the epidermis near the hair; some passes to the surface between the hair-shaft and the follicle-sheath. The cells from which the hair grows seem to resist the inflammatory process more than the other cells of the bulb. When permanent alopecia results both the follicle-sheaths and the base of the follicle are completely destroyed. If eczema is present the root-sheaths and follicle-sheaths are acted upon in their entire length at the same time.

"The sebaceous glands may also become affected, though not at so early a stage of the disease as the fundus of the hair, and the whole gland may be destroyed by a process of molecular retrograde degeneration. The sweat glands generally escape, but the epithelial cells may become detached or the glands destroyed."

GIOVANNINI (86 ap.) has found the affected hairs thicker than normal, with more or less notched contour, so that the hair presents from two to six irregular projections. In the middle of the hairs he has found broad and irregular cavities filled with pigment, fat, and detritus. UNNA (312 ap.) teaches that sycosis may be caused by the common pus cocci gaining access to the hair follicles by traumatism or mechanical friction. The hair remains unaltered for some time after the invasion of the follicle. Any pustular inflammation may start this form of sycosis. TOMMASOLI (311 ap.) describes a bacillary form due to short, rod-shaped, somewhat thick, and elliptical bacilli with rounded ends, the *bacillus sycosiferus fœtidus*. This he believes to be of tubercular nature.

DIAGNOSIS.—The distinguishing characteristic of sycosis is the presence of pustules pierced by hairs. It must be diagnosed from trichophytosis barbæ, eczema barbæ, the small pustular syphiloderm, acne, and lupus.

Differential diagnosis from trichophytosis barbæ :

Trichophytosis barbæ.

Begins as a small scaly spot, a superficial ringworm, and gradually involves the deeper parts of the hair.

Has its favorite seat upon the chin and the sub-maxillary region; rarely attacks the upper lip. Often asymmetrical.

The eruption consists of tubercles and nodules which tend to group and are studded with a number of hairs. The internodular portions of the skin often remain unaffected.

Is a deep inflammatory process as soon as the hairs become affected.

Hair is diseased primarily, and is twisted, split and broken.

May readily be removed by slight traction and without pain. Its root is often dry.

Subjective symptoms slight, may be only slight pruritus.

Patches of ringworm often present on other parts of the body, and sometimes the disease extends upon the neck or face.

Hairs and scales loaded with the tricophyton fungus.

Is a progressive disease, and when cured not liable to relapse.

Sycosis.

Begins suddenly with an outbreak of papules which soon become pustules, each of which at the start involves a hair.

Its favorite seat is the upper lip, and sometimes it alone is involved. Involves the hairy portions of the face more generally and often symmetrically.

The eruption consists of papules and pustules, each of which is pierced by a single hair, and they show no disposition to group. The intervening skin is generally reddened, and may be diffusely infiltrated; and abcesses may form.

Is a more superficial inflammation.

Hair diseased secondarily and comes away at first with difficulty, causing much pain. Later is easily removed and its root is swollen with pus.

Subjective symptoms of pricking, burning, and tension of the part. These are often intense and attended with swelling of the face.

Limited in most cases to hairy parts of the face. No tendency to extend on non-hairy parts of face or neck.

No fungus present.

The course of the disease made up of a number of acute outbreaks. Liable to relapse.

The differential diagnosis from *eczema* of the beard cannot be made with so much certitude, and often we must remain for a while in doubt as to the true nature of the case. At times the sycosis is a legacy left by a preceding eczema, and we may meet with a case in the transition stage when a sure diagnosis would, manifestly, be impossible. A typical case of pustular eczema is attended by a far greater amount of crusting than is sycosis, and the crust is of a more greenish or blackish color. Upon removing the crust in eczema a moist and oozing surface will be exposed, while in sycosis we will do no more than remove the tops from a number of pustules. In eczema the pustules break down more readily than in sycosis, and they are not so accurately located about the hairs. In eczema the whole surface of the skin is involved and the process tends to extend upon non-hairy parts of the face. While exceptionally eczema is confined to the hairy portions of the face this is always so in sycosis. The duration of the disease will at times help us to a diagnosis, sycosis being far more chronic than is eczema.

In *syphilis* when the beard is involved we will find pustules upon other portions of the body, and the history will help us to a correct conclusion. The pustules or papules of syphilis are grouped in circles or segments of circles, of peculiar color, and their development is painless and comparatively slow. Pustular syphilis more often causes permanent baldness than does sycosis.

Acne is scattered about the whole face, is usually met with in young persons, comedones are present and its papules, pustules, or tubercles have no definite relation to the hair.

The course and history of *lupus* are so different from that of sycosis, that it is hardly possible for them to be confused. In lupus vulgaris we have the characteristic brown tubercles which do not contain pus, are not confined to the hairy portions of the face, generally begin in early life, and tend to ulcerate or to be absorbed and leave behind cicatrices.

TREATMENT.—The treatment of sycosis is both general and local. While many cases will yield to local treatment alone, there are quite as many, if not more, which require general treatment. The surroundings of the patient must be inquired into, and also his mode of life. He should be urged to take exercise in the daylight; a powerful means for good if the case happen to fall upon one constantly employed in badly ventilated rooms, such as tailors and the like. He should be advised against exposing himself to dust and wind, and even against smoking, especially in the wind where the smoke blows against the face. The proper regulation of the diet is important. Many cases will improve if we stop their tea, coffee, hot drinks of all sorts, ale, beer, and spirits. If the digestive process seem at all embarrassed, it is well to put the patient on a light diet for morning and evening, and direct him to take his principal meal at noon, eating meat only at that time. Anything that is known to him to be indigestible must of course be prohibited. In a word the diet and hygiene of the patient should be regulated.

What medicines we should administer will depend upon the stage of the disease. In the acute stage, when there is much swelling and inflammation, a good dose of blue pill, calomel, or some other active cathartic is to be ordered, to be followed by an alkaline diuretic. When pustulation is active the sulphide of calcium or calx sulphurata will do good. PIFFARD (399) recommends this very highly, giving one-tenth of a grain two or three times a day. Care must be exercised that our patient obtain the drug fresh. I have found the exhibition of the drug in the form of the tablet triturate by far the most active way. Small doses of calomel, as one-tenth of a grain three times a day for two or three days at a time, are useful in relieving the congestion of the skin. In chronic cases, iron, cod-liver oil, and other tonics are indicated if there is a state of debility. Arsenic is advised in very obstinate cases. If indigestion is present we must address our remedies to its relief before we give calcium, arsenic, or other remedy for the disease proper.

The local treatment is more important than the general, and is required in every case. It must vary with the condition found, whether it be acute or sub-acute. When the disease attacks the upper lip the nose must be examined for evidences of catarrh, and that condition treated if found. I have had, at times, good results from the subnitrate of bismuth or powdered cubebs, used as a snuff in this condition, but it is best for each physician to use for this, that which experience has taught him to be most useful.

In the management of an acute case of sycosis soothing remedies are needed. Hot water should be sopped on the part for some five or ten minutes once or twice a day, and this should be followed, if the beard is growing, by the use of a simple oil such as olive oil or sweet almond oil, or if the face is shaved the zinc oxide ointment or cold cream may be used; or better still, LASSAR's paste, as follows:

    Amyli,
    Zinci oxidi. āā . . . . $\fiveZ$ ij—8.
    Vaseline, ad . . . . $\fiveZ$ j—32
M.

Powdering the part with corn starch, or bismuth and talc, after smearing on a little vaseline, will at times give ease and comfort. If the process is attended by a good deal of œdema and the inflammatory symptoms are severe, warm poultices will relieve the disagreeable sensations of the patient and reduce the inflammation. In some cases cold starch poultices will be better borne. DEVERGIE (387) recommends steaming the inflamed parts every second day, and covering the affected parts constantly with cold or almost cold thin flaxseed poultices. Even in the early stage, if the inflammatory symptoms are not very intense, a mild white precipitate ointment will sometimes check the disease. DUHRING (10) recommends bathing the face with "black wash" followed by zinc oxide ointment with a drachm of alcohol or half a drachm of camphor to the ounce, spread on cloths and bound on; and speaks well of the oxide of zinc ointment with fifteen to thirty grains of calomel to the ounce. Tumenol oil has yielded surprising results in some cases. The chief objection to it is its color. It may be used pure or diluted with vaseline or other oil. When the disease has reached the pustular stage, and there is more or less crusting, the crusts are to be removed by the free use of olive oil, or oil of sweet almonds, letting it soak in thoroughly over night and washing the part with soap and warm water the next morning. If the crusts are thick, it is a good plan to tie up the bearded face in a towel after anointing it with oil. A poultice may be used for the purpose of removing the crusts. After the crusts are gotten rid of, pull the hairs out of the pustules, and insist upon the patient shaving himself every second day. At first he may rebel against the use of the razor, but if plenty of warm water and soap is used, and a good lather formed, the shaving will not be very painful, and it is only the first shave that is painful. Epilation of the hair from all the pustules and papules is to be continued until they cease to form. Shaving is to be continued until some months after the skin is apparently well. It is possible to cure a case without shaving, but the cure will be more difficult to effect. The patient must be made to understand that epilation is necessary, both for the cure of the affection and the salvation of the hair. After epilating, the oxide of zinc ointment, Lassar's paste, or diachylon ointment is to be used. Sulphur in the form of an ointment, half a drachm to a drachm to the ounce, or in powder, will sometimes do good, but often will prove too irritating. TILBURY FOX (390) recommends the use of the following ointment after shaving:

Zinc oxide, ...
Zinc carbonate āā . . . ʒj— 4.
Rose ointment ad . . . ʒj—32.
M.

Instead of an ointment we may use oxide of zinc one drachm to the ounce of linseed or other oil. SHOEMAKER (402) advises the application of equal parts of oleate of mercury and olive oil.

In sub-acute and chronic cases a more active treatment is necessary. Here our aim is not so much to allay inflammation as to stimulate the skin. To this end we may use the soap and salve treatment of HEBRA, which renders such good service in chronic cases of eczema. It consists of frictions with green soap, soft soap, or better the tincture of green soap, composed of two parts of the soap and one part of alcohol, followed by a soothing ointment, such as the oxide of zinc ointment. Some of the soap is poured or placed upon a piece of flannel, this dipped in hot water, and then rubbed actively upon the part to remove all the tops of the pustules or papules, and leave the surface a little raw. Then the soap is all washed off, and the part covered with the ointment spread thickly upon old linen or cheese-cloth. The dressing is firmly bound down with a roller bandage. The ointment is to be changed two or three times a day, but the soap is to be used but once a day or every other day. In some cases better results will be attained by the use of diachylon ointment, or Lassar's paste, with ten or fifteen grains of salicylic acid to the ounce. In very obstinate cases where there is much thickening of the skin, the soap may be kept applied to the part like an ointment. When sufficient inflammatory reaction is produced, emollient measures, as in the acute stage, should be used. The use of a strong tincture of tar after Pick's formula of forty parts of tar and twenty parts of alcohol sometimes answers well. In some cases where there is a good deal of pustulation and the patient is shaven, it is a good plan to curette the patches, tearing off the tops of the pustules and letting out the pus. This procedure is to be followed by the application of Lassar's paste with salicylic acid.

Our success in treating these cases, will vary with the thoroughness with which the dressings are applied. All ointments must be spread on cloths, not on the skin, and the dressings must be kept continuously in close contact with the affected part. Sometimes a sulphur ointment, one half a drachm to two drachms to the ounce; an ointment of iodide of sulphur; the ointment of the ammoniate (gr. xv-xxx. ad ℥ j), or the red oxide (gr. v-xv. ad ℥ j) of mercury will prove useful. ROBINSON (33) recommends the following ointment:

  Ungt. diachyli (Hebra) .
  Ungt. zinci oxidi, . . āā ℥ iss—50.
  Ungt. hydrarg. ammon. . ʒ iij—10.
  Bismuth subnitrat. . ʒ iss— 5.
M.

He has found cod-liver oil the best local application in strumous subjects. VEIEL (40)* advises painting the affected parts twice a day with a two per cent. solution of pyrogallol in alcohol, and applying during the night:

  Sulphur. lact. . . . .
  Alcohol, ...
  Aquæ rosæ . . . . āā 30.
  Mucilag. gum acaciæ, . . ℳ xx-xxxx.
M.

HANS VON HEBRA's (394) plan of treatment is to epilate and shave, and then with a stiff brush to rub in once or twice a day some of the following ointment:

---

* Ziemssen's *Handbuch des Speciellen Path. u. Therap.* p. 235.

Ol. fagi,
Flor. sulph., ... āā 10.
Pulv. cretæ alb., ... 5.
Adeps, ...
Sapo. viridis., ... āā 20.
M.

and cover with flannel. DEVERGIE (387) recommends painting the part every fourth or fifth day with a solution of nitrate of silver, one part, in five of water by weight.

BEHREND (42) has obtained good results by scraping the affected parts with the dermal curette, and dressing with a simple ointment or oil. All abscesses must be opened. BROOKE (295 ap.) recommends the application, after epilation, of an ointment of

℞ Hydrarg. oleat. (2½ per cent), . ℥ i.
Ichthyol. s. ammon., . . . ℳ xx.
Ac. salicyl., ... gr. x.
Ol. lavandulæ, ... gtt. ij.
M.

which is to be kept applied on strips of linen, or mixed with enough zinc oxide and Armenian bole to make a skin-colored paste, and smeared on.

Our rule of treatment then is in acute cases to allay inflammation by emollient dressings; in chronic cases to stimulate. In all cases to epilate, at least from pustules, and where possible to have the patient shave. The dry and reddened skin sometimes left after the disease has run its course is to be treated with hot water sopped on for five minutes once or twice a day, and some simple ointment to protect the skin from dust and exposure to the weather.

Sycosis affecting other locations than the beard is to be treated by epilation and emollient ointments. When the hairs within the nose are affected HARDA-

Way (299 ap.) has found the best plan of treatment to be to foment the parts several times a day, to apply glycerin both to the inside and outside of the nose, and to pluck the hair from the follicles. When pain and tension have subsided he uses the following:

℞ Squire's glycerol. plumb. sub-
  acetat., ... ℨss.   2.
  Glycerin., ... ℨiss. 6.
  Ungt. aquæ rosæ, . . . ℥i. 32.
  Ceræ albæ, ... q. s.
M.

Prognosis.—This is one of the most obstinate of diseases, specially in such cases as are due to occupation. Left to itself, when once under headway, it shows no tendency to get well, and has been known to last twenty or thirty years. Even under the most judicious treatment it is an obstinate disease, taking weeks or months before a cure is effected. Relapses are exceedingly liable to occur, and these sometimes show a disposition to recur at certain seasons. Unless the hair is carefully withdrawn from the inflamed follicles permanent baldness may be caused. But the disease is not dangerous to life, and it is curable.

FOLLICULITIS DECALVANS.

Within the past few years there has been described by French writers a group of diseases of hairy parts that is characterized by, 1, a follicular and perifollicular inflammatory process; 2, a complete destruction of the hair-papillæ, causing absolute baldness; 3, the formation of apparently cicatricial tissue; and 4, a tendency of the lesions to agminate or group. This group of diseases has been named by Brocq "folliculitis et perifolliculitis decalvans."

The disease has been described under many names applied by different observers to different phases or locations of the malady. It has been called "lupoid sycosis" by MILTON; "alopécie cicatricielle innominée" and "acné pilaire cicatricielle dépilante" by BESNIER; "folliculite epilante" by QUINQUAUD; "Acné décalvante" by LAILLER and ROBERT; "Ulerythema sycosiforme" by UNNA; sycosis chronique. BROCQ would also include under it Kaposi's "dermatitis papillaris capillitii."

The disease may show itself either as a sycotic affection of the beard, pubic and axillary hairs, passing over at times to the scalp; or as a bald area upon the scalp, which, upon more careful examination, will be found to be due to a follicular and perifollicular inflammation of the hair-follicles.

Folliculitis decalvans of the bearded portion of the face has its seat of predilection upon the cheeks, from whence it may invade the temporal region of the scalp. It begins as a redness of the skin, which is soon followed by the appearance of little vesico-pustules at the mouths of the hair follicles, forming isolated islets of disease. The patches soon become crusted and appear eczematous. When the acute process subsides the pustulation ceases, and the surface of the patches becomes red and scaly. It will then be noted that the skin is cicatricial and the hair is permanently destroyed. There may be but one patch or a number of them. The patches may be symmetrical or non-symmetrical. They tend to spread slowly, serpiginously, and peripherally. The disease is exceedingly chronic in its course, and after it has lasted a number of years, unless it has been of very limited extent, the bearded portion of the face, and perhaps the temporal regions of the scalp, will be found sown over with cicatricial spots which may be depressed, bridled, or keloidal

Folliculitis decalvans of the scalp alone simulates alopecia areata so closely as often to be mistaken for that disease. Usually the first thing noted is the appearance of one or more bald spots, and, attention being thus attracted to the scalp, further search will reveal evidences of folliculitis. It assumes one of two forms: 1. The *alopécie innominée* of Besnier, in which we find on the scalp irregular, ill-defined bald patches bounded by bouquets of sound hairs, or by tufts of hair which has partially fallen out, or by neighboring bald patches. The scalp appears cicatricial, thinned, slightly depressed, smooth or stippled over with the follicular mouth-openings, and with or without pigmentation. There may be no evidences of dermatitis or traces of it about the follicular orifices. Often there is a slight, diffused, ill-defined redness with furfuraceous desquamation; or some very small and superficial pustules occupying the infundibula of the hair, which in a short time are transformed into depressions in the epidermis, out of which the hair deprived of its sheaths falls or is readily plucked. As soon as the hair falls the inflammation subsides, but the hair has been permanently destroyed. The disease spreads in a very erratic manner, and there frequently are many isolated spots scattered over the scalp.

2. This is the *folliculite épilante* of Quinquaud. While usually affecting the scalp, it may affect the beard, pubes, and axillæ. It resembles the first variety in producing bald, smooth, irregular-shaped cicatricial patches, but is preceded or attended by a more marked folliculitis. The patches are disseminated; about the size of a silver quarter of a dollar or a franc piece; pale, with a few red points in them, while about their peripheries and in the hair of neighboring parts are various evidences of folliculitis, such as purulent points, punctiform miliary abscesses, with hair in their centres. When these hairs fall or are plucked they leave a red,

scarcely moist point. Instead of pustules there may be only punctiform redness, with or without secondary desquamation, or red follicular papules. The successive loss of a number of hairs produces bald patches of large size, which are separated from each other by tufts of sound hair. There is never any appearance simulating that of eczema.

ETIOLOGY AND PATHOLOGY.—We do not know anything positive about the cause of the disease. QUINQUAUD describes a micrococcus as the cause of his folliculite epilante, which occurs as a monococcus, diplococcus, and in series of four in the follicles, the blood, and in the inflamed skin. The fluid from the cultivation, when rubbed into animals and man, produced a disease apparently identical with the parent disease. In all forms of folliculitis decalvans there is atrophy of the hair-follicles and sebaceous glands. It has been surmised that some of the cases are due to syphilis.

DIAGNOSIS.—Folliculitis decalvans affecting the bearded portion of the face differs from sycosis chiefly in causing cicatricial destruction of the skin and permanent loss of hair. Moreover, its pustules are more superficial and its surface is never so crusted. It occurs in patches, and not so disseminated through the hair, and affects primarily the skin between the hairs rather than the hair-follicles themselves. Folliculitis decalvans affecting the scalp produces bald spots that are to be distinguished from those of alopecia areata by the presence of inflammatory symptoms. Without a well-marked history of the occurrence of favic crusts it would be impossible to distinguish alopecia from old favus from an alopecia of this disease in quiescence.

TREATMENT thus far has been inadequate to the cure of the disease. The scalp should be kept clean and some mild antiseptic lotion or ointment used.

The PROGNOSIS is bad. The course of the disease is slow, and permanent baldness follows. After a time the disease may reach a quiescent stage.

PART III.

PARASITIC DISEASES OF THE HAIR.

Trichophytosis.—Favus.—Pediculosis.—Beigel's Disease.—Trichomycosis Nodosa.

CHAPTER XII.

TRICHOPHYTOSIS CAPITIS.

SYNONYMS.—Herpes tonsurans; Herpes circinatus; Herpes squamosus (Cazenave); Tinea tonsurans; Tinea tondens (Mahon); Trichonosis furfuracea (Devergie); Porrigo furfurans (Devergie); Porrigo tonsoria (Alibert); Trichosis tonsurans (Wilson); Trichosis pityriasica seu furfuracea (Wilson); Trichomykosis (Gruby); Dermatomykosis trichophytina; Phytoalopecia seu Trichomyces tonsurans (Malmsten); Rhizophyto-alopecia (Gruby); Dermatomykosis tonsurans (Köbner); Squarus tondens (Mahon); *French*, Herpes tonsurante, Teigne tondante (Mahon); Teigne tonsurante, Teigne annulaire (Rayer); Teigne herpétique furfuracée (Gibert); Trichophytie tonsurante (Hardy); Trichophytie (Gruby); Porrigine tonsurantie (Alibert); Dartre furfuracée arrondie (Alibert); L'herpes circiné parasitaire; *German*, Scherende Flechte; *English*, Ringworm of the scalp; *Slav*. Ringskurv.

DEFINITION.—A contagious parasitic disease of the hairy scalp, due to its invasion by the Trichophyton fungus; and characterized by the formation of partially bald, scaly, more or less circular patches, in which "stumps" of broken-off hair will be found. It is a disease peculiar to children and runs a chronic course.

SYMPTOMS.—Ringworm of the hairy scalp begins, like ringworm of the body, by the formation of a small, round erythematous spot upon which ephemeral vesicles and pustules soon form, which rapidly go on to desquamation. Or the spot may become covered with furfuraceous scales without the appearance of vesicles or pustules. This stage is so rapid in its course, and gives so little annoyance to the patient, that it is seldom brought to the notice of the physician. The patch spreads, the hairs become early affected, and then we have the typical patch of the disease before us. This is circular in shape; denuded of hair, though not completely bald; covered with a greater or lesser amount of scales; and more or less raised above the surface of

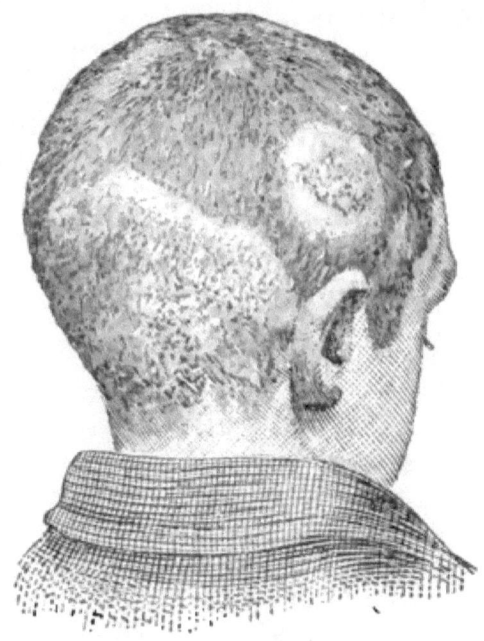

Ringworm.

the scalp. There may be only one patch upon the head, or there may be a number of them. The size of the individual patch may be quite small or it may be as large as a silver dollar. If several patches occur close to each other, they may coalesce and form a huge patch, which may involve the whole top of the head. The color of the patch varies somewhat; it may be reddish, gray, slate, greenish, bluish or even blackish. The color depends upon the amount of scaling present, upon the complexion of the individual, upon the extent to which the inflammatory process has gone, and upon the admixture of dirt or foreign matter. If the inflammation is but slight, and the complexion is medium, the patch will be grayish or reddish. In dark-skinned subjects the color is apt to be slate. If in a strumous subject, and pustulation has taken place, we will meet with greenish or blackish patches. A. R. Robinson (33) draws attention to the fact that sometimes the central part of a patch may be gray or slate color, and the periphery yellowish or blackish brown from the drying up of the vesicles at the margin. Probably "slaty-gray" will best describe the color of the majority of the patches.

The amount of scaling is rarely excessive, and, though at times abundant, never reaches to the formation of thick mortar-like crusts such as we meet with in favus. In a case complicated with eczema, arising either spontaneously or from over treatment, thick purulent crusts may form, but, of course, quite independently of the ringworm. Upon the removal of the scales, the exposed scalp will be found reddened, swollen, and, it may be, œdematous and tender. In chronic cases the scalp will be merely reddened and scaly, and sometimes of goose-flesh appearance.

The condition of the hairs is characteristic. They are affected very early in the course of the disease, becoming dry, lustreless, opaque, brittle, twisted, and readily breaking off upon the slightest traction, or of their own accord. They lose their elasticity, as shown by taking a hair and pressing it with the nail, when it will readily bend at an angle which it will retain. If a healthy hair is subjected to the same treatment it will soon regain its usual form. Another proof of the loss of elasticity, is that when the hair is combed the wrong way upon the head, while the healthy hairs will immediately fall into their former position, the affected ones will stand up for a moment, and then slowly fall. The hairs breaking off leave their roots and a small portion of their shafts, it may be only one or two lines in length, in the scalp. These are known as "stumps," and are pathognomonic of the disease. A stump, then, is the broken-off shaft and root of a ringworm hair, a few lines in length, with a ragged, nibbled-off-looking end, which is split and frayed out, and laden with spores. It may present itself as only a minute dark-colored dot on the scalp. They are met with in this disease alone, and must not be confounded with the ends of cut or accidentally broken-off healthy hairs, which are not split or ragged, and do not contain spores. The amount of hair present in a patch of ringworm varies. Sometimes there will be quite a number of long though diseased hairs; sometimes there will be only stumps; and sometimes both are present, the stumps being superabundant. In a typical patch there will be mostly stumps, which give to it the appearance of having had the hair cut off clumsily with a dull pair of scissors. In some cases, as the result of treatment, or when the disease has taken an exceptional course, the hairs will entirely fall out, and thus there will be formed a completely bald spot.

Ringworm affects by preference the vertex and the parietal regions, though it may occur anywhere upon the head, and at times may pass over upon the skin of the face or neck. The only subjective symptom which it presents is itching, which is often the first thing noticed, and leads to investigation of the scalp. It is usually slight. The disease, though chronic in its course, and obstinate to treatment, is yet self-limited, and does not of itself cause baldness.

Besides this typical form of trichophytosis capitis, the one which we meet with in the vast majority of cases, there are several other forms or varieties. These are pustular ringworm, disseminated ringworm, and kerion. The last differs very much from the other varieties, and is not always due to the trichophyton fungus; therefore it will be reserved for special treatment.

The *pustular* form of ringworm occurs chiefly in ill-nourished or scrofulous children. Instead of a scurfy place forming, we have pustules produced, and greenish crusts. Sometimes this form may be caused by treatment, too strong remedies being used, or remedies being improperly applied, and it is especially prone to occur in eczematous subjects. It is indeed an impetiginous eczema, complicating a ringworm of the scalp, and may involve a large portion of the scalp. The process is superficial, and if the crusts are removed, under them will be found the characteristic stumps. ALDER SMITH (79) regards this form as especially liable to spread amongst schools. It is not the same disease as is the deep inflammatory process called kerion. Sometimes we meet with a chronic pustular ringworm, which presents pustules pierced by hairs.

Disseminated ringworm is that form in which, instead of a single patch or a number of patches occurring on the scalp, the disease involves more or less of the whole scalp, not in the form of patches, but diffused throughout the hair. The hair may seem to be growing well, but when the scalp is inspected, we will find it scurfy, as in eczema or pityriasis; most of the hair will be of normal length and appearance, but here and there will be found a number of stumps, either isolated or in groups, or there may be only black dots on the surface, the roots of broken-off stumps. The long hair may be firmly fixed or may come out easily. This form is seen most frequently in chronic cases; and is often overlooked. It will sometimes last in this sluggish condition for years after the well-defined patches have disappeared, and the case is considered by the physician and the friends as cured.

Chronic squamous ringworm is that form in which we meet with a patch or patches of the disease which do not spread, and which are partially covered with apparently healthy hair. The patch is still scurfy, and the hair may look a little dry, as it is commonly met with in seborrhœa, but that may be all which arouses our suspicions. Upon careful examination stumps will be found close to the scalp, hidden by the long hair.

The incubation period of ringworm has been shown by experiment to be about three days. Its rate of growth is rapid, a spot as large as a ten-cent piece may develop in forty-eight hours, and attain the size of a fifty-cent piece in twenty-four hours more. When it has reached the size of a silver dollar, it, in most cases, ceases to enlarge. The life of the fungus is also limited. Though the disease may have lasted many years in a child, it tends to get well as the child reaches the age of puberty.

The scalp is not the only hairy region affected by the trichophyton fungus. The beard is its frequent habitat. It may also occur upon the pubes and in the axilla, and give rise to appearances somewhat similar to those met with in ringworm of the beard.

ETIOLOGY.—Trichophytosis capitis is due to a single cause, the implantation and growth of the trichophyton fungus. This view is one now accepted by all dermatologists, though up to quite recent years, there were some eminent ones who believed it to be a disease of nutritive debility. Thus WILSON (33) taught "that it was essentially an arrest of development of the hair-cells and the cells of the rete mucosum. That the cells retained their primitive molecular character, and the granules taking on a proliferous growth are converted into a tissue closely resembling a mucedinous vegetation."

The fungus gains lodgment in the upper layers of the epidermis, after the most superficial layers are removed in some way, and from its point of entrance, spreads. LIVEING (452) thinks that the fungus is not the essence of the disease, but plays a secondary, though important part, in its development. His reasons are the following: First.—The food of this kind of vegetable parasite is dead or dying structures. Secondly.—The development of the fungus is not always in proportion to the changes present in the skin and hair, showing that other causes are at work. Thirdly.—In many cases the comparatively healthy hair of the whole scalp loses its lustre and becomes harsh, dry, brittle and more opaque than in health, without the growth of the fungus beyond the ringworm patches, and this condition may persist for months after the parasitic growth has disappeared. Fourthly.—If the fungus were the essence of the disease we should expect the malady to be less capricious in its nature.

The disease is very contagious, much more so than is favus. It is nearly always endemic and sometimes epidemic. When it gains entrance into a school or children's hospital or asylum, it spreads with great rapidity, and such institutions are the most important agents in keeping it alive. BERGERON (45) has shown that in France it is more common in cities than in the country. It attacks children almost exclusively. It is rare to meet with it after puberty, excessively rare to see it on the head of an adult, and very infrequent in infants. This shows that it requires, like other parasites, some peculiar condition of the soil for its growth, though what that condition may be is not yet determined. It attacks all classes of children, the rich and the poor, the clean and the uncared for. It occurs often amongst strumous children, and those who are badly nourished; but as these children are found most frequently in those classes which live under other conditions favorable to infection, it is difficult to determine the exact predisposing force of the diathesis. When the disease gains entrance into asylums and schools it shows no disposition to spare the healthy and robust children. TILBURY FOX (12) taught that children of lymphatic temperament were prone to the disease.

The means of infection are mediate and intermediate. Thus, it is readily conveyed directly from the body or head of one infected person to the head of another, or from the body of a child to its own head, ringworms sometimes passing over from the non-hairy to the hairy parts. It may also be communicated from animals, the disease being common in cats and dogs, and it is met with in cows, horses, rabbits, squirrels and other domestic or pet animals. In epidemics of ringworm in children's hospitals, the air of the room has been found loaded with floating spores. The most common means of mediate contagion are hats, caps, brushes, and combs.

PATHOLOGY.—The disease is caused by the vegetable fungus called *tricophyton tonsurans* or *achorion Lebertii*. This consists of mycelia and conidia, which bear a close resemblance to those of the penicillium glaucum. It is, without doubt, a distinct species of vegetable growth, resembling (though not the same as) the achorion Schœnleinii, as inoculations made with pure cultures produce ringworm alone. On the other hand inoculations with pure cultures of the achorion Schœnleinii give rise to favus alone. It is aërobic.

In the hair the conidia are far more numerous than the mycelia, and sometimes are present in such numbers as apparently to burst the hair. The conidia are often found arranged in rows parallel with the long axis of the hair. Sometimes they are scattered irregularly through the hair; usually they are so numerous about the bulb and root as to appeared crammed together. They are round, highly refractive bodies, of a grayish or pale green color, and a diameter of from .0021 mm. to .0035 mm.

The mycelia are often absent from specimens examined. When present they run through the hair in its long diameter, and are long, jointed and wavy. Their diameter varies from .0018 mm. to .0026 mm.

Bacteriologists, by their more advanced methods of staining and cultivation, are striving to advance our knowledge of the micro-organisms of the skin. FURTHMANN and NEEBE (329 ap.) believe that they have found no less than four different fungi as causes of ringworm, resembling each other very closely, and

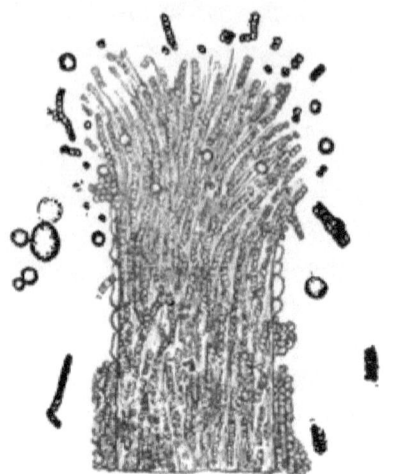

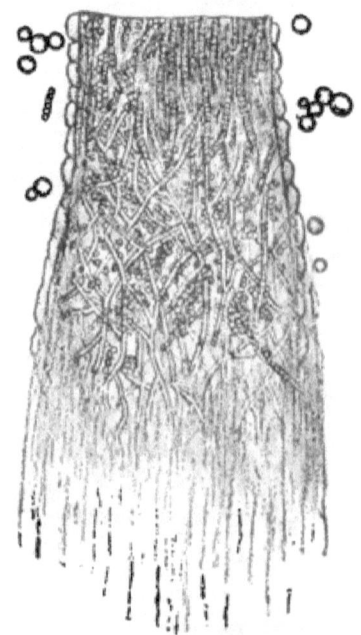

Trichophyton (Smith).

distinguished one from the other mostly by their cultures. It is probable that the fungus of ringworm, like other fungi, varies in form and manner of growth according to the physical and chemical properties of the soil in which it grows, and that after all there is but *one* micro-organism in ringworm. QUINCKE (248 ap.) has found that the fungus of ringworm has many points of resemblance to his α-fungus of favus, but differed also in many respects. It grows slower on gelatin, with greater tendency to grow downward; and more rapidly on agar-agar. Under the microscope its mycelia were simply rounded and not tapering, and were straighter. It would not grow on potato.

In 1892 SABOURAUD (354 ap.) published what seems to me to be one of the most important contributions to our knowledge of ringworm. He believes that the fungus belongs to the botanical species of Botrytis, of which there are probably a number of distinct species capable of causing trichophytosis in different animals. In the human there are two principal varieties, viz.: 1. This has small spores and is found only on the scalp. It is the constant cause of the disease on the infant's scalp and of all the obstinate cases. Under the microscope it has a spore of 3 μ in diameter, with no mycelia. The hairs are full of the spores, and they escape from it to form a sheath about it. 2. The other species has large spores. It may cause trichophytosis capitis, and is found in thirty-five per cent. of the cases, and these are easy to cure. It is the common cause of trichophytosis barbæ, and, with another special large-spore species, is the cause of trichophytosis corporis, specially when that follows upon the disease in the beard. It has a spore of 7–8 μ diameter, with visible mycelia.

A case with one species of spores produces by contagion a case with the same species, and no case presents both varieties at the same time. The patches caused by the small spores are round or oblong, and usually not more than 5 cm. in diameter. At the beginning they are raised and the scalp feels thickened and infiltrated. The affected hairs are usually fine, and they are almost constantly broken off at more than 3 mm. from the mouth of the follicle. The patches caused by the large spores are more often large than small, irregular in shape, with tufts of healthy hair that encroach on the circumference of the patch. It looks almost completely bald because the diseased hairs are cut off very short. The hairs are often of large size and appear as black points at the follicular orifices. Besides these common species SABOURAUD has found two more rare types. One of these has large spores and occurs in trichophytosis corporis. It differs from the other large spore species in its more vigorous and rapid growth with a fluffy centre in cultures. The other species presents large and unequally-sized spores. He also found in one cultivation black, and in another rose-colored, spores that are thought to be forms found only in animals.

He found that he would have to make many attempts before succeeding in inoculating some subjects, and that the patient must have an alkaline reaction to his sweat before success can be obtained.

The hair in ringworm of the scalp is early affected, the first point attacked being, according to TAYLOR (486) that portion of the shaft immediately on a level with the surface of the skin, from which point it spreads up and down. The cortical substance in its peripheral part is the most frequent and earliest seat of the fungus; but the whole hair is frequently involved. According to most observers the bulb is invaded to only a slight extent and the papilla and root-sheaths are spared. A. R. ROBINSON (33) has met with the spores and mycelia in the root-sheath and even in the perifollicular tissue. However, as a rule it may be stated that the part of the hair most infected is above the neck of the follicle. It grows up to a long distance in the shaft, but seldom if ever to the point of the hair. When present in the hair, in small amount, the hair preserves its cuticle entire, and looks scarcely altered. When present to such an extent as to cause fracture of the hair and the formation of stumps, the cuticle will be broken, the whole stump will be disorganized, and its end frayed out. Often the hair under the microscope seems as if it had burst at many points and allowed the spores to escape. In such a case the spores will be found lying along the outside of the hair-shaft, and grouped and scattered about the fractured portion. In some cases some hairs in a patch will escape for some time, but eventually all will become involved.

The amount of irritation caused by the fungus will vary with the amount of the fungus and its seat. When only a few spores are present, and these are superficially seated, the scalp will be only slightly reddened and scaly, or there may be some vesiculation. When the spores are present in greater number, and have penetrated into the hair-follicles, they will cause more redness of the scalp, a greater or less amount of perifollicular inflammation, and tumefaction of the scalp. The extreme degree of irritation is that met with in kerion, as we shall learn in another place. BEHREND (3) well points out the difference between the growth of the fungus in ringworm and favus, when he says: "The achorion remains for a long time confined to the superficial layers of the hair, growing quite high up in the shaft, while the trichophyton involves in a few days the whole thickness of the hair and makes it brittle, so that it breaks upon the slightest traction. On this account the hairs of favus preserve their normal lustre and consistence, while those of ringworm very soon lose the same."

DIAGNOSIS.—The diagnosis of a typical patch presents no difficulty, as there is no other disease which occurs in the form of round, partially bald, scaly patches, with disorganized hairs and "stumps" growing in them. But at times cases do occur which are not so easily made out. Seborrhœa, psoriasis, favus, lupus erythematosus, and eczema occur upon the scalp in the form of scaly patches; alopecia areata causes circular bald patches; and the other forms of alopecia denude the scalp of hair. From these, then, under certain conditions, ringworm of the scalp must be differentiated.

1. *From seborrhœa sicca capitis.*—Seborrhœa may appear at any time of life. Ringworm is almost exclusively confined to childhood. Seborrhœa has no history of contagion, and is variable in its course, getting better and worse of itself. In ringworm it is generally easy to trace the case back to its source of contagion, and the disease once started is progressive, showing little tendency to get well of itself until puberty is reached, when it generally disappears completely, never to return. Seborrhœa involves the scalp pretty generally and uniformly; if it form patches they are irregular and not sharply defined. Ringworm usually occurs in one or more isolated sharply defined patches, affecting by preference the vertex. The scales of seborrhœa are prone to heap up into thick masses, and are tenacious and greasy to the feel. In ringworm they are seldom heaped up, are readily detached, and are not so greasy. Seborrhœa may cause baldness, this condition being usually preceded by a progressive thinning of the hair-calibre over a series of years, and affecting the top of the head. With the increase

of the baldness there will be a decrease of the seborrhœa, and the baldness will be complete and permanent. In seborrhœa there is a complete absence of the tichophyton fungus, and "stumps."

2. *From psoriasis.*—Psoriasis occurs in the form of thick, crusted patches scattered more or less over the whole scalp, and tending to form a row of characteristic lesions along the margin of the hair upon the forehead. Ringworm is more apt to be confined to the vertex, its patches are scaly, and if it encroaches upon the non-hairy parts, it will form a characteristic patch of ringworm of the body. Psoriasis does not affect the hair; ringworm causes partial alopecia. In psoriasis there will be a history of relapses in many cases; this is entirely absent in ringworm. In psoriasis there is no history of contagion; in ringworm there is such a history in the majority of cases. Psoriasis rarely if ever occurs on the scalp alone, and therefore characteristic patches of the disease will be found upon the arms or elsewhere; ringworm is quite commonly confined to the scalp.

3. *From favus.*—Favus is an imported disease in this country in the vast majority of cases, and hence is seen mostly in foreigners; ringworm is endemic and often epidemic. Favus presents either the pathognomonic cupped crusts, or else thick masses of mortar or asbestos-like, grayish, friable crusts; ringworm has no cupped crusts, and the scaling is but slight. Favus causes distinct, irregular, perfectly smooth, atrophic-looking, red, bald patches, scattered over the whole scalp; ringworm causes only partially bald, slightly scaly, circular, grayish patches, confined most often to the vertex or side of the head, and the scalp is not atrophic. In favus the hair is affected secondarily, and readily pulls out entire with its root; in ringworm the hair is early affected, and when pulled

on it breaks easily and leaves its root behind in the scalp, forming the "stump." Favus is a very chronic disease, shows little tendency to get well of itself, and often lasts until late in life; ringworm is not so chronic, is most often seen in children, tends to get well of itself as its subject reaches puberty, and is rarely met with in adults. Under the microscope the mycelia and conidia of favus scales are seen to be slightly larger than those of ringworm. Its conidia are more manifold in shape, being ovoid, often elongated, and sometimes dumb-bell shaped, while the conidia of ringworm are uniformly round. In favus the hairs are mostly invaded by mycelia which may be seen as long filaments in the hair; in ringworm the conidia are found in superabundance, often exclusively, and so numerous at times as to burst the hair. It must, however, be said that it is very difficult for any but the most expert microscopist to always distinguish between the fungi of the two diseases, as they at times resemble each other very closely. Cultivation of the fungus is the only sure method of diagnosis.

4. *From lupus erythematosus.*—The only resemblance it has to ringworm is the formation of a scaly bald patch; but there is no history of contagion, the course of the disease is slower, the hair is affected secondarily, the patch has a cicatricial depression in the centre, and is of irregular outline. There is no fungus to be found.

5. *From eczema.*—Squamous eczema is the form which is most apt to be confounded with ringworm. At times a pustular eczema may complicate a case of ringworm, or a case of disseminated ringworm may simulate a pustular eczema.

Squamous eczema.—Squamous eczema has no history of contagion; in ringworm the history of contagion can generally be made out. Squamous eczema attacks all ages from infancy to old age; ringworm is usually met with in childhood alone, sparing infants and not lasting after puberty is reached. Squamous eczema often is diffused over the whole scalp, and when it does occur in patches they are not sharply defined; ringworm occurs in more or less circular, sharply defined patches. Squamous eczema is very itchy and the scalp shows scratch-marks; ringworm is but slightly itchy and scratch-marks are not commonly found. The scales of squamous eczema can be removed in plates though they are thin; the scales of ringworm are more powdery and bulky. The hair in squamous eczema is not affected, is firmly implanted in the scalp, and when epilation is attempted, it is painful; in ringworm the hair is early affected, dry, lustreless, and either comes out readily and painlessly on slight traction, or breaks off. In squamous eczema there are no "stumps;" in ringworm they are always present, and give the diagnosis, even if eczema occurs as a complication of the ringworm.

Pustular eczema. — A pustular eczema will only need to be differentiated from that rare form of disseminated and pustular ringworm. Here the condition is one of ringworm plus eczema. The diagnosis is made by the presence or absence of the characteristic hairs and stumps of ringworm. In doubtful cases the microscope will decide.

6. *From alopecia areata.*—In typical examples of this disease there should be no mistake in diagnosis. The perfectly smooth, non-scaly, non-pruriginous, perfectly bald patch of alopecia areata is in strong contrast to the rough, scaly, more or less itchy, partially bald patch of ringworm in which are stumps and diseased hairs. Alopecia areata comes on suddenly, often with antecedent symptoms of headache and pain in the scalp, and the patch is formed at once. Ringworm comes on with comparative slowness, and without antecedent

symptoms. Alopecia areata occurs at a later period of life than does ringworm, as a rule; and if a circular, circumscribed bald spot occur in an adult, it is far more likely to be one of alopecia areata than of ringworm. Sometimes a patch of alopecia areata will present a number of black dots, the unfallen roots of hair, which may bear some resemblance to the stumps of ringworm, but if they are examined under the microscope, the root will appear shrunken and atrophied, and there will be an entire absence of the spores and mycelia of ringworm. Sometimes a patch of ringworm will be perfectly smooth and without stumps. This is usually the result of treatment. Stumps will be found elsewhere on the scalp, or diseased hairs, if the disease be ringworm, and the fungus will be found in the scales from the border of the patch.

7. *From other forms of alopecia.*—The history and course of other forms of baldness, as well as the time of their appearance, are so different from what obtains in ringworm, as hardly to give rise to any difficulty.

The recognition of the presence of the fungus in the hair by the aid of the microscope is easy after a little practice. All that is needed is a microscope with a lense magnifying some 250 diameters, a slide and covering glass, and a drop of liquor potassæ with or without glycerine. After dropping the liq. potass. upon the hair, (a stump is the best one to examine,) and putting on the covering glass, wait a few minutes before looking at the specimen. The mycelia will be recognized as long, branched, jointed threads running up the hair-shaft, and the spores will be seen as small, round, refracting bodies in rows or closely packed together. Care must be exercised not to mistake air-bubbles for spores. Air-bubbles are recognized by their change of color upon changing the focus of the microscope. It is also possible to mistake the striated condition of the hair for the mycelia, but a little care and practice will prevent this.

Dyce Duckworth (421) has pointed out a ready means of determining the presence of fungus in hair, which is by laying the suspected hair in chloroform, and allowing the chloroform to evaporate. If fungus is present the hair will become white or primrose yellow where the fungus is. But this reaction is not peculiar to ringworm, as it is also found in favus and pityriasis versicolor.

The recognition of the *disseminated form* of ringworm is often a most difficult task, but it is most important for the physician to recognize it, as one such unsuspected case may be the cause of fresh epidemics. To examine these cases the child should be placed with its face to the light and its back to the physician. Then the hair is to be turned back on the head in the opposite direction to its growth. By carefully watching the scalp, some stumps will be found here and there; and Tilbury Fox has shown that, if there are any diseased hairs present, they will stand out from the head after the normal hairs have fallen into their places.

Treatment.—That nothing is easier to cure than a recent case of ringworm of the body all will agree. That a chronic case of ringworm of the scalp is very difficult to cure all writers and observers attest. Our treatment must vary with the stage of the disease. External applications are far more valuable than internal medication; and indeed the former may be relied on in the vast majority of cases for the cure of the disease without recourse to the latter.

General Treatment.—If the patient is evidently strumous or any way out of health, of course he should be given the internal treatment best fitted to his case, such as cod-liver oil, iron, arsenic, *etc.* Attention to the patient's general surroundings is in most cases of more service than giving medicines. An infected child should be isolated as much as possible, and by all means kept out of school. Isolation can and ought to be rigidly practiced in all children's asylums, and the attending physician should see to it that the infected children do not come in contact with the healthy ones. In private families isolation cannot be so readily accomplished, but much may be done to prevent the disease spreading to other children in the family by having the infected child sleep by itself; by providing towels, brush and comb for its special use; and letting it wear a close-fitting linen cap. Other children should not wear the infected one's clothing, for not only do caps carry the contagion, but also the collars of jackets and coats. The patient's head must be kept saturated with an antiparasitic oil, lotion, or ointment during the whole course of treatment, to kill the loose spores and prevent them from being carried to other heads. For this purpose we may use either a two per cent. salicylated oil, that is salicylic acid in castor oil; a saturated solution of boracic acid; a two to five per cent. carbolized oil; a solution of hyposulphite of soda, two drachms to the ounce; or an ointment of sulphur, one drachm to the ounce; or one of the ammoniate of mercury. Of course this does not exhaust the list of useful applications, but the ones mentioned are probably the best. My own preference is for the salicylated oil, as it is odorless and efficient.

Local Treatment.—In the local treatment of the disease our chief reliance is upon the use of parasiticides, to which in some cases epilation must be added. The remedies that we call parasiticides, such as mercury, chrysarobin, pyrogallol, and the like, have been supposed by some observers to do good, not by any specific action, but by the production of inflammation and scaling. (We know that the fungus cannot live in the presence of pus, and for that reason we employ, in very chronic cases, croton oil to produce acute pustulation. Here the good done is due to the inflammation produced by the remedy.) It is true that many of them do have this action, but as ringworm can be cured by these remedies, without the production of inflammation, and as no exfoliation of the epidermis caused by them could account for the cure of trichophytosis pilaris, we must hold that they do act by virtue of their specific action upon the fungus.

The first step in the treatment of all cases is to have the head well washed with soap and water, and all crusts removed. After washing, the parasiticide is to be applied at once, unless epilation is practised. The frequency with which shampooing of the scalp is to be repeated will vary with the remedial application used. As a rule it may be stated that the longer a parasiticide ointment or oil is kept in contact with the scalp the better, and it is only to be washed off when there is an accumulation of scales, or for purposes of cleanliness. But, as we shall see, some plans of treatment require daily washing of the scalp.

Epilation.—Epilation is unnecessary in recent cases, but serviceable in chronic cases, and should always be practised in the pustular form. It is by no means as effectual in ringworm as in favus, because the hair is so brittle that it breaks off when pulled on and leaves its spore-laden root in the scalp. Still, a certain number of the roots will be extracted, and this will have two good effects, namely; 1. The removal of a certain amount of fungus from the scalp; and, 2. The mouth of the follicle will be open so that the parasiticide may gain more ready access to its deeper portion. In pustular ringworm the hair comes away readily, and thus relief is afforded to the perifolliculitis. Epilation should be immediately followed by the application of the parasiticide.

Treatment of recent cases.—If we are so fortunate as to see the case in its early stage, when it is still superficial, it will be easy to effect a speedy cure. It is in such cases that many vaunted remedies have made their reputations. One of the most reliable parasiticides is the bichloride of mercury in alcohol, gr. i—iii. ad ℨ j; or even stronger, if used by the physician, sopped on three or four times a day. Here as elsewhere the best means of applying the solution is a little absorbent cotton on a wooden toothpick, or any small stick. Caution must be had to use a fresh swab each time, otherwise a remnant from the previous application will dry on the swab, and we soon will have a much stronger solution than we want. LEVISEUR (336 ap.) uses the bichloride in tincture of benzoin, 1 in 300, rubbing it in with a tooth-brush, after epilation. As soon as irritation subsides he applies a ten-per-cent salicylic acid ointment. Epilation and the bichloride are to be used once a week, and the ointment daily, the strength of each to be slowly increased to double that used at first. KERLEY (335 ap.) uses two grains of the bichloride dissolved in alcohol and added to a half ounce each of kerosene and olive oil, followed by vaseline or simple ointment, and repeated when irritation subsides. The scalp is to be frequently washed with soap and water. More rapid results were obtained by alternating a saturated tincture of iodine with the bichloride solution. These methods are also useful in the more chronic cases. Various other remedies are useful, such as sulphurous acid in full strength; sulphuret of potassium, ℨ ss-j, water ℨ j; hyposulphite of soda ℨ iij, water ℨ j; sulphite of soda ℨ j-ii, lard ℨ i; salicylic acid, two to five per cent.

in castor oil; a saturated solution of boracic acid; carbolic acid, ten to twenty per cent. in glycerine, varying with the age of the child; sulphur ointment; tinct. of iodine, and others. When the disease is still recent, but the hairs have become invaded and a characteristic patch has formed, the hair is to be cut from the patches and a slight area around them. In this stage any of the just mentioned remedies may be used. It is in such cases that chrysarobin (chrysophanic acid) effects its most brilliant cures. FAYRER (426) was the first, in 1874, to draw the attention of European physicians to its use in the treatment of ringworm as occurring in India, where it was used under the name of Goa powder, araroba, or po'dê Bahia. It was first used in the form of an ointment, and DA SILVA LIMA (474) in 1875, reported excellent results from one composed of Goa powder, 20 grains, acetic acid, 10 drops, ointment of benzoin, one ounce. It has since then been used successfully in the form of an ointment of five to twenty per cent. strength. But this manner of applying the drug on the scalp is objectionable, as it is exceedingly irritating, and frequently causes intense œdema and cellulitis of the scalp and face. The best method of using it is in the form of a "pigment" composed of

Chrysarobin,	10 parts.
Flexible collodion,	90 "
Castor oil,	3 drops.

M.

This is to be painted on the scalp with a stiff brush, and renewed as often as the film loosens or scales off.

The oleates of mercury or copper in five to ten per cent. strengths according to the age of the child are often useful, especially the first. The ointments of the ammoniate and red oxide of mercury are favorites with many. Tincture of iodine will sometimes act brilliantly in these cases, and the combination of equal parts of iodine, carbolic acid, and chloral hydrate, as recommended by CUTLER (166 ap.), is excellent both in this stage and in the more chronic ones.

Treatment of chronic cases.—The chronic cases are exceedingly hard to cure, and medical literature is rich in methods of treatment. The first requisite to success is patience both on the part of physician and patient. As there frequently is need for changing from one method of treatment to another before the case is cured, it has seemed advisable to give below a number of plans which have been found useful by competent observers.

TILBURY FOX (473 and 12) recommends epilation followed by

Ol. cadini.	ʒ iij,
Sulphur,	ʒ iij,
Lard	ad ℥ j.
M.	

He directs that the parasiticide be well rubbed in for fifteen or twenty minutes every morning and night. He also recommends the following ointments:

Sulphate of copper,		
Ammoniate of mercury,	āā gr.xx	= 1.50
Oil of cade,		
Sulphur,	āā ʒ iij	= 12
Lard,	ad ℥ j	= 30
M.		

If this is too strong the strength is to be reduced by adding more lard. Another ointment is:

Oil of cade,		
Sulphur,		
Tinct. iodine,	āā ʒ iij	= 12.
Carbolic acid,	gr.xl	= 3.
Lard,	℥ j	= 30.
M.		

He aims at producing irritation almost to the point of suppuration, and if the above ointments do not effect this, he prescribes either:

Corrosive sublimate, gr. iv-vi.	=	.25-.565
Acetic acid, 3 j	=	4.
Lard, ad. ℥ j	=	30

M.

or:

Corrosive sublimate, gr. vj	=	.565
Tinct. of cantharides, ℥ ss	=	16.
Nitric acid, ʒ j	=	4.
Distilled water, ad ℥ vj	=	200.

M.

When there are but a few small spots, he blisters them, and continues the treatment until the growth of diseased hairs is checked. For blistering fluids he uses acetic acid, or Costar's paste (iodine ʒ iij, colorless oil of tar ad ℥ j), applied sparingly to small surfaces and blotted off when they begin to smart. If much irritation, pain and swelling follow their use, he uses a poultice for one or two hours. Blistering may be repeated every four to six days. Then epilation and one of the above ointments, or the lotion is continued until the disease is cured. During the treatment the hair is kept either well greased, or soaked with dilute sulphurous acid, and covered with a silk cap. In the wards of children's hospitals he directs that sulphur be burned to disinfect the air.

E. A. BROWN (414) treats his cases by rubbing the patches with rectified oil of tar and covering them with a layer, one eighth of an inch thick, of a paste composed of tannin, iodine, gum arabic and a few drops of oil. This is to be left on for three or four days, then scraped off, and re-applied.

SAML. GEE (434) gives the following directions for

treatment. 1. Cut the hair everywhere quite close, and keep it cropped close. 2. Wash the scalp twice a day with warm water and soap. 3. After drying rub well into the scalp.

   Sulphocyanide of potassium, ℥ ss = 15.
   Glycerine, . . . ℥ j = 30.
   Water, ... ad ℥ vij = 200.
 M.

4. Keep a piece of lint soaked in the same lotion on the scalp day and night, covering it with oiled silk and a calico cap.

LADREIT DE LACHARRIÈRE (446) uses in chronic cases croton oil followed by poultices, only a small place at a time being treated. He employs a cosmetic pencil composed of equal parts of croton oil and white wax; or of equal parts of cocoa butter and white wax with fifty per cent. of croton oil. The mass is melted and poured into hollow cylinders with a diameter of half an inch. A cure is said to be effected in from six to eight weeks.

LIVEING (453) applies the tincture of iodine in double strength every day in extensive cases, and follows it with the ointment of the nitrate of mercury; or the red or white precipitate ointment with sulphur; or a ten per cent. oleate of mercury.

STARTIN (481) advises washing the part with soap and water, drying and applying a blistering fluid. After inflammation has subsided, apply equal parts of oil of cade, creosote, and tincture of iodine, and a lotion of hyposulphite of soda (ℨ ij ad ℥ j). If the skin is sore from the use of the above, apply equal parts of white precipitate ointment and vaseline.

COTTLE (416) has had most success by using "liniment crotonis" to produce pustulation, after which an ointment or lotion of salicylic acid, ten to forty grains to the ounce, is applied two or three times a day. Sometimes he combines oil of cade or carbolic acid with the salicylic acid. If an eczematous condition is present, he combines one of the soothing salts of zinc or mercury with the salicylic acid.

RICHARDSON (466) has used ethylate of sodium successfully in a case of chronic ringworm.

MORRIS (460) has been very successful with the following:

Thymol,	ʒ ss	=	2.
Chloroform,	ʒ ij	=	8.
Olive oil,	ʒ vj	=	ad 30.

M.

The amount both of the thymol and chloroform is to be reduced in young children and when the disease tends to become pustular. The oil is to be rubbed in gently two or three times a day. The part is to be washed with soap and water before the first application, but not subsequently. If at any time there is the slightest irritation the rubbing is to be stopped and the oil merely smeared on. No cap should be worn in the house, as the head is to be kept cool.

HARRISON (436) by a series of experiments upon the comparative power of various substances to penetrate the scalp, found that the most powerful combination of remedies was to apply to a small part of the diseased scalp for a few days Solution No. 1., as follows;

Iodide of potassium,	ʒ ss	=	2.
Liquor potassæ,	℥ j	=	30.

M.

on pledgets of cotton. This is followed by Solution No. 2.

Corrosive sublimate,	gr. iij	=	.2
Sweet spirits of nitre, vel.			
Water,	℥ j	=	30.

M.

In this way the head is gone over several times. He has treated thirty cases by this plan, curing them in about two months. His theory is that the liquor potassæ softens the hair and conveys the iodide of potassium to its root and bulb; then the mercury penetrates to the same place, meets the iodide of potassium and forms the biniodide of mercury just where it is wanted. In 1889 HARRISON modified his procedure on account of its being a troublesome one, and recommended that the patches be rubbed every night and morning with the following:

    ℞ Potass. caustic.,   .   gr ix,   say 2.
       Ac. carbolici,   .   gr. xxiv, " 5.
       Lanolini,
       Ol. cocois,   .   . āā ℥ss,   " 50.
    M.

And the whole scalp anointed with:

    ℞ Ungt. boricis,
       Ungt. eucalypti,   . āā ℥ij, say 33.33
       Ol. caryophilli,   . ℥ss, " 1.
       Ol. cocois,   .   . ad ℥vi, " 100.
    M.

FOULIS (428) claims to cure his cases in seven days by cutting the hair short over the affected parts, or off the whole head, if the disease is extensive, and then rubbing in a liberal quantity of spirits of turpentine. When the scalp begins to smart it is to be washed with warm water and a ten per cent. carbolic soap It is then to be dried and painted with two or three coats of the tincture of iodine. When the scalp is dry the whole head is to be anointed with carbolized oil, 1 in 20. The treatment is to be repeated once or twice a day. In very severe cases it is well to use a solution of ten grains of iodine in one ounce of turpentine.

GEO. T. ELLIOT (424) recommends painting the part with the following:

Pyrogallol (pyrogallic acid), ᴈ ss-ij = 2-8.
Salicylic acid, . . ᴈ ss = 2.
Flexible collodion, . ℥ ij = 60.
M.

LESSER (451) has found a ten per cent. solution of corrosive sublimate applied twice a day the most useful parasiticide.

KAPOSI (19) regards as specially useful the following:

Oil of birch, ... 15.
Tinct. of green soap, . . . 25.
Precipitated sulphur, . . . 10.
Spirits of lavender, . . . 50.
Balsam of peru, ... 1.50
M.

GAMBERINI (59) recommends,

Flowers of sulphur, . . . 10.
Camphor, ... 10.
Lard, ... 30.
M.

ALDER SMITH (79) has often been successful in treating ringworm with,

Boracic acid, . ᴈ iv vel q.s. = 16.
Common ether, ℥ v . = 150.
Alcohol, . ad ℥ xx . = 600.
M.

which forms a saturated solution of boracic acid. The hair is to be cut, and all scurf and sebaceous matter washed from the patches with hot water and soap. This washing is to be repeated every morning. After drying the head, the solution is to be well dabbed and pressed into the hair follicles with a small fine sponge for ten minutes. Repeat three to six times a day. During the first few weeks of treatment the solution is to be applied over the whole head. No pomade should be used during this treatment. Sometimes under this plan the scalp becomes so dry that the hair falls out of itself, and leaves bald smooth patches, similar to what is seen in alopecia areata. In very chronic cases, and especially in disseminated ringworm, his chief reliance is upon croton oil. In the disseminated variety, a minute drop is placed upon every stump, or black dot. If there are only a limited number of these, the oil is to be pressed into the follicles by means of a fine, blunt gold pin. A pustule results, the hair loosens and can be readily pulled out. If the hair breaks off, the oil is to be re-applied when the pustule has healed. A large thin poultice worn day and night under an oil-skin cap will hasten suppuration and aid in extracting the hair. In a chronic case, which has resisted all other forms of treatment, croton oil is to be applied to a small place, not much larger than a ten cent piece, at a time and followed by a poultice. If one application does not give rise to suppuration, repeat until artificial kerion is produced, the scalp being swollen, tender, puffy, and pustular. The hairs are now loosened and are to be removed, and soothing remedies applied, as in kerion. A second patch is to be attacked as soon as the first one is progressing favorably, and so all diseased portions are to be treated. This plan is to be kept for a last resort. If a bald spot remain after the use of croton oil, a stimulating lotion is to be ordered.

The same author recommends the removal of obstinate disseminated diseased hairs by electrolysis, which of course destroys the follicle.

LEFTWICH (450) advises cutting the hair close from and around the patch, and painting the scalp with an alcoholic solution of the iodide of mercury, made by adding calomel to tincture of iodine, and using the supernatant colorless fluid. When the soreness caused by it has passed off, an iodine plaster is to be applied, and left on for a week or so. This plaster is made by adding a half drachm of solid iodine to an ounce of plaster mass, and spreading it on kid. By this treatment the doctor expects to cure his cases in a month.

CHARON and GEVAERT (321 ap.) commend the treatment by galvanism proposed by REYNOLDS, of Chicago—that is, by saturating the electrode of the positive pole with a three- to five-per cent. solution of bichloride of mercury, and applying it to the diseased spot for ten or fifteen minutes. PURDON has used with benefit an ethereal tincture of the seeds of croton tiglium with salicylic acid. A neat and often reliable method of treatment is that devised by DOCKRELL (322 ap.). The patch is to be shaved and washed with a five-per cent. hydronapthol soap and hot water. After drying the patch it is to be covered with narrow strips of ten-per-cent. hydronapthol plaster, overlapping at the edges and going beyond the periphery of the patch. This is to be covered with ten per cent. hydronapthol gelatin. After four days the plaster is to be removed, the washing and drying repeated, and a twenty-per-cent. plaster used in the same way. This is to be left on for one week, and followed by a ten-per-cent plaster for ten days.

HUTCHINSON (332 ap.) has the scalp washed twice a week with one drachm of liquor carbonis detergens in one pint of hot water. He shaves the patch, or has the hair cut close, and applies every morning and evening, or only in the evening:

℞ Chrysarobin, . . ℨi, say 12.
   Hydrarg. ammon., . gr. xx, " 4.
   Lanolini, . . ℨi, " 12.
   Adipis benzoat., . ad ℨij, " 100.
   Liq. carbonis detergens, ♏x, " 2.
M.

When the case seems to be cured the same ointment, of less strength, is to be continued for six months.

BUTTE (320 ap.) has the scalp washed every second day with a spray of hot water, and then applies protochloride of iodine (ten per cent.) in lanoline. If epilation is thought to be necessary he paints the patch with many layers of the following:

℞ Alcohol at 95°, ... 12.
    Iodine, ...75
Dissolve, and add:
    Collodion, ... 35.
    Venetian turpentine, . . . 1.50
    Castor oil, ... 2.
M.

This is to be repeated every day for three or four days; then the border of the artificial skin is to be loosened and the whole gently torn away. Then parasiticides are to be used until another epilation is necessary.

QUINQUAUD (350, 351 ap.) recommends lightly scraping the patches with a curette at intervals of five to eight days, and applying twice a day:

℞ Hydrarg. biniod., . . . 20.
    Hydrarg. bichlorid., . . . 1.
    Alcohol at 90°, ... 40.
    Aquæ dest., ... 250.
M.

On the fifth or sixth day, after scraping, apply to patches and all over the scalp:

℞ Chrysarobin,
    Ac. salicylici,
    Ac. boricis, ... āā 2.
    Vaselini, ... 100.
M.

The head is to be kept covered with a rubber or cloth cap held in place by a zinc or bismuth gelatin paste. The ointment is to be used for one or two days, and then the lotion as before.

Vidal's method, according to ELOY (327 ap.), is to bathe with spirits of turpentine, and then apply tincture of iodine to a limited portion. The scalp is to be kept anointed with vaseline every other day, and covered with a closely fitting cap.

CARRERE (50 ap.) gives Besnier's treatment as consisting in epilating for six to eight millimetres about the patch, curetting it, washing with tar, salicylic acid, or sulphur soap, and covering with emplastrum Vigo. HALLOPEAU has the patches washed with sapo viridis, and then rubs in

℞ Alcohol, ... 125.
  Spts. turpentine, . . . 25.
  Liq. ammon., ... 5.
M.

and applies half an hour afterward vaseline with one per cent. of iodine. A rubber cap is to be constantly worn, and the vaseline is to be renewed at night.

D'AUDRAIN (35 ap.) applies tincture of iodine once a day, or twice if not too irritating. Every second day he removes the destroyed epithelium. When the parts have become smooth he epilates and scrapes them, and makes intradermic injections of the following:

  Hydrarg. bichlor., ...01
  Ac. tartaric., ...40
  Cocaini hydrochlorat., . . 1.
  Alcohol,
  Aquæ destill., ... āā 30.
M.

The syringe needle is to be directed obliquely and deep under the skin. One drop of the fluid is to be pressed out at each insertion, and the insertions are to be made close together. He has made as many as fifty at one sitting. Twelve days after the injection the patches are white, smooth, and absolutely bald, and he expects the hair to appear within from three to eight weeks.

The rapidity of cure will depend largely upon the thoroughness with which directions are followed, and the physician should make the applications himself during the first few days, until some one of the family becomes properly trained. The results are often best in hospitals because of the constant supervision of the house physicians and the skillful manipulations of the trained nurses.

After a case has recovered from trichophytosis a dry and scaly condition of the scalp may be left. This condition may readily be cured by

   Hydrarg. ammon . . . ℈ij = 3.
   Hydrarg. chlor. mitis . . ℈iv = 6.
   Vaseline . . . . ad ℥i = 30.
M.

a favorite formula of Dr. E. B. Bronson of New York. Or a sulphur ointment may be used of the strength of one drachm to the ounce.

When to stop treatment is a very important matter to determine. The mistake is often made of stopping as soon as the hair is growing fairly. We should suspend treatment as soon as the scalp is no longer scurfy, the hair is growing healthily, the microscope no longer shows the presence of fungus in the hair, and there are no stumps to be found in the scalp. The patient should be kept under observation for a few months, and if these same conditions are preserved he may be discharged as cured.

Over-treated cases are sometimes met with, either too strong remedies having been employed, or proper remedies continued too long. The condition present is usually one of eczema, the original disease perhaps having been cured. By suspending all treatment from time to time we will easily avoid this.

PROGNOSIS.—The prognosis is good, though the disease is often very rebellious to treatment. A too speedy cure should not be promised, and three to six months must often be given to the treatment of a chronic case. It must be borne in mind that the disease is in most cases self-limited, and the most inveterate cases tend towards spontaneous recovery with the approach of adult years, rarely lasting later than the fifteenth or sixteenth year. This should prevent us from making use of such remedies as croton oil except when other things have failed, as it produces at times permanent baldness if not carefully employed. Cases have been known to spontaneouly heal in from one to three years. Baldness rarely follows the disease excepting as the result of treatment.

Granuloma trichophyticum.—Under this title MAJOCCHI (458), in 1883, described a form of ringworm attacking hairy regions which differed from sycosis and kerion, and consisted of round tumors of normal skin-color, painless, non-scaly, surrounded by a colorless areola, of the size of a half walnut, at first elastic, then soft and fluctuating like an abscess. In the middle of each is a trichophytic hair, or a filament of fungus. Histologically the tumors have the characters of a subcutaneous granuloma, young granulation cells with blood-vessels and giant cells. MAJOCCHI believes that in these cases the fungus penetrated into the corium through the hair follicle and sebaceous gland.

CHAPTER XIII.

KERION.

Synonyms:—Trichomykosis capillitii (Auspitz); Vespajo del Capillizio, Vespajo tricofitico del Capillizio (Ital). Tinea kerion; Kerion Celsi.

Definition.—A more or less chronic inflammation of the hairy scalp, characterized by the formation of a prominent, boggy, uneven swelling, studded with numerous foramina out of which oozes a sticky, mucoid substance. The tumor at times undergoes suppuration, and generally follows upon ringworm of the scalp.

Symptoms.—Kerion derives its name from a Greek word meaning a honeycomb. The Italian name signifies a wasp nest. It has generally been regarded as a stage of ringworm of the scalp, but it is better to look upon it as a form of that disease rather than a stage, as it may be produced artificially and independently of trichophytosis. It is analogous to the nodular swellings met with in trichophytosis barbæ. Tilbury Fox (502), in 1866, first identified kerion as a form of ringworm.

As ordinarily met with the disease or condition follows upon a simple patch of ringworm. The affected part becomes red, œdematous, swollen, and boggy; it may be purplish in color. Its surface is glazed, uneven, and studded with a number of yellowish suppurating points, or with foramina out of which oozes a sticky, gelatinous, viscid, transparent fluid. At times if the inflammatory process is more intense, the swelling may reach considerable size, and instead of a mucoid fluid escaping from the foramina, true suppuration may take place attended with a sero-purulent discharge. The amount of the discharge is in proportion to the amount of inflammation present, and the depth to which the process goes. The swelling is rounded or

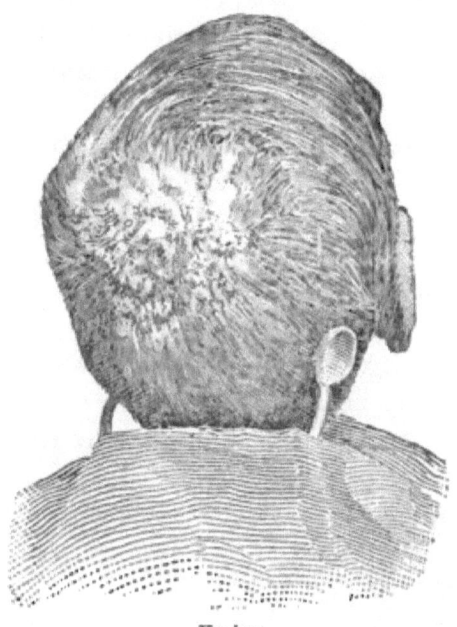

Kerion.

oval in shape, and varies in size; usually it is one or two inches in diameter, but it may become as large as a turkey's egg.

The hair on the affected part at first has the characteristic appearances of that of ringworm, when kerion follows that disease, being broken off, and presenting stumps. The pustules of the early stage of the disease form about the hairs at their exit from the scalp. Later the hairs loosen, and are easily plucked; as the inflammation progresses, they fall, and from the openings of the hair-follicles the mucoid or sero-purulent discharge takes place. If the disease is not properly managed, or if the inflammation is very intense, permanent baldness may result from destruction of the hair follicles.

The subjective symptoms are more or less pain; tenderness on pressure; at times itching and burning. The course of the disease is chronic, and it may last a very long time. At times the posterior cervical glands are enlarged, as is common in inflammatory diseases of the scalp.

ETIOLOGY.—The disease is rare. I have met with it only three times in six thousand cases. It occurs in all classes of society, but affects children especially. The scrofulous habit or a poor constitution favor this form of inflammation, though it may occur in healthy subjects. The exciting cause is, in most cases, the trichophyton fungus passing deep down into the hair-follicles. It may be produced by over treatment of a case of ringworm of the scalp; or by the application of irritants to the scalp quite independently of ringworm; or it may follow eczema or sycosis of the scalp. According to MAJOCCHI (503) this condition is sometimes met with in favus.

PATHOLOGY.—When due to the trichophyton tonsurans, the fungus penetrates deeply into the hair-follicles and there sets up an inflammation. This will vary in intensity with the irritation produced. According to ATKINSON (499), if the irritation goes only to a certain extent there will result a purely catarrhal inflammation of the hair-follicles, and the production of a mucoid secretion. If the irritation is greater, a suppurative inflammation will be established, and there will be a sero-purulent discharge. ROBINSON (33) says, that "in tinea kerion the glands of the skin seem to be affected as well as the hair-follicles, and pour out a mucoid secretion. In this form, though there is no true

suppurative process, the inflammation in the given area is so general, deep, and long-continued that the follicles are destroyed, and permanent alopecia results."
MAJOCCHI (503) found the parasite in the hair-follicle, along the hair-shaft, and in the connective tissue around the hair-follicle; the follicles filled with epithelial cells, pus corpuscles, spores, and mycelia, and the sebaceous glands and skin in the neighborhood inflamed.

We need not here describe the trichophyton fungus, as that has been done in the preceding chapter. If hairs are plucked from a non-suppurative patch of kerion the fungus will be found in abundance in them and their sheaths. If suppuration is active in the patch the fungus may be destroyed, and it may not be found in the hairs.

DIAGNOSIS.—Kerion is most apt to be mistaken for a subcutaneous abscess. It must also be diagnosed from a papilloma of the scalp, from a gummatous tumor, a sebaceous cyst and a fatty tumor.

An *abscess* is not preceded by ringworm, has no history of any irritation directly applied to the part, and may arise without any antecedent disease of the scalp. Kerion is commonly preceded by a ringworm, or there is a history of some antecedent disease, or the application of some irritant to the part. An abscess as a rule is very painful, and the patient experiences a sensation of throbbing in the part. Kerion is much less painful and sometimes itchy. Abscesses occur in subjects of lowered vitality. Kerion often occurs in the otherwise robust. The formation of an abscess is accompanied by chilliness, fever, and general malaise. These symptoms are absent in kerion. An abscess when ripe shows fluctuation and contains pus. Kerion is boggy to the feel and generally does not contain pus. There is no discharge from an abscess unless it has been opened either naturally or by the knife, and then it gives exit to pus. Kerion pours out a mucoid secretion from numerous foramina. In the hairs pulled from over an abscess the trichophyton fungus is wanting. In the hairs pulled from a kerion the fungus is usually present, or it will be found in hairs from other parts of the head.

The other diseases mentioned above should not be confounded with kerion. A papilloma is non-inflammatory, exceedingly chronic, firm to the touch, and has no discharge. A gumma is usually accompanied by other signs of syphilis, and tends to break down and ulcerate. A sebaceous cyst is slow in its growth; the skin over it is normal; it shows no great tendency to break down, and if opened it gives vent to a fetid, cheesy mass. A fatty tumor is a chronic swelling, freely movable, rather elastic to the touch, and the skin over it is normal.

PROGNOSIS.—The disease is curable, although sometimes with difficulty. The chief thing to be feared is the occurrence of permanent baldness, and this will occur in some cases even with the greatest care.

TREATMENT.—The first thing to which we should give attention is to the thorough epilation of the affected part. This will sometimes save the hair from destruction and prevent baldness. It will remove a certain amount of the fungus from the scalp, and open up the follicles for the escape of the mucoid or seropurulent secretion.

The subsequent treatment will depend upon the causation of the case. If it is due to the application of an irritant, such irritant must be stopped, and a poultice, hot water, or some mild emollient dressing applied. Mild antiphlogistic remedies are also indicated in cases complicating eczema and sycosis. But as most cases are due to the trichophyton tonsurans we should at once apply antiparasitics, just as in scabies, for instance, we use sulphur, no matter how irritated the skin may be. This is advised against by some authorities, but most are in its favor. The antiparasitics mentioned in the chapter on trichophytosis capitis are useful in kerion. Of them dilute sulphurous acid; a solution of bichloride of mercury, a grain to the ounce; hyposulphite of soda, one or two drachms to the ounce of water; carbolic acid, twenty to thirty grains to the ounce of water, are perhaps the best. These, joined to epilation, will generally result in a speedy cure.

CHAPTER XIV.

TRICHOPHYTOSIS BARBÆ.

SYNONYMS.—Tinea sycosis; Sycosis parasitica, seu parasitaria, seu contagiosa, seu menti; Tinea barbæ; Trichomykosis barbæ (Auspitz); Dermatomykosis barbæ nodosa; Mentagra; Herpes tonsurans barbæ; Trichophytie sycosique, Sycosis parasitaire (Fr.); Teignementagra, (Bazin); Parasitische Bartfinne, (Ger); Parasitic mentagra, Ringworm of the beard, Barber's itch (Eng.); Sicosi parasitaria (It.).

DEFINITION.—A contagious parasitic disease of the hair of the face and neck, caused by the trichophyton tonsurans, which invades the hair follicles, disintegrates the hair, sets up a peri-folliculitis, with inflammation of the skin and subcutaneous tissues, and gives rise to the formation of pustules, tubercles and nodular swellings. It runs a chronic course, is rebellious to treatment, and may cause permanent baldness.

SYMPTOMS.—This is the barber's itch proper, and presents different appearances in accordance with the depth to which the parasite has penetrated and the amount of irritation it causes. It begins as an ordinary ringworm of the body, a reddish, more or less circular scaly patch, appearing on the bearded portion of the face, which is either scarcely raised above the surface of the skin, or has its circumference markedly raised, and, it may be, vesicular or pustular, while its centre is scaly. In some cases, under appropriate treatment, the process may go no further; and sometimes it stops here spontaneously.

In most untreated cases the parasite penetrates the hair-follicles, sets up a folliculitis, and peri-folliculitis, and more or less inflammation of the skin and subcutaneous tissues, giving rise to the formation of pustules, papules, tubercles, nodular swellings, and, rarely, abscesses. According to BEHREND (33) it takes about fourteen days for the fungus to penetrate the deeper parts of the skin. The hair is early affected like as in the

Trichophytosis Barbæ.

other forms of ringworm, becoming dry, brittle, twisted, and broken off. Over the tubercles and nodules the hair may be extracted with the greatest ease and without pain to the patient, and it may fall spontaneously. The hair roots may be dry, or they may be swollen and boggy.

A characteristic case of the disease presents the following features: Upon the chin, neck, and submaxillary regions, we find a number of tubercles and nodules, varying in size from a split pea to a half cherry. These are irregularly shaped, for the most part rounded; are prominently raised above the surface of the skin, it may be to the height of half an inch; and show a marked tendency to group in segments of circles and to form patches. There may be one group of nodules or there may be half-a-dozen or more. The nodules themselves have a congested, purplish look. They are either hard and scaly, or they discharge a thick sticky fluid from many follicular openings, or they suppurate. The hair over them is broken or stubbed, or it has fallen out so that the affected parts are more or less bald. The skin between the separate groups is usually unaffected, as also may be the case between the individual nodules. But very often the skin over the patches is reddened and crusted, and there are a number of pustules about the hairs at their exit from their follicles. In some cases the amount of pustulation is so great that the appearances are very like those found in sycosis, and when the crusts are removed the affected part presents that moist, raw surface, studded with numerous points discharging a glutinous material, which suggested the name of σῦκον, a fig.

Subjective symptoms may be wanting; commonly some itching and burning are experienced; sometimes, if the inflammation runs high, there will be more or less pain, and even some constitutional disturbance in the form of chills, slight fever and loss of appetite. Is the vast majority of cases, the patients are more troubled by the unsightliness of the disease than by any physical discomfort.

The disease commonly affects the chin, neck, and submaxillary regions. In most cases the upper parts of the cheeks are spared, and the upper lip is very rarely invaded even in the worst cases. The malady may be limited to a single patch, or to one side of the face, or may involve the whole bearded portion of the face. It is very chronic in its course when it once becomes deep-seated. At times it may pass over from the hairy to the non-hairy contiguous parts.

ETIOLOGY.—The trichophyton fungus is solely responsible for this disease. As it affects the bearded portion of the face, men are naturally its victims. It is most frequently met with in men of early and middle manhood, say from the twentieth to the forty-fifth year. The same idiosyncrasy is shown in the susceptibility to this as to other forms of ringworm, not every man being capable of taking it. HYDE (18) has met with it more often in men with light hair and eyes, and light brown, reddish or sandy beard. It occurs most often among those who shave, and especially those who are shaved by barbers. All classes are attacked by it, but it is more common amongst the poor.

The barber's damp towels, and, may be, his fingers, are the most active agents in spreading the disease. The shaving brush and razor may convey the fungus, though if the water used in shaving is hot, the danger is reduced. The razor strop and mug may be mediate carriers of infection. The disease may also be acquired directly from animals, from cases occurring on other individuals, or may be conveyed from other parts of the body to the beard. The health of the person has no influence upon the disease.

Trichophytosis barbæ is not very common. In Boston, WHITE met with it thirty-eight times in 5,000 cases. In New York, BULKLEY had twenty-four cases in 8,000. In Glasgow, ANDERSON saw it but eighteen times in 10,000 cases. It is said by DUHRING (10) to be more common in France, and rare in Vienna. In Germany it is quite common, because, as MCCALL ANDERSON (504) says, "the men there kiss each other, and go daily to be shaved."

PATHOLOGY.—We need not say anything here about the appearances of the fungus in the hair and scales, as they are the same as in ringworm of the scalp. ROBINSON, (33) has found it in the matrix of the hair and between the hair-sheaths. If there is much suppuration it will destroy the fungus, and none may be found in many hairs examined. There is, as a rule, more mycelia in the hairs of ringworm of the beard than in the same disease of the scalp. At times little, round, glistening bodies are seen in the hairs, in their frayed-out ends, and in the remains of the root-sheath, which are regarded by LANG (520) as the product of disorganization, but not of the fungus. He affirms that the proper fungus remnants are cubical, large, glistening masses, either simple or forked. It is curious to note that before the identity of the parasite of tinea sycosis with that of trichophyton tonsurans was proven, it was called the microsporon mentographytes.

The severity of the symptoms varies with the amount and seat of the fungus, and the manner in which the tissues react to it as a foreign body, which is somewhat a matter of idiosyncrasy. That tubercles and nodular swellings are found here and not in the same disease as it occurs upon the scalp, is due to the amount and looseness of the subcutaneous connective tissue. LANG (520) believes that their presence is accounted for by the inflammation and suppuration of the sebaceous glands, which, in the bearded portion of the face, lie in very loose connective tissue, and are freely supplied with blood-vessels which anastomose around them. TILBURY FOX (12) has drawn attention to the appearances of trichorrhexis nodosa which the hair sometimes presents in this disease. This, however, is not peculiar to trichophytosis barbæ.

DIAGNOSIS.—The differential diagnosis is, mainly, between trichophytosis barbæ, and sycosis and pustular eczema. Sometimes a large papular or tubercular syphilide, an indurated acne, or an epithelioma will need to be differentiated. If the characteristics of the disease as already given are borne in mind, there should be little difficulty in making a diagnosis. The finding of the fungus in the hair is decisive against any of the diseases mentioned above.

Sycosis affects all parts of the face, notably the upper lip, and is entirely devoid of nodular swellings. Trichophytosis barbæ affects by preference the chin and submaxillary regions, spares the upper lip as a rule, and has large tubercles and nodules which tend to group. Sycosis is an active inflammation, and presents many pustules pierced by hairs. Trichophytosis barbæ is a more sluggish inflammation, and presents few if any pustules. The hair in sycosis is only affected secondarily, and is firmly planted in the skin, giving rise to pain on extraction. In trichophytosis barbæ it is primarily affected, split, twisted, broken, and is readily extracted without pain. In sycosis there is no fungus; in trichophytosis it is often abundantly present. Sycosis relapses when apparently cured, and often seems to get better and worse with the condition of the patient's health. Trichophytosis barbæ is not so prone to relapse, and is not influenced by the condition of the patient's health.

Eczema barbæ is an active inflammation of the skin of the bearded portion of the face; *tinea barbæ* is an inflammation of the hair-follicles. Eczema develops rapidly; involves large portions of the bearded face, or all of it; is devoid of nodular swellings; and presents a large amount of crusting. Tinea barbæ is gradual in its advance, affects the chin and submaxillary regions, spares the skin between the tubercles, nodules, and groups of the same, and, as a rule, is not crusted. In eczema the hairs are firmly fixed and free of disease; in tinea barbæ they are evidently diseased, easily extracted, and often wanting. Eczema is accompanied with far more itching and burning than is tinea barbæ, and is not contagious.

The large papular or tubercular *syphilides* tend to group in circles or segments of circles, as do the lesions of trichophytosis barbæ. But there is generally no more than one group of lesions in syphilis; it has a characteristic color, and a different history; other evidence of syphilis may be found, and the lesions tend to ulcerate.

Acne indurata occurs not only on the hairy but also on the non-hairy parts of the face; does not involve the hair; and if the nodules, which here are cutaneous abscesses, are opened, they give vent to a large amount of purulent sebaceous matter. A parasite is wanting.

Epithelioma should hardly be mistaken for trichophytosis barbæ. It occurs in the form of a circumscribed lesion with waxy margins over which course fine blood-vessels. The whole history of its origin and progress is different from what pertains to tinea barbæ, and when the almost inevitable ulceration takes place all doubt as to its nature vanishes.

The superficial form of tinea barbæ presents the same appearances and has the same symptoms as met with in tinea corporis, *viz.:* a superficial, more or less circular scaly patch, with a vesicular or pustular advancing edge.

TREATMENT.—The treatment of trichophytosis barbæ is prophylactic and curative. Prophylaxis consists in not being shaved by a barber, or, better, in not shaving at all, as the disease rarely affects those who do not shave. If you go to a barber's shop, owning your own brush, cup, and razor will not save you. You do not know that you can always trust your barber not to use your apparatus on other men's faces. Then too, as has been said, the damp towels, the razor strop, and perhaps the barber's fingers, may convey the contagion. If you do not own your own utensils, of course, your chances are all the worse. Above all things, the cheaper class of barber's shops are to be avoided.

The curative treatment follows very much the same line as given in the two preceding chapters. In the early stage, and before the hair-follicles have become involved, we can often succeed in stopping the progress of the disease by painting the affected part with a solution of four or five grains of the bichloride of mercury to the ounce of alcohol, to which may be added a little glycerine. This is to be painted on twice a day with a cotton swab, a fresh one being used for each application. IHLE (515) extols resorcin for this stage, exhibited in the following paste:

Pure resorcin,	10.
White vaseline,	50.
Oxide of zinc,	
Starch,	āā 25.

M

Apply two or three times a day.

Painting with the tincture of iodine; the application of a chrysarobin pigment, (Chrysarobin 10 per cent. in flexible collodion); the various mercurial ointments, are all good at this stage. When the hair-follicles have been invaded, and tubercles formed, a more active treatment must be instituted. Epilation now forms an essential part of the treatment. The hair is to be pulled not only from over the lesions but also for a small zone about them. Shaving is also to be practised, and it is often well to shave one day and epilate the second or third day, according to the rapidity with which the hair grows. IHLE claims that by using his paste, as given above, epilation is not necessary, as the diseased hairs come out of themselves. He advises cutting of the beard and applying the paste two or three times a day, its strength being gradually increased if well borne, up to twenty-five or fifty per cent. When pustulation and inflammation begin to lessen, the strength of the resorcin is to be reduced. When apparently well, a three per cent. ointment is to be continued for some time.

The rule, however, is to epilate and shave, and to apply your parasiticide after either, and re-apply it two or three times a day. A bichloride of mercury solution, gr. i to ij, in water or alcohol ʒ j; a solution of hyposulphite of soda ʒ j, to water ʒ j; dilute sulphurous acid; a two to four per cent. carbolized oil; a five to ten per cent salicylated oil (castor oil preferred); a five to ten per cent. oleate of mercury; a saturated solution of boracic acid; tincture of iodine; an ointment of sulphur, or of yellow sulphate of mercury, thymol, or napthol; these are all good parasitics. TILBURY FOX (512) advised the use of an ointment composed of

Hydrarg. Ammoniate.	gr. v	=	.3
Hydrarg. oxidat. nitrici,	gr. v	=	.3
Acidi carbolici,	gr. x	=	.6
M Adepis,	ʒ j	=	30.

HARDY (514) directs that the part be epilated and then bathed with a sublimate solution 1 in 500. The next day an ointment composed of

Hydrarg. sulph. flavæ,	2 parts.
Pulv. camphoris,	1 part.
M Adepis,	27 parts.

or of

Sulphur. sublimat.	2 parts.
M Adepis,	28 "

is to be constantly worn. When there is much inflammation he uses emollient dressings until it has subsided, and then employs the above plan. NEUMANN (27) advises green soap frictions, the opening of pustules and multiple scarifications of the patches. When there is deep infiltration he uses a plaster composed of equal parts of mercurial and diachylon plaster, with enough olive oil to make it soft.

BEHREND (3) speaks highly of scraping the patch or patches with the sharp spoon. GAMBERINI (59) paints the affected parts with a solution of oil of juniper in tincture of iodine. In the *Algemein. Wien. Med. Zeit.* (506) for 1884, it is recommended to apply after epilation and every morning and evening, the following:

   Hydrarg. bichlor. 01.–02
   Sapo-viridis,
M Ol. cade, . . . āā 5.

Epilating, shaving, and the careful use of parasiticides will effect a cure.

MORROW * epilates and applies an ointment of iodide of sulphur, thirty or forty grains to the ounce. UNNA in superficial cases recommends a ten-per-cent. resorcin ointment, or a spray of

 ℞ Resorcin, ... 5.
  Hydrarg. bichlor., ...01–.05
  Aquæ Colognien.,
  Alcohol, ... 50.
  Ol. ricini, ... 1.
 M.

followed by powdering. In the nodular form he binds on plaster-mulls made either of hydrarg. 20, ac. carbol. 7.5; or hydrarg. 10, ac. carbol. 10, corrosive

---

* Journal of Cutaneous and Venereal Diseases, 1886, iv., 32.

sublimate 1; or the last with two parts of sublimate and ten of oxide of zinc.

PROGNOSIS.—When left to itself the disease is very chronic. It may end spontaneously, though with permanent loss of hair. If properly treated it is perfectly curable, and leaves no traces. The prognosis of ringworm of the beard is indeed the same as that of the head, though it is somewhat easier to cure.

CHAPTER XV.

FAVUS.

Derivation: Favus (Lat.) a honeycomb.

SYNONYMS.—Porrigo lupinosa (WILLAN;) Porrigo favosa, Porrigo lavalis; Porrigo scutalata (LEBERT); Porrigophyta (GRUBY); Tinea favosa; Tinea vera; Tinea ficosa (PARÈ); Tinea lupinosa (GUI DE CHALIAC; Tinea maligna; Trichomykosis favosa (AUSPITZ); Dermatomycosis favosa; Kerion; Teigne faveuse (ALIBERT); Teigne du pauvre; Crusted or honeycomb ringworm, scall head, true porrigo (Eng.); Erbgrind (Ger.); Kopskurv. (Slav).

DEFINITION.— A contagious vegetable parasitic disease due to the invasion of the hairy scalp by the *Achorion Schoenleinii*. It is characterised by the presence of discrete or confluent, circular, pale, sulphur-yellow cupped crusts perforated by hair, or by asbestos-like masses of grayish crusts; by loss of hair producing irregularly shaped, disseminated, red bald patches; by running a chronic course; and by causing permanent atrophy of the scalp.

SYMPTOMS.—Favus begins either as one or more scaly erythematous spots, as minute yellowish puncta or as a group of vesicles smaller than those met with in ringworm. But though this is the mode of origin, the physician seldom sees a case in that stage, excepting as he may find new points developing upon a scalp already bearing the disease in a pronounced form. As usually met with, we find that the hair is dry and lustreless, and in places it has fallen out, leaving irregularly shaped bald patches, of all sizes, and of rather brilliant red color. Upon separating the hairs, and examining the scalp more closely, we will find both upon the bald patches and upon parts of the scalp covered with hair, little, sulphur-yellow, cup or saucer-shaped crusts, with raised or rounded edges, out of the middle of which one or several hairs will be growing. Besides these cups there will be more or less scaling, and if the disease is of some age, thick, mortar-like

Favus.

crusts of grayish color. If we approach near enough to the patient, we will appreciate a peculiar odor from the scalp.

The characteristic features of favus are: 1. Sulphur-yellow cupped crusts. 2. Baldness occurring in irregular patches. 3. Dryness and loss of lustre of the hair. 4. A peculiar odor. In a typical case all these symptoms are present. Let us examine each one by itself.

1. The cupped crusts. These are present very shortly after the commencement of the disease, and

are to be found at some period in all cases. They are situated about the hair-follicles. At their beginning they are small, about the size of a pin head, but growing rapidly they attain the size of a split pea. These are called favi or scutula. Though usually described as cup-shaped, they seem to me to be more like a saucer in form. Their edges are rounded and several lines in thickness. The depression in their centres is well marked. They are round and concavo-convex, the concavity looking upwards. At first they are covered with a thin layer of epidermis, but later the edges are free. When they are picked off from the scalp, which can readily be done, they leave a moist depression, which soon fills up. Or the scalp, under them, may be dry, red and atrophied; or it may be pustular. The color of the crust is pale or sulphur yellow: if of long standing it may become a dirty or greenish yellow from extraneous matter. The surface of the crust is uneven, and its centre is pierced by one or more hairs. These crusts occur discretely and disseminated; sometimes in groups. They are firm in consistence, and when crushed between the fingers they impart a feeling of crumbling, somewhat like mortar. Around them is a slight zone of redness.

In old cases these cupped crusts may not be present. We will then find thick mortar or asbestos-like grayish crumbly masses of scales, or perhaps these may be greenish yellow as if they were composed of dried pus. But if they are cleared off and the scalp left to itself, the scutula will again develop.

2. Baldness occurring in irregular patches. This is almost as characteristic of favus as are the scutula. The patches are of all shapes, though they show little inclination to become round or oval. There may be only one or two bald spots, or the whole scalp may be so denuded of hair that the condition may be better de-

scribed as a bald scalp with irregularly shaped and sized patches of hair. In the active stage of the disease the bald patches are covered more or less with scutula and crusts, and here and there will be solitary hairs, or little tufts of hair. Their color is an inflammatory red in the active stage, which pales with time. They have a cicatricial look, due to the atrophy which always takes place. They are wanting in glands and hair-follicles, and are permanent.

3. Dryness and loss of lustre of the hair are always present. The hair, unlike what obtains in ringworm, is affected secondarily, and grows, though with impaired vigor, for some time after the scutula have formed. Eventually, the growth of the hair is interfered with by the general atrophy of the skin, and by invasion of their roots and shafts by the parasite. They then become dry and brittle and, may be, split longitudinally and fall out of themselves, or on account of traction from brushing, scratching, and the like. They are easily pulled out when their roots are invaded and do not break so readily as in ringworm.

4. The odor is always present in a well-marked untreated case, and has been variously described as "menagerie like," or "mousey" or, like the urine of cats, or as "stale straw." In doubtful cases it may be an aid in diagnosis as it is quite unlike the smell of pustular eczema or syphilis.

Itching is the only subjective symptom. The disease may be complicated by eczema, syphilis, pediculosis, or any other disease of the scalp, and these may somewhat alter the clinical appearances. In this way we may have a decidedly pustular element added. In Randall's Island Hospitals there are always some forty or more ringworm and favus cases in the same wards. During four years I have never seen a case of ringworm and favus on the same scalp, though the children mingle freely with each other. As in other inflammatory diseases of the scalp, enlargement of the glands of the neck are quite commonly encountered. KAPOSI (19) says, that in the course of such fevers as typhus, variola, and pneumonia, the favus growth is checked only to begin again in convalescence.

Various names have been applied to designate the different clinical pictures presented by favus. Thus we have: F. discretus, when confined to one spot; F. confertus, when it extends over large surfaces; F. scutiformis, when in moderate oval patches; F. cohœrens, when a number of cups join; F. granulatus, when in mortar-like masses; F. urceolaris or dispersus, when disseminated. CHARPY (539) has described one form in which the crusts are millet seed sized and disseminated throughout the hair, to which he has given the name of F. miliaire.

ETIOLOGY.—The disease is due to the implantation and growth of the *achorion Schœnleinii* primarily in the scalp and secondarily in the hair. This vegetable fungus was first described by SCHÖNLEIN in 1839, in Mueller's *Archiv für Anatomie und Physiologie*. In 1841, GRUBY also described it, and at the time he was unaware of the discovery of Schönlein.

Favus is highly contagious, though not so much so as is ringworm. Its victims are mostly children of the poorer classes, such as Hungarians, Poles, and other foreigners, who neglect the most ordinary rules of hygiene. It is exceedingly rare to see favus of the scalp in our own people, and in this country it is one of the rarer skin diseases; the statistics of the American Dermatological Association in 1885 record only 32 cases in 16,863. But while it is not so contagious as is ringworm, it is more persistent. While ringworm tends to spontaneous recovery as puberty is reached, and it is exceedingly uncommon to see it in the adult, favus persists indefinitely. I have seen it in a woman well on in her twenties, and in a man in his thirties, not only in the form of permanent bald spots, but with scutula and asbestos crusts. The disease may be acquired either mediately or immediately; that is, directly from another individual suffering with the disease, or by wearing the cap or using the brush or comb of a favus patient. Hebra and Kaposi (15) say it is more common in males than females. Bergeron (45) found that it was more prevalent in the country than in the city. It does not affect all individuals, but seems to require some undefined peculiarity of soil for its lodgement and growth. How great a rôle the strumous or other diathesis plays in etiology it is difficult to determine. Many of the children with favus are strumous, but that is a very common condition in this class of children. Aubert (533) asserts that the disease is prone to follow injuries to the scalp, and Kaposi (19) says that the spores must land upon macerated epidermis or in a hair-follicle, and lie there for some little time before they will grow.

Animals are subject to favus, and from them it may be acquired by man. It occurs in mice, rabbits, dogs, fowls and cats. When occurring in the mouse the pressure of the fungus may cause not only atrophy of the skin but also of the bones of the skull, and kill the animal by exposure of the brain, the bones of the skull being completely destroyed.

Pathology.—A. R. Robinson (33), who has carefully investigated the parasitic diseases of the skin, thus describes the appearances of favus: "The mass (favus cup) is composed almost wholly of the luxuriant vegetable growth in various stages of development. The most apparent is the mycelium in the shape of flat, narrow threads branching and inosculating with one another in various directions. Their diameter is about the $\frac{1}{800}$ part of an inch, and their color is pale-gray,

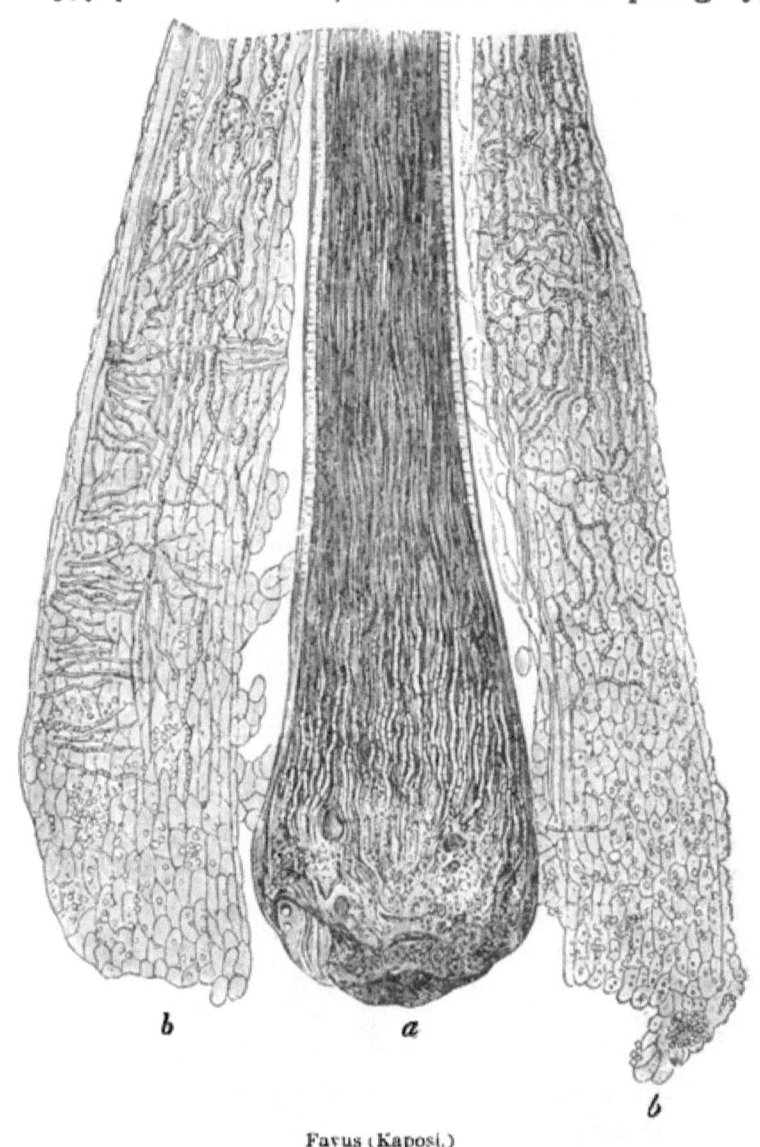

Favus (Kaposi.)

sometimes tinged with green. When in a state of fructification these tubes are divided into numerous small compartments by delicate cross lines, sometimes with constrictions, giving a chain-like appearance; and in each compartment are seen young spores in various stages of growth. The spores or conidia are present in abundance amid the meshes of the apparent growth. They are very small, of varying form, round, oval, flask or dumb-bell shaped, and of a pale greenish color. Intermediate forms between the spores and mycelia are always present, and fungoid growths of various kinds, as well as micrococci and bacteria, are often accidentally in the field of view.

" The parasite first obtains a lodgement in the funnel-shaped depression in the epidermis, through which the hair-shaft emerges upon the surface. It grows luxuriantly in the upper part of the hair-sac, and insinuates itself on all sides between the superficial layers of the epidermis. When it reaches a short distance on all sides of the follicle-mouth, it breaks the looser layers and appears on the surface, giving us the familiar cup-shaped bodies. It also invades the hair-shaft itself, though not to the extent that the trichophyton does. It penetrates between the cellular layers of the root-sheath, and multiplies in the cortical substance of the hair. The nutrition of the hair is interfered with by the mechanical pressure of the growth upon the papillæ. The hair falls out, and eventually, in many cases, the papilla atrophies, and a new growth becomes impossible. In cases of any standing the parasite may be demonstrated not only in the cortical, but in the medullary substance of the hair. Splitting of the hair may occur, as in tinea tonsurans, but as a usual thing the hair falls out before that occurs.

" In the skin itself the parasite usually confines itself to the upper corneous cells, and does not extend to the living tissues. In cases where the surface is covered by irregular, mortar-like masses of parasite, the entire upper layer of the epidermis will be found infiltrated with the achorion.

"The corium itself is usually in a state of chronic inflammation, and suppuration, which may be quite abundant, often occurs under the crusts. Even where no pus is found, the pressure of the parasite causes atrophy of the skin, and at last pit-like depressions, or more extensive reddened scars are left. When the granular structures are entirely destroyed, the achorion no longer finds a suitable nidus, and the disease at that spot is at an end."

The fungus of favus does not affect the hairs so readily as does that of ringworm. UNNA (579) has found that the hair is sometimes diseased only in its upper one-third, while the inner root-sheath is diseased throughout. Sometimes, on the other hand, the hair may be diseased throughout and the inner root-sheath unaffected. He has found the hair-bulb uniformly free from fungus, and the cortex generally undiseased till to about the upper border of the lower fourth of the hair-follicle. The usual point of entrance of the fungus into the hair is at that part of the hair follicle where the sebaceous glands find entrance, the fungus making its way through the cuticle of the hair. From the point of the entrance into the hair, the fungus grows up and down. The achorion furnishes a cement to the corneous cells in which it lives, so that the corneous layers do not scale off but form cups, and, later, mouldy or brittle masses, and the hairs do not break up as in trichophytosis.

The formation of the yellow cup is accounted for in several ways. 1. The parasite gains entrance into the hair-follicle and grows out in all directions from it as a centre. The outer rings are the newest and more succulent, while those nearer the hair or point of departure are the oldest and drier. Hence the latter will be depressed by the pressure of the atmosphere. 2. The epidermic cells in the immediate neighborhood of the hair are more firmly attached to the hair than are those further away, hence will not so readily give way and bow out under the pressure of the parasitic growth. 3. ROBINSON (33) has shown that the peripheral portion of the cup consists of a dense collection of mycelia, imbedded in a granular *débris*, while the central portion is composed almost entirely of spores which are not very closely packed together. From this it results that the peripheral portion of the cup is much firmer and more resistant to external pressure than the central part, and does not so readily sink in.

The objective symptoms of favus are due to the growth of the achorion. The skin is atrophied on account of the pressure of the growing fungus upon the constituents of the skin, squeezing, as it were, the life out of them. Whether the *achorion Schœnleinii* is the only parasite causing favus is a question that is still under discussion. The majority of investigators believe that it is, and inoculation experiments support them in their belief. GRAWITZ (435) has recently inoculated several subjects with pure cultures of favus and ringworm, and each has produced the disease peculiar to itself alone. QUINCKE (568) has found in his cultures of favus crusts, that at least three different fungi are capable of producing the clinical picture of favus. In each of the examined cases, one form was found. The three forms he designates only as α, β, γ fungi. They show marked difference under cultivation. As yet QUINCKE has not been able to classify them botanically. UNNA (402 ap.) has outdone Quincke and describes no less than nine sorts of favus fungus. On the other hand, PICK, supported by KRAL (382, 383 ap.), and MIBELLI have found only one fungus. At present the question is far from settled, and it is not possible to give here the details of the investigations of the bacteriologists. It is probable that the so-called "forms" are due to the different methods of the observers in making cultures, and to the different reactions of different skins to the favus irritation.

DIAGNOSIS.—Most cases of favus are easy of diagnosis, their features of sulphur-yellow cupped crusts; asbestos-like grayish masses; red, atrophic bald spots; and peculiar odor, being so well marked. It is to be differentiated from ringworm, eczema, seborrhœa sicca, psoriasis, lupus erythematosus, and baldness arising from various other causes.

1. From *Ringworm.* Ringworm is met with chiefly in children, rarely persisting to the age of puberty. Favus usually begins in early life, but often continues into the period of adult life. Ringworm is indigenous; favus is most often an imported disease, and is met with in Hungarians, Poles, and other foreign races which are uncleanly in their habits. Ringworm occurs in the form of circular, circumscribed spots, partially denuded of hair, and covered with grayish scales in moderate amount. It has no cupped crusts, and entire baldness is rare except from the results of treatment. Favus is more multiform and presents to view either discrete or grouped yellow-cupped crusts; or gray mortar-like masses of parasitic *débris*; or irregularly shaped bald spots scattered through the whole head, which are atrophic, devoid of follicles, and either red or white in color, according to their age. In some cases there will be a mixture of all these features. In ringworm the hairs look as if nibbled off, and we find many "stumps;" in favus stumps are not met with, and the hair though dry and cracked is not broken off. When we attempt to epilate in ringworm the hair breaks and leaves its root behind; in favus the hair with its root will come out entire. Ringworm has

no characteristic odor. In a well-marked case of favus the odor of stale straw or mice will be readily appreciated. In doubtful cases we may often decide whether the disease is ringworm or favus by the microscope. The conidia of favus are more manifold in shape than those of ringworm and slightly larger. In the hair, in favus, we have chiefly mycelia, which sometimes are exceedingly long; while in ringworm the conidia are found in greater abundance. But it is by no means easy to accurately differentiate between the two diseases by the microscope alone, and we must trust chiefly to the symptoms and course of the disease in our diagnosis.

2. From *eczema* favus needs to be differentiated when it is of long standing and there are no cupped crusts present. Sometimes the scalp of a patient with favus may be irritated by treatment to such an extent that it may become eczematous; but this is an accident that need not detain us. We would speak here of the diagnosis from pustular eczema in a doubtful case. In favus we have often a history of contagion, which is not found in eczema. Favus is usually confined to the hairy scalp. Eczema of the scalp is usually accompanied by eczema behind the ears, and quite commonly it extends upon the neck and forehead. The grayish, dirty-looking, thick, friable crusts of favus are in strong contrast with the greenish, tenacious crusts of eczema. On removing a favic crust we leave a dry, red surface. When an eczematous crust is removed, a moist, exuding surface is exposed. In favus the hair is dry, lustreless, and more or less split longitudinally. In eczema the hair is matted together, and may be dry, but it is not otherwise altered. Favus causes permanent bald spots of red and atrophic appearance. Eczema as a rule does not cause baldness, and never gives rise to permanent alopecia. The mousey odor

of favus is vastly different from the sickening smell of pustular eczema.

The pustules of a discrete pustular eczema or impetigo are rounded, non-umbilicated, whitish or grayish in color, and contain pus. These should not be confounded with the straw-colored or yellow cups of favus, which are umbilicated and firm and do not contain pus.

3. From *seborrhœa sicca*. Favus affects all parts of the scalp indifferently and irregularly; seborrhœa is most frequently confined to the upper portions of the head, and when present to a sufficient degree to require differentiation from favus, it will form a continuous patch covering the whole top of the head with the hair dry and more or less matted. The crust of favus is dry and brittle and gritty to the feel; that of seborrhœa is friable, but greasy to the feel, and when removed leaves a normal or pale skin. Seborrhœa has no characteristic odor, no cups, and if it causes baldness it will be at first a general thinning of the hair, and with it the amount of seborrhœa will lessen. The baldness caused by favus is permanent, and the skin is smooth and atrophic.

4. From *Psoriasis*. Psoriasis does not occur upon the scalp alone; when found there, other lesions will be found elsewhere on the body. Favus is quite generally found on the scalp alone. The crusts of psoriasis are scattered about through the whole scalp, are circumscribed and discrete, and when removed the skin underneath will be found reddened but not atrophic. Psoriasis presents no lesion like the cupped crust of favus, and does not cause permanent baldness.

5. *Lupus erythematosus* resembles favus only in forming red cicatricial bald areas, which are crusted. Its bald patches are really cicatricial, while those of favus are atrophic. The cicatrix of lupus is often de-

forming. Lupus is much slower than favus in its course, is generally more limited in its distribution, has no cupped crusts, and when crusted the crust is thin and adherent, and never forms the thick mortar-like masses of favus.

6. From *Alopecia*. Alopecia areata presents perfectly bald, smooth, non-scaly, white, circular, circumscribed patches, the skin of which is normal in appearance, though it may be somewhat pale and thin. Its history and course are distinct from what obtains in favus. The baldness arising from *syphilis*, as seen in the early stages of the disease, and due to the general hydræmia, is more like the baldness arising from favus than is any other form of alopecia. But the hair in syphilis is less affected than in favus, and when it falls it gives the head a ragged appearance, as if the hair had been cut off in an irregular manner with a pair of dull shears. There are no crusts, and there is a history of an initial lesion and a general eruption. When the baldness is due to a late ulcerating lesion the history will be distinct, and the cicatrix well marked. The other forms of baldness will be sufficiently distinguished by the history of their onset and course, and the absence of all the other characteristics of favus.

In doubtful cases the finding of mycelia and conidia in crusts and hair will positively exclude eczema, seborrhœa, psoriasis, lupus, and alopecia. If a case presents itself with thick crusts; or, if by reason of cleanliness, it has only red, bald patches, and we are told that it has lasted for some time and is very scaly, we can determine the presence or absence of favus by letting the disease follow its own course for a time, and watching it. Of course all the crusts that may be present must be removed. In the course of two or three weeks or less, if the case be one of favus, we will notice, at first, little red spots upon the scalp, and later the development of cupped crusts.

PROGNOSIS.—Though the disease is **obstinate to treatment**, still it is perfectly curable when handled with intelligence and perseverance. A promise of speedy cure should never be given, as it is always a matter of months, and sometimes of years. Even after the disease is apparently cured, we should have the patient present himself for inspection at intervals of a few weeks during a year. Unfortunately we can do nothing to remedy the damage done to the scalp in the formation of bald patches. These are permanent, though they will become less disfiguring by time, as the redness gradually fades.

TREATMENT.—In the treatment of favus patience and method are of more value than any special medication. Without perseverance on the part both of physician and patient it will be impossible to cure a case. Epilation, cleanliness and parasiticides are the means at our command for combating the disease.

The first thing to be done is to clean the scalp of all crusts. This is accomplished by keeping the whole scalp soaked in oil for a day or two, according to the thickness of the crust, and then washing with an abundance of soap and water. For an oil we may choose either sweet or olive oil, or oil of sweet almonds, and it is useful to add some parasiticide to the oil, such as carbolic acid gr. xv ad ℥ j; or salicylic acid, three per cent. Or we may use a poultice to remove the crusts, though this is a more disagreeable method. When the crusts have been got rid of we must epilate, and prevent the new formation of crusts by the use of our chosen parasiticide, and by washing, at intervals, with soft soap or the tincture of green soap. As a rule it is best to allow the parasiticide to remain undisturbed upon the scalp, and not to wash the head more than once or twice a week.

The most important means of cure is epilation, which may be accomplished by one of three methods: 1. By the pincette or epilating forceps; 2. By KAPOSI's (19) method; and 3. By the "calotte," or by the epilating stick, which is a modification of the calotte.

The first method, or that of the forceps, is the one most commonly employed, and when systematically carried out, is thoroughly reliable. Care must be given to the selection of the forceps. They should have an easy spring and their edges should accurately coapt. I prefer to have the blades grooved transversely on their inner face, as the hair is more firmly held by a roughened than by a smooth blade. Each hair must be plucked out from the diseased patch and from a little area around it. The operator should begin at one portion of the patch and clear off a small part every day, each hair whether sound or unsound being pulled out by a rapid jerk of the hand in the direction in which it grows. The chosen parasiticide should be applied immediately to the part epilated.

KAPOSI's (19) method has for its object the epilation of the diseased hairs alone. When the forceps is used, both the sound and the unsound hairs are pulled out. Of course it is desirable to spare the sound hairs, and Kaposi maintains that if little bunches of the hair are pulled between the thumb and some firm object, such as a straight spatula, held within the grasp of the fingers enough traction will be exercised to extract the loosened and diseased hairs, but not enough to disturb the healthy and firmly seated hairs. This is KAPOSI's method, which has given him satisfaction. It is rapid, and less painful than any other mode of epilating.

The "*calotte*," or pitch-cap, is the most rapid means of epilating, but is painful and sometimes dangerous. The "calotte" is composed of,

Vinegar,	150	parts.
Wheat flour,	25	"
Black pitch,	25	"
White pitch,	25	"

made into a mass and spread on leather. This is applied, while soft, to the whole head, and when it has set, it is pulled off suddenly by taking hold of the part over the forehead, and removing it from before backwards. Its use was abandoned on account of some unfortunate accidents. Since then various epilating sticks have been invented to take its place. One of these much used by BULKLEY (538) is composed of

Ceræ flavæ,	ʒ iij	say	10.
Laccæ in tabulis,	ʒ iv	"	15.
Picis burgundicæ,	ʒ x	"	35.
Gummi damar,	℥ iss	"	40.

These are made into a mass and cast in sticks from half an inch to an inch in diameter and two inches in length. One end is heated and applied with a sort of boring movement to the part to be epilated. When cold it is removed by a sudden twisting motion. The hair will be found sticking to the end, and must be removed by burning before the stick can be used again. Several sticks may be employed at once to a large patch. For hospital or dispensary work these epilating sticks are serviceable, but are rather harsh for private practice.

Whatever method is used it must be faithfully carried out. A skilled nurse should be employed for the work, and where such is not attainable, the physician must make it his duty either to do the epilating himself, or to train some member of the family for the work.

Parasiticides are next in importance to epilation, and should be applied after that procedure, not only to the part epilated but to the whole scalp; the latter in order to prevent the spread of the disease. There are many excellent ones from which to choose, such as a half per cent. solution of bichloride of mercury in ether or alcohol; the oleate of mercury or copper; the essential oils; tar; oil of cade; creosote in ethereal or alcoholic solution, or in oil; sulphurous acid in full strength; carbolic or salicylic acid, 3 to 5 per cent. in oil; the ointments of the ammoniate or yellow sulphate of mercury; and various others. Ointments or solutions of thymol, napthol, pyrogallol, or chrysarobin in five to ten per cent. strength, are among the newer remedies. HYDE (18) quotes Lenzberg as saying that he had never failed to cure a case of favus without epilation by generating sulphur fumes, carrying them to the head by means of a paper cap, and continuing the fumigation for five or ten minutes. PERONI (390 ap.) recommends spraying the scalp with acetic acid. If this causes excoriations they are to be treated with diachylon ointment, and when healed the scalp is to be washed with warm water and sublimate soap.

SAWICKI (572) recommends cutting the hair short and covering the whole scalp, crusts and all, with a paste made out of powdered chalk or gypsum with five or ten per cent. of carbolic acid. The paste is to be laid on 0.5 cm. thick, and the head bound with a damp cloth. In three days the whole is removed, the scalp washed with a potash soap, and the dressing reapplied. Three or four applications are said by him to effect a cure. The oil of naptha has been used from time to time, applied morning and night, after washing the scalp with soap and water. This often proves irritating. UNNA (493a) has recently recommended icthyol in the form of spray for favus. IHLE (554) cured one case in two months by means of a five per cent. resorcin ointment. I have seen as satisfactory results from epilation and the use either of a three to five per cent. solution of salicylic acid in castor oil, kept constantly applied, or from the constant and persistent use of sulphur ointment, as from any other plans of treatment.

The treatment of favus consists then, in 1st. Cleaning the scalp; 2nd. Epilating; 3d. Applying a parasiticide. These procedures must be repeated again and again, and the whole diseased portion of the scalp worked over and over, until the hair is apparently growing in a healthy manner and the scalp is free from reddish points freshly cropping out. Then the scalp should be left entirely alone for a few weeks. At the end of that time it should be inspected, some hairs examined under the microscope, and if there is no sign of the disease present, the case may be discharged. Even then the patient should be examined from month to month for some months, to make sure of the durability of the cure.

CHAPTER XVI.

PEDICULOSIS CAPITIS.

SYNONYMS:—Phthiriasis; Phtheiriasis; Morbus pediculars; Pedicularia; Malis pediculi; Pediculosis capillitii; Lousiness (Eng.); Lausesucht (Ger.); Phthiriase, ou Maladie pediculaire (Fr.).

DEFINITION.—A contagious disease of the hairy scalp due to its invasion by lice. It is characterized by the presence of the lice and their ova upon the scalp and hair; by the wounds inflicted by their proboscies; by pruritus; and by lesions consequent upon scratching.

SYMPTOMS.—The attention of the patient is first drawn to his disease by itching of the scalp, due to the irritation produced by the louse in its endeavor to obtain its nourishment from the skin, which it does by inserting its proboscis into the follicles; and by its moving about on the scalp and hair. The amount of itching varies with the susceptibility of the individual, with the number of the lice present, and with the extent and duration of the disease. One louse will cause as much itching in one individual as a whole army of them will give rise to in another. In susceptible individuals, and in aggravated cases, the itching may be so intense as to cause loss of sleep, and consequent loss of health, emaciation, and such constitutional symptoms as to constitute a grave disease. The itching always induces scratching, and the rubbing and tearing of the scalp by the nails give rise to the lesions of the disease. These are either merely excoriations, or isolated pustules, or a veritable pustular eczema, varying with the individual. People who are in bad hygienic surroundings, poorly nourished, uncleanly, and of strumous habit present the most pronounced lesions.

The part of the scalp most frequently affected is the occipital region. The parietal regions stand next in order of invasion, but always in conjunction with the occipital region. In the vast majority of cases the disease is limited to these localities, but sometimes the whole head is affected, and in bedridden, uncleanly individuals the head-lice may be found on the trunk.

The picture presented by a well-marked case of pediculosis capitis is the following: On lifting the hair from the back of the head a reddened, oozing, excoriated, crusted patch of varying size is uncovered, with outlying pustules, and cutaneous abscesses or furuncles, a veritable pustular eczema of high grade. The hair is matted together by the exudation, and myriads of lice will be seen running around amongst it and crawling along the hairs. There will be multitudes of glistening white or yellowish ova firmly attached to the hair shafts, sometimes quite near the scalp, sometimes at long distances from it. Over the whole scalp we will find isolated pustules, furuncles, excoriations, and patches of eczema. The post-cervical glands will be enlarged, there will often be patches of eczema behind the ears and upon the face and neck, and, according to KAPOSI (19), pemphigus-like blebs may occur upon the face.

Pediculosis may occur alone or it may complicate other diseases of the scalp, such as eczema, favus, and trichophytosis

ETIOLOGY.—The head louse is the only cause of the disease, and this is always derived from some other individual suffering with phthiriasis. It used to be thought that lice were the product of an eczema or other disease of the scalp, but this view is no longer tenable. It gained its support from the presence of an eczema in the cases observed. But though eczema does not produce lice, an existing eczema may favor their invasion, providing for them a good and convenient feeding ground.

The disease is met with most frequently amongst the poor, but the better classes, even those who are most cleanly, are not exempt. Thus Bulkley met with 228 cases of the disease in 8,000 cases of skin diseases, of which 7 occurred in private practice. It is more frequent than is pediculosis corporis, as Bulkley's tables show. He met with but 147 cases of the latter against 228 of the former. Children suffer from the disease more than adults, and, among adults, women more frequently harbor the vermin than do men. Infants at the breast are, usually, exempt from the disease; I have seen but few cases in them. The reason for the exemption of infants is found in the scantiness of their hair, and the greater care that is bestowed upon them. Women wear their hair long and their skin is more tender than is that of men; they therefore form a more favorable lodgment for the lice than men do, and are more often affected.

The cause of the lesions of the scalp in phthiriasis is the scratching, nature's plan for allaying the itching. The swollen glands are sympathetic with the inflammation of the scalp; they are commonly met with in all inflammatory diseases of the scalp.

The louse does not bite, as it has neither mouth nor mandibles, but simply inserts his haustellum into the follicles of the skin and sucks its food from the deeper parts. Thus it is improper to speak of louse-bites.

PATHOLOGY.—The pathology of this disease is concerned mainly with the louse itself. The pathology of the lesions of the disease is the same as that of artificial eczema of the scalp, or a common traumatic dermatitis.

There are three distinct varieties of lice which infest the human body. 1. The pediculus vestimenti seu corporis. 2. The pediculus capitis. 3. The pediculus pubis. Each has its own province, out of which it rarely passes.

The body louse resides in the clothing, which it leaves only to draw its food from the wearer of the same. This genus of louse does not concern us here. The pubic louse is found mostly on the pubes, rarely on the scalp, and will be described in the next chapter. We have here to do with the head louse alone. The male pediculus capitis is from 1 to 2 mm. long (according to Neumann (27) it is 3 to 5 mm. long)- and from 0.6 to 1 mm. broad: the female is larger by a few millimetres. It is smaller than the body louse, and larger than the pubic louse. Its head is triangular in shape; its body forms an elongated oval whose outline is scalloped with seven deep notches. It has three pairs of legs, situated well forward at the sides of the thorax, and these have powerful joints and strong claws. The legs are covered sparsely with bristly hairs. Two fine-jointed antennæ come off from the head. The color of the louse is gray with black outlines. The color varies with the race. Thus in the Esquimaux it is white, in the Negro black, and in the Chinese yellowish brown. The ova or "nits" are about one quarter of a line long, are of oval or pear-shape, and of gray color. They are found in great abundance glued to the hair-shaft, one hair bearing anywhere from one to four or more, the lower one being the youngest. They are first laid by the female along the lower part of the hair-shaft, and then are carried further away from the scalp by the growing hair. They are difficult to re-

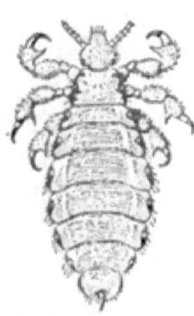

Pediculus capitis.

move and are arranged along the hair like grapes upon a stem.

The female is very prolific, laying from fifty to sixty eggs. The young hatch out in six days, and within seventeen to twenty days are capable of propagating. This accounts for the rapidity of spread of the disease, one louse being capable of producing 8000 lice in eight weeks.

DIAGNOSIS.—Pediculosis capitis is most apt to be mistaken for an eczema of the scalp, and is not infrequently treated with ointments, such as that of the oxide of zinc, which only add the element of rancid lard to the already disgusting condition present. This mistake need never occur if we bear in mind that an eczema of the occipital region is due, in an overwhelming majority of cases, to lice. In every such case then, look for the vermin, and you will find them or their nits, if the case is one of pediculosis. In some cases where only itching of the scalp is complained of and the patient is very careful of his scalp, we may find neither eczema nor lice, but simply some little grayish bodies fastened on the hairs. These may be epithelial scales or sebaceous masses perforated by hairs, a not uncommon condition; or they may be nits. The doubt is readily settled by attempting to remove them. If they are epithelial or sebaceous particles they will be easily removed by brushing. If they are ova they will resist the brush, and sometimes the comb, and often will require the application of some agent, such as acetic acid, to dissolve their connection with the hair.

PROGNOSIS.—The disease is rapidly and completely cured by appropriate treatment.

TREATMENT.—The promptest, cheapest, and most readily obtainable remedy for pediculosis capitis is crude petroleum or common kerosene oil. The head is to be saturated with this for a day or two, and then well washed with an abundance of soap and water. This will destroy the lice, but will not prevent the hatching out of the ova. These are to be removed by the fine comb, care being taken not to touch the scalp, and by pulling the hair through a towel saturated with vinegar or dilute acetic acid. By patience all the nits may be removed. As a precautionary measure the scalp should be well wetted with kerosene for a few days, after it is apparently well, to insure the death of any louse which may have hatched from a missed nit. The objections to this plan are the danger of fire, and the unpleasant odor. The first objection may be met by cautioning the patient to keep away from the fire; or the danger may be reduced by mixing the petroleum with some other oil, such as olive oil. The second objection is overcome by adding some perfume to the oil. But this is only partially effective, and the plan with all its excellencies cannot be used in private practice.

It is always preferable to use lotions rather than ointments in treating pediculosis capitis. Ointments are apt to mat the hair together and to become rancid. There are many drugs and chemicals that destroy lice. One of the most desirable for use in private practice, and in adults, is a solution of bichloride of mercury, two or five grains to the ounce of cologne water or bay rum, sopped on several times a day. This is poisonous and had best not be used on the heads of children, and is contraindicated where there is eczema. In children a good substitute is carbolic acid, ten, twenty or more drops to the ounce of alcohol, to which a little glycerine may be added. Larkspur seed (*staphisagria*) can be readily obtained, and a strong decoction of it will be found a quick and sure destroyer of lice. Instead of the decoction the tincture may be employed. Napthol, five per cent. in oil, is another good remedy. The essential oils will kill lice. The decoction or tincture of cocculus indicus enjoys a reputation of the same sort. A five or ten per cent. salicylated oil may be used. Instead of a lotion we may use an ointment, especially where there is considerable irritation of the scalp and an artificial eczema. One of the best is that of the ammoniate of mercury made with vaseline in twenty per cent. strength. Where eczema is present, we want to kill the lice first, when it will be an easy matter in most cases to cure the eczema. These eczematous heads with lice will often bear kerosene very well. After the vermin has been destroyed by the kerosene, the application of oil of cade, one drachm to olive oil one ounce, will act admirably upon the eczema and at the same time be a parasiticide.

An ointment of sulphur is efficient, as is also one of sabadilla. The latter should not be used upon a sore head as it may cause dangerous symptoms of poisoning. All the ointments of mercury will render good service. An ointment of tobacco will destroy the pediculi, but it is too disagreeable for use except when nothing else is available. Chrisma, a derivative of petroleum, has been indorsed by CRANE as an active parasiticide. (587.)

Instead of ointments some recommend powdering the hair with either calomel, pyrethrum roseum (Persian insect powder), powdered seeds of wormwood, rue, or parsley. But in my judgment ointments and powders are all objectionable. If an ointment must be used, then it is better to have it made with vaseline rather than with lard, as it is less liable to mat the hair. In all cases soap and water should be used for washing, and this should be done daily, excepting where eczema is present, when water is contraindicated. In all cases the nits are to be removed with the greatest care, and we must impress upon the patient or the attendant, that as long as there is a single nit, there is danger of a new outbreak. The nits may be removed, as already stated, by means of vinegar, or acetic acid, which is the best. To the same end alcohol, and alkaline solutions, as of borax or soda or the tincture of green soap, may be used.

CHAPTER XVII.

PEDICULOSIS PUBIS ET PALPEBRARUM.

THESE two diseases are caused by the same species of louse, the Phthirius inguinalis, or, as commonly called, crab louse. We will first consider,

PEDICULOSIS PUBIS.

SYMPTOMS.—When the louse has taken up its habitation upon the pubes it begins to propagate in the same energetic way we have learned of in the preceding chapter; and to insert its haustellum into the follicles of the skin and suck its nourishment therefrom. The movements of the lice upon the hair, and the irritation caused by their puncturing the skin, give rise to itching, which the patient tries to allay by scratching. The itching is at times intense, specially at night, but is usually not so violent as in pediculosis capitis, though it is as persistent. As a rule there is no great amount of dermatitis or eczema present, and it is rare to see such a picture presented as we constantly meet with in pediculosis capitis. This may be accounted for by the fact that the pubic region must be scratched for the greater part of the time through the clothes, while the scalp is readily accessible to the nails. If the lice are present in very great number, or the patient's skin is very vulnerable, we may have a well-marked eczema. The patient usually applies to the physician on account of a pruritus of the genital region, and in the majority of cases, is not conscious of having lice on him. In all cases of itching about the genitals it is imperative to make an examination of the parts. If lice are present we will find them as small, flat, translucent, brownish or reddish specks close down to the skin and clinging to the roots of the hair. Their position is characteristic. They usually are found with their heads buried in a skin-follicle, their claws firmly fastened about a hair, and their hinder parts sticking up in the air. At times one may be found moving about upon the hair, but they are far less active than the lice which infest the head and clothing. Besides the lice, we will find their excrement lying amongst the hair-roots in the form of minute reddish particles; and their ova upon the hair. As has already been said, we may find a well marked eczema of the pubic region. This is exceptional. But we constantly see red, punctate macules or papules scattered about, which are scratch marks. Now and again, but not constantly, we will see dull or slaty gray, or peculiar pale blue macules, varying in size from that of a lentil to that of a split pea or larger, scattered over the regions infested by the lice, as over the pubes, abdomen, and inside of the thighs. These do not disappear on pressure; they last for a few days and then vanish of themselves. They are known as *maculæ ceruleæ* or, in French, *tâches ombrées*. They are unaccompanied by any subjective symptoms.

The pubic louse is more nomadic than any of its congeners. While it is most frequently found, as its name indicates, upon the pubic region, it is not infrequently met with upon other parts of the body supplied with hair, as upon the thighs, about the anus, on the abdomen, thorax, and arms, in the axillæ, occasionally upon the eye-brows, eyelashes and beard, and sometimes on the arms. In all these situations it gives rise to the same symptoms as when on the pubes, and we must be on the watch for it in all cases of pruritus in those regions. It is not found on the scalp. Adults are the subjects of the disease in the vast majority of cases.

ETIOLOGY.—The cause of the disease is the infesting of the part with the pubic louse. The infection most commonly is effected during sexual intercourse. It is also acquired by sleeping with an infected individual or in a bed already infected by some one; and by wearing infected clothing. It is quite possible for the vermin to infest a woman in public conveyances, on account of the peculiar make of her clothing. A man's clothing affords better protection. PIFFARD(29a) thinks that it is possible to acquire the disease in water-closets. It is much rarer than are the other forms of pediculosis. BULKLEY records but 8 cases in 8,000 cases of skin diseases. The tables would indicate that it is proportionately more common in private practice than are the other forms, as of the 8 cases met with, three were in private practice and five were in public practice.

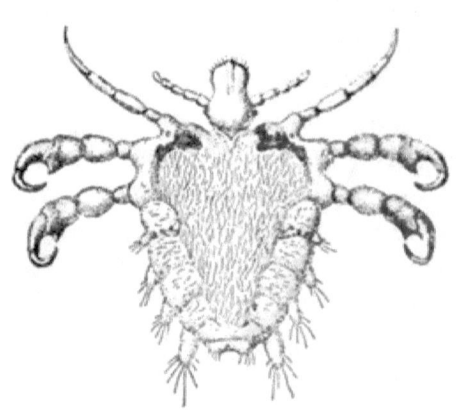

Pediculus pubis.

PATHOLOGY.—The pathology of this affection is the same as that of pediculosis capitis with but two exceptions, and those are the characteristics of the louse, and the maculæ ceruleæ. These alone will require notice here.

The Pubic Louse,—synonyms: pediculus pubis seu ferox, phthirius inguinalis, morpio, feralis pediculus; morpion (Fr.); die Filzlaus (Ger.); Crabs (Eng.),—has a less triangular and more broad head than the other species of louse, and a well-marked neck. Its body is heart-shaped, and its eight segments are not clearly indicated by notches in its outline, as is the case with the head louse. It has three pairs of legs which come off from its thorax, of which the anterior pair are proportionately delicate, and bear only a small claw at their ends. The two other pairs have a stout hook-shaped claw. The male is from 0.8 mm. to 1.0 mm. long, and from 0.5 mm. to 0.7 mm. wide; and the female is from 1.0 mm. to 1.5 mm. larger. The female has a triangular-shaped notch at the termination of the abdomen. From the head of the louse project two fine-jointed antennæ. The color of the louse is yellowish gray; and it is more or less transparent.

Maculæ ceruleæ have long been known, but until very recent times their significance has been misunderstood, and they have been considered as symptomatic of grave fevers, such as typhus fever.

MOURSON (591) was the first, in 1878, to draw attention to the connection between these spots and the presence of lice. He showed that while everyone who has pediculosis pubis does not have the maculæ, yet everyone upon whom the maculæ are found has the pediculi, or has had them shortly before presenting for examination. They were found more commonly in those who had fine skins and did not wash often. They are further proved to be due to the lice by the fact that if only one axilla was affected with lice, they would be found in that axilla alone, the other being free. Most of the cases have been found in patients suffering from some debilitating disease, and their great rarity,—GIBIER (589) failed to find a single case in two hundred subjects of pediculosis,—would indicate that some predisposition on the part of the skin is a requisite for their formation. The causal connection between the pediculi and the maculæ having been established, it next remained to determine how they were produced. DUGUET (588), in 1880, obtained the same appearances by pricking the skin with a lancet point charged with a paste made by rubbing up the bodies of twenty-five lice. But he could produce the maculæ only in individuals already bearing them; another indication of the part idiosyncrasy plays in the disease. He believes that the maculæ are caused by the emptying of the contents of the salivary glands of the louse beneath the epidermis of the human, unconnected with any further alterations in the skin. These spots spontaneously disappear in about ten days.

DIAGNOSIS.—The detection of the louse or its nits will at once give the diagnosis. It is for us only to look for the evidence of pediculi in every case of pruritus cutaneous, especially when such pruritus is limited to the pubic or axillary regions. We should suspect and carefully look for lice in all cases of eczema limited to the pubis, and even in eczema of the genitals and thighs. The pediculus pubis is less easy of detection than are the other species of louse on account of its small size, light color, translucency, greater quiescence, and more or less perpendicular position as it lies deep down among the hair-roots. Still if one is but alive to the possibility of the vermin being present, he will have no difficulty in detecting them when present.

It is important to determine whether we must deal with the pediculus vestimentorum, the so-called body louse, or with the pubic louse, as the two species of louse demand different treatment. The body louse inhabits the clothes alone, and if we search carefully we will find either the lice crawling about the clothing or will see its eggs deposited in groups along the seams of the clothing. The pubic louse dwells upon the hairy skin alone, and neither it nor its egg is to be found in the clothing. The scratch marks of pediculosis vestimentorum are found over the shoulders posteriorly, about the waist, and along the outer side of the limbs where the seams of the clothing come. The long parallel scratch marks over the shoulders are pathognomonic of this form of lice. The scratch marks of pediculosis pubis are limited more to hairy parts, and therefore are seen more commonly on the anterior face of the trunk, and in the axillæ. Maculæ ceruleæ are met with in this form alone.

PROGNOSIS.—There will be no difficulty in curing the disease if we have once made the diagnosis.

TREATMENT.—The quickest and neatest method of treatment, when the patient will allow of it, is to shave the affected parts. By this means we at once destroy the lice, nor need we wait until any complicating eczema is cured. One of the most frequently employed remedies is mercurial ointment. It is efficacious, but very often proves irritating and sets up a pustular eczema. A lotion of the bichloride of mercury is better. If there is much excoriation mercurials are to be avoided. Any of the remedies used in pediculosis capitis will be useful here, and need not now be detailed. Chloroform may be used to destroy the lice where the skin is uninjured, using the plan proposed by HAMAL (590) in 1857, namely, after washing the part with soap and water, and then with clear water, and drying, pour chloroform on drop by drop and rub in. Then cover with a folded handkerchief for a half hour, and wash again to remove the debris of pediculi. As the chloroform is irritating it is advisable to protect the skin of neighboring parts with powder. Ether will act in the same way as the chloroform. Hot baths with the free use of soap, and subsequent bathing with carbolized water, is a good plan of treatment for a generalized pediculosis caused by the pubic louse, care being taken to get rid of the ova.

Mourson caused the maculæ ceruleæ to disappear promptly with a solution of hypochlorite of soda.

Pediculosis Palpebrarum is a very rare form of pediculosis, which occurs most frequently in children. It is caused by the pubic louse and is communicated in some cases from the beard of a man in the act of fondling a child. As a rule there is but little itching in this form of pediculosis, though there is enough to cause the patient to rub the eyelids. The irritation by the lice and by the rubbing gives rise to redness of the lids and even eczema. Usually the eyelids will be found reddened, crusted, and scaly, and sometimes some of the eyelashes are broken off. Close examination will show the lice either at the roots of the eyelashes in the characteristic position, or wandering about, and the eyelashes will have ova upon them. Minute reddish specks of excrement are to be seen on the edge of the lids, especially the lower one. One or both lids may be affected, and the disease is symmetrical or unilateral. The whole of a lid may be diseased, or only places here and there. The eyebrows may also be involved with the eyelashes, or be affected independently of them.

The *treatment* consists either in pulling out all diseased hairs, and then smearing on mercurial ointment, which is a rapid and radical plan of treatment; or the mercurial ointment may be smeared on, and soap and water used to wash off the lice and remove the ova. As the first plan results in a temporary deformity, and the second is quite as effectual though slower, the latter is to be preferred.

CHAPTER XVIII.

BEIGEL'S DISEASE, OR THE CHIGNON FUNGUS; PIEDRA; TRICHOMYCOSIS NODOSA; AND OTHER PARASITIC DISEASES.

BESIDES favus, ringworm, and pediculosis, there have been described from time to time other parasitic diseases which involve the hair. In this chapter will be found Beigel's disease, Piedra, Tinea nodosa, Trichomycosis nodosa, and some unclassified parasitic diseases.

BEIGEL'S DISEASE.

In 1866, Professor LINDEMANN of Petersburg first described a parasite that he found in artificial hair, and which he considered a species of gregarine, or low form of animal life on the border line between the vegetable and the animal world. Shortly after this BEIGEL (44) examined a number of samples of artificial hair, and in one, which had been discarded by the hair-workers as unfit for use, he found upon each hair a number of dirty brown knots which adhered very closely to it. The hair itself was unaffected, and the knot formed a sort of sheath to it. With considerable care it could be stripped off without damaging the hair. Microscopical examination showed that these knots were composed of micrococci in chains, and of large round cells containing two to four large nuclei. The fungus growth was pronounced by KUCHENMEISTER as a new species of pleurococcus. HALLIER (44) subsequently determined that they were a new species of Sclerotium, and a stage in the development of the well-known penicillium. BEHREND (3) considers the chignon fungus as being a species of Piedra.

BEIGEL determined by experiment that the chignon fungus was incapable of producing the slightest ill effect upon the skin. It is therefore simply a scientific curiosity.

PIEDRA.

The best description of this is given by MALCOLM MORRIS (293) in his paper upon the subject reported in the *Transactions of the Pathological Society* of London, 1879, vol. xxx., page 441: also in the *Lancet*, 1879, x., 407. It is upon this paper I mainly rely in the following account of the disease, as from its rarity I have not had an opportunity to study it.

DESCRIPTION.—Piedra is said to occur only in Cauca, one of the United States of Colombia, and was first described in 1874 by DR. N. OSORIO of the University of Bogota. It consists in the occurrence, along the shaft of the hair, of from one to ten small dark-colored nodes which are very hard and gritty, and rattle like stones when the hair is combed or shaken. The stony hardness of the nodes gave the disease its name, which in the Spanish language means "stone." These nodes are always placed at irregular intervals along the hair-shaft, and are first met with at about half an inch from the point of exit of the hair, the root being unaffected. The disease occurs most frequently in women, men being but rarely affected, and it is the scalp-hair alone which exhibits the disease. It is non-contagious and seems to be met with only in warm valleys. The hair has an acid odor.

ETIOLOGY.—DR. OSORIO thought that the nodes were produced by an agglomeration of epithelium in certain parts of the hair. MORRIS believes that it is a fungoid growth, and due to the use of a peculiar mucilaginous linseed-like oil by the natives, especially by the women, to keep the hair smooth and shiny. Another theory is, that it depends upon the use of the waters of certain stagnant rivers, which are very mucilaginous. Heat seems essential for its production, as the use of either the oil or the water fails to cause the disease in cold climates.

PATHOLOGY.—The hair is found to be dark, weak, and flaccid. The nodes are very hard to cut, and when considerable force is used they break. Under the microscope the appearance is that of a honey-comb mass consisting of spore-like bodies, deeply pigmented on their surface. The mass in its early stage seems to originate from one cell that grows by budding in every direction, forming radiating columns of spore-like bodies. As soon as the mass has grown to a certain size, the surface cells seem to alter in shape, become darker in color, and form a pseudo-epidermis. MORRIS, (293).

DIAGNOSIS.—It is differentiated from trichorrhexis nodosa by the stony hardness of the nodes, by its occurring principally upon the scalp-hair, by its probable etiology and by the microscopical appearances it presents.

The third nodular disease of the hair is

TINEA NODOSA.

This name is selected by CHEADLE and MORRIS (278) to designate a condition of the hair which differs from trichorrhexis nodosa, in the presence of a parasitic growth resembling, though larger than, that of tinea tonsurans; in the marked incrustation of the hair-shaft by this growth; and in the absence of multiform symmetrical nodosities. It also lacks the stony hardness of Piedra.

According to their description the hair follicles and the skin were unaffected, and the hairs were firmly fixed in their follicles. The affected hairs looked as if incrusted by a granular material around and external to the shaft. In some places splitting of the hair and incrustation occurred together; in other places incrustation was seen alone, the hair-shaft being intact though somewhat cloudy and opaque. The incrustation increases toward and reaches its maximum at the free extremity. With high powers the incrustation was seen to consist of an agglomeration of minute, spherical, light-refracting bodies of uniform size, and having all the characters of a vegetable parasite. The sporules adhered, as a rule, in masses like fish-roe; but a few were found scattered about in small groups. As a rule none were found within the hair-shaft. When splitting and fibrillar disintegration were well advanced, a few spore-like bodies were seen clinging to the fibrils, or more rarely lodged in the interstices between them. The spores were larger than those of tinea tonsurans.

LEPOTHRIX.

This disease was first described by PAXTON (295), and received its name from WILSON. PATTESON (418, 419 ap.) has studied it carefully and proposed for it the name of Trichomycosis nodosa.

It affects the hair of the axillæ and scrotum, and assumes two forms, a diffuse and a nodular form. In the first the hairs appear dry and dull, and look as if they had been steamed. They feel rough and knotted, on account of the projection from the side of the shafts of numerous minute concretions. The whole hair may be involved, or there may be intervals of sound hair. The nodular form consists of small, rounded masses, generally most thickly placed on the terminal third of the shaft. Both forms may be present on the same hair. They are firmly adherent, and cannot be separated from the hair without injury to it. The hairs themselves are usually unaffected and end with fine points. Occasionally they break through a node and their ends split up into fine fibres. The follicles are not implicated.

The disease is due to the growth of bacilli upon the hairs of individuals who sweat freely. The sweating loosens the epidermic scales of hair and allows of the lodgment of the bacilli. At first these lie in small pits or depressions on the hair, but by their growth they break the cortical fibres. They also produce a hard, homogeneous, granular substance, which lies between and around the bacteria and forms the nodes, at the same time acting like a cement to prevent the breaking of the hair. The bacilli are short, fine rods with slightly rounded ends, two or three times as long as broad, and about one-fourth the diameter of a red blood corpuscle. They stain readily with aniline dyes, but best by Gram's method. They are sometimes joined together, but do not form threads. They are not cultivatable by ordinary methods.

It will be seen that this disease bears a close resemblance to tinea nodosa. Treatment has not been very successful, but an antiparasitic lotion is indicated and the use of soap and water.

DISEASES OF THE HAIR CONNECTED WITH EXCESSIVE SWEATING.

Various colored nodes surrounding the hair are found in subjects who sweat profusely. These are met with most often in the axillæ, then upon the chest, genital regions, and inside of thighs. They are of large and small size, completely invest the hair sometimes to a considerable length, and adhere closely. BEHREND (3) says that they are found in twenty per cent. of all individuals in Germany. They certainly are much less frequent in this country. They occur most often in Summer and in those who do not bathe frequently, and, like BEIGEL's chignon fungus, are of no pathological significance, excepting that they may be the cause of chromidrosis staining the underclothing red.

The color of these nodes is grayish, yellowish red or brown. When the hair is dry they are hard, and appear under the microscope as sharply contoured bodies. When plucked from the sweating skin they appear as slimy or gelatinous masses. They are composed entirely of micrococci, of which several different forms have been described. EBERTH (598) has met with them in the form of colonies composed of round, oval and biscuit-shaped bacteria; as ball-shaped, somewhat larger, bacteria; and as bacteria in chains.

The sweat macerates the hair and loosens its cuticle in places. In these the micrococci gain lodgment, and their growth being favored by the heat and moisture, they rapidly increase, still further lift up the cuticle, and form nodes. It is possible that they may even penetrate between the cortical fibres, though Waldeyer is inclined to doubt this.

Under the title of "*Dermatomykosis Palmellina,*" PICK (600), in 1875, described similar micrococci due to sweating and occurring in the same situations. But he found in his case that the hair was fragile, and in places broken off so as to leave bald spots. MARTIN (599), in 1862, reported a case in which a patch of hair on the occiput of a girl recovering from typhoid fever became golden or yellowish red, and looked as if smeared with a yellowish red pomade. In one part of the patch the hair had dropped out or was broken off, and in other parts it looked as if it had been singed. The pomade-like substance was composed of epiphytes, the *Zooglea Capillorum* of Bühl.

UNCLASSIFIED PARASITIC DISEASES OF THE HAIR.

DUHRING (597), in 1876, reported a case of parasitic disease of the hair of the head in which ova were found upon the hair, and papules and pustules on the scalp. No insect could be detected about the scalp or person, and the ova, though watched for some time, did not undergo change. The ova were firmly attached to the shaft of the hair, close to the root, and were of elongated, elliptical shape. One end was glued to the hair, the other terminated in a rounded, somewhat tapering form. They were firm in consistence, half a line in length, and a twelfth of a line in width. Their color was straw yellow. From four to a dozen ova were fastened to a hair, arranged in rows, close to one another and pointing obliquely upwards away from the scalp.

THIN (601), in 1882, reported a case of parasitic disease of the moustache occurring in a man of good health, readily cured, but reappearing once a year for five years. It caused a bald strip one quarter of an inch broad, extending from the upper to the lower border of the moustache. The diseased surface had the characteristic appearances of ringworm. The hair contained spores similar to the trichophyton fungus, but were diseased at their free extremities first, instead of at their roots, as is the case in ringworm.

PART IV.

DISEASES OF THE HAIR SECONDARY TO DISEASES OF THE SKIN.

CHAPTER XIX.

DANDRUFF.

Synonyms:—Seborrhœa sicca capitis; Pityriasis capitis; Dandriff; Eczema seborrhoicum.

The term dandruff or dandriff has been used to designate at least four distinct diseases of the scalp, namely: pityriasis simplex, seborrhœa sicca, eczema erythematosum or squamosum, and psoriasis; and it is probable that a fifth disease, diffuse trichophytosis capitis, has been included under it. Properly speaking, its use should be limited to that scaly condition of the head which is due to seborrhœa sicca or pityriasis simplex—the seborrhœal eczema of Unna.

Whether these latter three diseases are identical or not, is still an unsettled question. By some authorities they are regarded as one and the same disease, but they present enough points of difference to entitle them to separate consideration. I have here placed them together for convenience, as they give rise to a somewhat similar condition of the scalp, and are amenable to the same treatment.

Seborrhœa sicca is, as commonly taught, a functional disease of the sebaceous glands, in which an abnormal amount of sebaceous matter of abnormal consistence is secreted by them. This dries upon the scalp, and either appears in the form of thin, fatty plates about the mouths of the hair-follicles, or adheres to the hairs in flakes, or, if very abundant, heaps up into thick, fatty masses or cakes, which cling with a good deal of tenacity to the scalp. This latter form is seen very frequently in children, during the early months of infancy, and is the remains of the vernix caseosa. If portions of these crusts or cakes are rubbed between the thumb and finger, they will impart an unctuous feeling. The scalp in this disease is usually pale or leaden-hued, and when the crusts are removed shows no tendency to moisture, or else exhibits a fatty, glistening surface upon which the crust is soon renewed. In some cases more activity is shown, and the scalp is slightly reddened. This affection runs a chronic course, is generally most pronounced on the vertex, but often is distributed quite uniformly over the whole head. Some pruritus at times is present, and in some cases, in consequence of scratching, there will be excoriations. Seborrhœa of the scalp may also take the oily form (seborrhœa oleosa), though this is unusual. In that case, instead of the scalp being covered with fatty plates and scales, it will be oily, and the hair shiny.

Pityriasis simplex, *seu capillitii*, is essentially an interference with the cornification of the upper cell-layers of the skin, on account of which, instead of the normally compact stratum corneum, we have a constant shaling off of imperfectly formed epithelial scales. The whole scalp may be quite uniformly affected, or the disease may be limited to the vertex, or it may occur in circumscribed patches. The scales are thin, easily detached from the scalp, sometimes so easily as to be readily blown off, and they do not pile up into crusts. When rubbed between the thumb and finger these scales do not impart the same unctuous feeling as do those of seborrhœa sicca, though there is usually a certain amount of sebaceous matter present, just as in seborrhœa sicca there is always an admixture of epithelial scales. Usually the scalp is of normal color, though often it is slightly reddened from hyperæmia.

There is never any moisture of the scalp. Pruritus often annoys the patient, especially when he is overheated or is using his brain actively, and this inviting scratching, excoriations are often met with.

Unna, of Hamburg, and Elliot, of New York, together with some other authorities, would do away with both of the names seborrhœa and pityriasis, and substitute that of *seborrhœal eczema*, as they regard both of the former as but degrees of the latter, and both of them as inflammatory diseases of the scalp.

These two diseases, differing mainly in their essential lesion and constituting dandruff, cause annoyance by the constant falling of the scales upon the shoulders of the patient, thus ruining the clothing, or giving it the appearance of being powdered; and by the pruritus which attends them. The hair, too, is generally dry, and will not stay smooth after brushing. It is for these reasons, in most cases, that the patients apply to us for relief. But dandruff is in many cases the forerunner of baldness, and the fact that a long continued seborrhœa sicca, or pityriasis, is the most frequent cause of premature alopecia, should stimulate us to use our best efforts to cure the disease.

Etiology.—Dandruff is one of the most common of skin diseases. Statistical tables would not show this, as it is seldom so bad as to demand medical aid. It frequently occurs in strumous individuals who are anæmic and have a sluggish circulation marked by cold hands and feet. Adolescence is its peculiar time of appearance, and chlorotic girls are apt to be annoyed by it. It is an attendant upon chronic debilitating diseases, such as rheumatism, syphilis, phthisis, and the like, and comes on after profound disturbances of the constitution, such as fevers and parturition. Dyspepsia and constipation are very common exciting causes or aggravants of the disease. Improper care of the scalp, the use of the fine-toothed comb, and of pomades, hair "tonics," and hair-dyes, will give rise to the disorder. In some cases there is apparently no cause for the disease, but careful inquiry, even in these cases, will usually bring out some latent cause, such as worry, overwork, mental or nervous strain, and the like. MALASSEZ, THIN, CHINCHOLLE, and some others, claim to have found a parasite as the origin of the trouble; and recent experiments by LASSAR and BISHOP would seem to prove that the disease, at least pityriasis simplex, is contagious.* The parasitic and contagious theory of its origin is gaining ground.

PATHOLOGY.—Seborrhœa sicca is a functional disease of the sebaceous glands, no structural derangement of them having been found. The normal change of the lining cells of the glands into oil globules, which escape through the gland ducts as an oily secretion, is imperfectly performed, and instead of an oily fluid we have an inspissated fat. Microscopical examination shows the crust of seborrhœa sicca to be composed of a granular mass of epithelial cells in various stages of fatty degeneration, and some oil globules.

Pityriasis is a consequence of imperfect cornification, a condition that sometimes follows a squamous eczema. By some authorities, as is stated under etiology, a parasite is maintained to be the reason of the desquamation. Both MALASSEZ (169) and CHINCHOLLE (604), in 1874, described parasites found in the scales from cases of pityriasis capitis, and asserted that they were the cause of the disease. They described them as oval or spherical in shape, and from 2 to 5 mm. in length, without mycelia, and forming more or less extended sheets upon the surface of the corneous layer of the

---

* The experiments of Lassar and Bishop will be found in Chapter VI., page 97, of this work.

skin, splitting it into layers. These spores of MALASSEZ and CHINCHOLLE are probably identical with the saccharomyces sphœricus et ovalis of BIZZOZERO (603), which according to him are found quite generally upon the normal human skin. The question of the parasitic or non-parasitic nature of pityriasis capitis cannot be considered settled as yet.

DIAGNOSIS.—Before we can intelligently treat a case of scurfiness of the scalp we must arrive at a correct diagnosis, and must differentiate between dandruff on the one hand and eczema, psoriasis, and diffuse trichophytosis capitis on the other.

Eczema is distinguished by the scales not being so abundant nor so greasy as in dandruff; by their being more parchment-like, as if formed rather of dried serum than inspissated fat, and more adherent to the scalp; by the disease not being so diffuse, but more limited to certain patches, or to one side of the head, and implicating contiguous non-hairy parts; by the greater amount of hyperæmia; by the moisture which is either present or readily induced by scratching; by its being far more pruriginous, and by its history. If thick crusts are present they will usually be of a greenish-yellow color and when removed will expose a reddened oozing surface. In eczema there will usually be a patch of the disease behind the ears.

Psoriasis rarely occurs upon the scalp without being found on other parts of the body. It occurs in the form of circumscribed, round, or oval, reddish, infiltrated patches, which if of large size are seen to be composed of a number of smaller round patches which have joined together at their edges. These patches are covered with a thick mass of grayish or white glistening scales which are not greasy, and on being removed expose a number of minute bleeding points or red dots; and they do not reform as quickly as those of seborrhœa. The disease tends to form a fringe under the hair on the forehead, and sometimes to push its white, glistening, scaly surface down upon it, and often presents a characteristic patch just in front of the ear.

Trichophytosis capitis (tinea tonsurans), when occurring as a "ring-worm," should offer no difficulty in diagnosis, its circular shape and the presence of broken and gnawed-off hairs being pathognomonic. The diffuse form is rare, and is to be diagnosed by its history of gradual spread from numerous reddish points or papules, by its scales not being greasy, by the hair being broken off and fragile, and by the microscopical examination of the hair and scales, which will reveal the trichophyton fungus in abundance.

Besides these three diseases, *lupus erythematosus* may sometimes call for differentiation. It is rarely met with upon the scalp, and then occurs in the form of a sharply defined patch, with an infiltrated reddened base covered by a thin adherent scale, which being raised shows on its under side a number of prolongations, the sebum plugs withdrawn from the follicles. It causes loss of hair and well-marked atrophic changes in the scalp.

TREATMENT.—A good deal in the way of preventive treatment of dandruff can be accomplished by the proper care of the scalp and of the general health. More care than is usual should be bestowed upon the operations of brushing and combing the hair, washing the scalp, and upon the selection of the brush and comb.*
Do not wash the head too much. I believe that the so-commonly practised daily sousing of the head in water is hurtful to the hair and scalp, especially if they are not carefully and thoroughly dried afterward, and

---

* Full directions as to these matters will be found in Chapter III., of this work.

a little oil or vaseline rubbed into the scalp. It is not the daily sousing which is objectionable, but the insufficient after-care. Water renders the hair dry, and the daily sousing only washes the head superficially. A good shampoo every week or ten days for those persons exposed to a good deal of dust, and every two or three weeks for other people, is sufficient for cleanliness. A shampoo composed of soap and water, borax and water, or of the yolk of an egg beaten up in lime-water, is simple and good, but it must not be forgotten to wash out these materials with plenty of clean water and to thoroughly dry the hair and scalp.

Patent hair "tonics," pomades, washes, and dyes are to be avoided. Those containing grease, the pomades, are, to use an Anglicism, "nasty," give the hair an unnatural lustre, smear the hat-band and whatever the hair touches, and, becoming rancid, act as local irritants. None of these dressings are needed by the healthy scalp, and the proper care of the scalp as above indicated will preserve the hair in better condition than they will.

The nearer the body can be kept to the standard of perfect health by means of bathing, exercise, and good diet, the less likely is dandruff to develop. When, therefore, the disease has appeared, and we are applied to for relief, one of our first inquiries should be concerning the general health, and our first efforts addressed to remedying anything found to be wrong. For, important as our local measures are in relieving the local disorder, in most cases we must depend upon internal treatment to render the cure permanent. The internal treatment must be along the lines marked out in works upon general medicine—tonics, as cod-liver oil and iron, for the debilitated; the acids and bitters for the neurotic and dyspeptic; mercurials, podophyllin, and the like for the bilious, etc. DUHRING (10) recommends sulphur and the sulphide of calcium as of special efficacy; and arsenic sometimes acts well. We should insist upon our patient obeying the laws of general hygiene and instruct him as to the proper care of the scalp.

Various substances, all of a more or less irritating nature, have been recommended for the local treatment of dandruff. Such are tincture of cantharides, ℨ j. to ℥ j.; tincture of capsicum, ℨ j. to ℥ j.; tincture of nux vomica, ℨ j. to ℥ j.; chloral, ℨ j. to ℥ j.; bichloride of mercury, gr. ij. to iij. to ℥ j.; the oleate and other mercurials in proportionate strength; sulphur, ℨ j. to ℥ j.; carbolic acid, gr. x. to xx. to ℥ j.; salicylic acid gr. x. to xx. ad ℥ j; tar or oil of cade ℨ j. to ℨ iv. ad ℥ j; quinia, strychnia, etc. These are used either in solution in alcohol, water, or the oil of olive, castor, rosemary, bene, etc.; or as ointments. A good menstruum for their exhibition is composed of glycerine, ℨ j. to ij., to dilute alcohol, ℥ j. Vaseline forms the best medium for their exhibition as ointments. Excepting where the hair is decidedly thin, so stiff an ointment as the ungt. zinci oxid. should not be used, and lard itself is apt to become rancid. Lanolin is too stiff a base unless very much diluted with vaseline or oil. The tincture of benzoin is made a constituent of many lotions for the scalp, and resorcin and icthyol have both been commended as remedies in seborrhœa. Thus FOURNIER (607) recommends the following lotion:

Beef marrow,	60.
Oil of sweet almonds,	20.
Flower of sulphur,	1.
Tinct. of benzoin,	6.

IHLE'S (159) formula for resorcin is:

DANDRUFF.

> Resorcin, ... 5 to 10.
> Castor oil, ... 45.
> Alcohol, ... 150.
> Balsam of Peru, ...5

Sig. Daily rubbed in with a piece of flannel.

UNNA has had good results from icthyol in ointment form, ten to twenty per cent strength.

HEITZMANN (608) expects a rapid cure of his cases by the application to the scalp of the following ointment:

> Oleum rusci . . . ℥ss = 15.
> Ungt. aquae rosae ad . . ℥iv = 100.
> Ol. rosæ . . . . gt. x to xx = 1.
> M.

This is to be used at night, and the superfluity is to be rubbed off with a dry rag in the morning. Twice a week the head is to be shampooed with castile soap and water.

Before using any remedy for the dandruff the scalp must be cleaned. If there is enough accumulation of fat scales to form crusts, the head must be saturated with oil, preferably sweet almond oil, for from twelve to twenty-four hours, and then washed with soap and water. After the hair and scalp have been well dried the chosen remedy is to be applied. In slight cases of dandruff the soaking with the oil may be omitted.

Of all the above remedies, I have been led by experience to place my main reliance upon sulphur and the mercurials, and would advise the following plan of local treatment. If the case presents itself with a decided accumulation of scales, or if crusts are present, direct the patient to saturate his head with oil, preferably sweet almond oil, before going to bed, and place over his head a flannel cloth soaked in the oil, and outside of all an oiled silk cap. The next morning he should shampoo his head thoroughly with soap and water, using by preference the tincture of green soap or the tincture of prepared olive soap, and wash out the soap with plenty of water. The scalp is then to be dried by vigorous rubbing with a coarse towel, and the hair by pulling it through a soft towel. If the crusts are not completely removed by this method, a little oil should be kept on during the day, the head soaked again at night and washed with soap and water in the morning. If the scalp should appear very hyperæmic after the crusts are removed, apply vaseline or simple ointment, such as rose ointment, until the hyperæmia is lessened. When the crusts are removed and the hyperæmia overcome, have an ointment composed of one drachm of sulphur loti to one ounce of vaseline, or, better, the sulphur cream whose formula is given under Alopecia furfuracea, applied once a day to the scalp. If the scales form rapidly, apply the oil every night and the sulphur ointment every morning, and wash the head every second or third day. As soon as scaling is lessened stop the use of the oil, but continue the ointment, at first using it every second morning, then gradually reducing its application to once a week. Throughout this plan of treatment the head should be shampooed about once a week with the tincture of green soap, borax and water, or the yolks of three eggs beaten up in one pint of lime-water, to which a half ounce of alcohol is added. Another excellent ointment for these cases, for the formula of which I am indebted to Prof. Bronson, of the New York Polyclinic, is composed as follows:

℞ Hydrarg. ammon., . . . gr. xx.
   Hydrarg. chlor. mitis, . . . gr. xl.
   Petrolati, ... ℥j.
M.

This applied once or twice a day has yielded most

admirable results in a number of cases in simple dandruff. Its consistence being that of a Mayonaise dressing, renders it an elegant pomade for private practice. Its use should be combined with the occasional shampoo, as directed above.

PROGNOSIS.—Dandruff is curable, though liable to relapse. In some cases the disease will disappear never to come back; but these are the exceptions. Quite commonly the condition will return whenever the health becomes deteriorated. We should inform our patients of this fact, and tell them that they will obtain a cure only by persistent, long-continued, and oft-repeated effort.

CHAPTER XX.

KERATOSIS PILARIS.

Synonyms.—Lichen pilaris; Pityriasis pilaris; Icthyosis seu hyperkeratosis follicularis; Cacotrophia folliculorum (T. Fox.)

Keratosis pilaris, as its name indicates, is a disorder of cornification. It is characterised by a heaping up of the corneous cells about the mouths of the hair-follicles in the form of small conical whitish or grayish elevations, the skin between them being normal in color and texture, though often of a grayish or brownish shade from lack of cleanliness and hyperpigmentation. It is met with principally upon the extensor surfaces of the limbs, the upper arm and thigh being most often affected, though it may occur almost anywhere on the body and not infrequently on the scalp. To the eye, the skin has the appearance of what is commonly called goose flesh, "*cutis anserina.*" It is seen to be dotted all over with little pinhead to small pea-sized, conical, whitish, grayish, blackish, or pinkish papules, each of which is either pierced by a hair or has at its summit a small black dot, indicating the mouth of a hair-follicle. These papules are often scaly, sometimes surmounted by a scale. Sometimes the hairs grow vigorously from the papules, sometimes they are broken off, and sometimes are to be found only by opening the papule, when they will be seen curled up inside of it. To the touch the skin feels dry, and harsh and somewhat like a fine nutmeg grater.

Subjective symptoms are wanting in most cases, but there often is more or less pruritus, especially in children; and in them there may be a slight degree of eczema on account of the scratching. Occurring on the scalp it sometimes produces baldness.

ETIOLOGY.—The disease is most often met with after puberty, though it may be congenital. It is a very common affection of the skin, but, as it gives little or no trouble, we are seldom called upon to treat it. It is the result of an inactive state of the skin, and is seen most often in those who do not bathe with sufficient frequence. It is met with in connection with icthyosis and prurigo, and is said to follow pityriasis rubra. It was formerly regarded as a species of papular eczema.

PATHOLOGY.—It is simply an anomaly of cornification, a thickening of the corneous layer of the skin about the mouths of the hair-follicles, by which the epithelial cells are heaped up into conical papules. It has nothing to do with any disorder of the sebaceous glands or of the hair-follicle itself. When this condition is congenital and constant in spite of treatment it has been named icthyosis follicularis, and corresponds to T. Fox's cacotrophia folliculorum.

DIAGNOSIS.—It is necessary for us to differentiate keratosis pilaris from cutis anserina; the miliary papular syphilide; lichen scrofulosus; papular eczema; lichen planus, and icthyosis.

Cutis anserina is a passing condition of the skin in which the arrectores pili muscles contract under the stimulus of cold, and raise up the hair and immediately contiguous parts into papules. Keratosis pilaris is constant and uninfluenced by temperature.

The *miliary papular syphilide* has its papules grouped; they are of a dark red or raw ham color, and deep seated, and are only slightly scaly. The syphilitic eruption is more apt to be a general one, and fades away of itself in a few weeks. The papules of keratosis are whitish, grayish or blackish, superficial so that they are readily removable by soap and water, scaly, and usually confined to the arms and thighs. It does not tend to get well of itself.

Lichen scrofulosus, as generally seen, is in the form of well-marked circular or crescentic patches of yellowish-brown papules upon the trunk, usually upon the abdomen. The papules are slightly scaly, and the subjects generally present other symptoms of struma. Keratosis pilaris does not have grouped papules; its papules are whitish or grayish, and its subjects are not necessarily strumous.

Papular eczema occurs as an eruption of bright red inflammatory papules which tend to run together and form patches. It is very itchy, and shows no predilection for the upper arm and thigh.

Lichen planus occurs especially on the anterior face of the wrists as a group of dull red or lilac-tinted, angular, slightly umbilicated, flat papules, and is attended with a good deal of pruritus.

Icthyosis is a congenital affection of the general integument, though most marked upon the arms and legs. The skin is dry and scaly, and marked off into polygonal spaces, and the disease is incurable. Keratosis pilaris rarely occurs before puberty, is limited to the mouths of the lanugo hair-follicles, and is perfectly curable.

TREATMENT.—The vigorous use of soap and water in an alkaline bath will promptly remove the papules. The best soap for the purpose is the sapo viridis or soft soap, and this may be used in the form of a tincture.

A vapor or Russian bath may be used for the same purpose. After the bath the skin should be anointed with oil, vaseline, lanolin, or any emollient. In some obstinate cases it may be necessary to use a mercurial such as:

℞ Hydrarg. ammon. . . . ℈j—ii
Hydrarg. chlor. mitis, . . . ℈ij—iv
Vaseline, . ad . . . . ℨj
M.

In congenital cases cod-liver oil should be given by the mouth and also rubbed into the affected skin.

PROGNOSIS.—The disease is perfectly and easily curable, but without treatment it may last indefinitely. It is subject to relapses.

CHAPTER XXI.

ECZEMA CAPITIS ET BARBÆ.

The hairy parts of the body are affected with eczema either in connection with or independently of eczema of other regions. Owing to the presence of the hair, and, in some places, as the scalp, to a difference in the structure of the under-lying tissues, the disease in hairy regions has some peculiar symptoms. We will describe eczema of the scalp, of the bearded portions of the face, and of the edges of the eyelids. The symptoms of the disease, as met with in these regions, include all that are to be encountered in any hairy region.

ECZEMA CAPITIS.

Synonyms:—Crusta lactea; Impetigo figurata, seu lactantia, seu mucosa, seu muciflua; Porrigo; Tinea amiantacea, seu furfuracea, seu granulata, seu asbestina; Melitagra; Achor; Eczema capillitii; Erythema ichorosum; Gourme (Fr.); Vesicular or running scall, scalled head, milk crust (Eng).

Symptoms.—By the above and a number of other names this very common disease of the hairy scalp has been designated. The vesicular, pustular and erythematous varieties of eczema occur upon the scalp as primary forms far more frequently than the papular variety, which is exceedingly rare: squamous eczema is met with as a sequella of the other forms. Eczema may affect the whole scalp or only a portion of it; and it may run an acute or chronic course. It may occur either in connection with eczema of other parts of the body, or independently.

Vesicular Eczema.—Eczema vesiculosum is an acute disease which breaks out either upon a part or the whole of the scalp. The vesicles are so short lived that the physician hardly ever sees them excepting upon newly formed patches. When the case presents itself, the scalp is seen to be swollen, at times so much so as to give a boggy sensation to the touch; it is moist, and in bad cases exuding so freely that it is covered with a sticky, yellowish, mucilaginous fluid. The hair is always stuck together in little bundles, and if it is long it will be matted. The head emits a sickening odor. The exudation dries into light-yellow adherent crusts; when these are removed a moist exuding surface is exposed that soon becomes again crusted. This form of eczema may change into the pustular form, or it may become squamous. It runs an acute course in most cases.

Pustular Eczema.—Eczema pustulosum occurs as an eruption of discrete pustules affecting a part or the whole of the scalp. The pustules soon become confluent, break down and discharge their purulent contents. The hair is matted together. The exudation dries into thick yellowish green or blackish adherent crusts and the head emits a foul odor. When the crusts are removed a moist exuding surface is left, which soon becomes covered with a fresh crust. After lasting an indefinite time the pustular form is exchanged for the squamous.

The pustular form differs from the vesicular form in having pustules rather than vesicles; in the character of the exudate; in its crusts being darker colored; and in having a more disgusting odor. Sometimes the pustules are located only about the hair-follicles and the disease will assume the form of sycosis. At times there will be marked swelling of the cervical glands; and, especially in children, abscesses of the scalp may form.

Erythematous Eczema.—Eczema erythematosum usually occurs in patches upon one side of the head but may affect the whole scalp. The patches are of irregular form, bright red in color, perfectly dry and slightly scaly. Sometimes the skin may be a little swollen or thickened. If the scales are removed with soap and water or slight friction, the skin will become moist and exude serum from innumerable pores. This form of eczema either gets well rapidly or changes to the squamous form.

Squamous Eczema. — Eczema squamosum on the scalp as elsewhere is the final stage through which the other varieties of eczema pass on their way to recovery. The whole scalp may be affected or only a patch here and there, depending upon the nature and location of the preceding primary form. It also presents several forms. It may be of only slight intensity when the scalp will be but slightly thickened, of light-red color, and covered with whitish or grayish scales, which come off readily in the form of flakes. Or it may be more severe, when the scalp will be decidedly thickened, of deep-red color, and covered with more or less adherent white or gray scales in the form of plates. Or the thickening of the skin and the inflammation may be yet more severe, and the scalp will be red, scaly and cracked. This form of eczema may last indefinitely as a chronic eczema of the scalp.

These are the four varieties of eczema as commonly met with on the scalp. The pustular and the squamous forms are the most frequent, the former especially in children, where it constitutes the "milk crust." The erythematous form is seen chiefly in adults. At times there will be several varieties of eczema upon the scalp at once—for example, the pustular variety in one place and the squamous in another. With eczema of the scalp we usually will find a patch of the same disease behind or upon the ears, upon the back of the neck, or on the face. Sometimes other hairy regions will be affected at the same time, as the pubes or axilla.

The hair is unaffected as a rule, excepting that it is glued together by the serous or sero-purulent exudation of the vesicular and pustular forms, and is dry in the erythematous and squamous forms. It is only when the disease is very chronic that there is baldness, which is not permanent; the hair in most all cases growing again when the eczema is cured. It is not uncommon to meet with baldness on the back of the head in children with eczema, the hair being rubbed off by the constant scratching of the head against the pillow or the nurse's arm.

Itching is present in greater or less degree in all cases. It is often intense in the erythematous and squamous varieties; less marked in the pustular and vesicular forms. In the acute stages of the disease the scalp will feel drawn, and, may be, painful, and there will be more or less of a burning sensation.

The sebaceous glands are usually functionally diseased in a case of eczema giving rise to a seborrhœa. In the crusts of eczema of the scalp oil globules are generally found, and in a chronic eczema the dry and lustreless appearance of the hair is in part due to a deficiency of the oily secretion of the sebaceous glands.

In neglected cases among people who are unclean in their habits pediculi find lodgment in the hair; and in some aggravated cases of uncleanliness we have developed that condition which is described in the next chapter as Plica Polonica.

ETIOLOGY.—The etiology of eczema of the scalp is no more settled than is the etiology of eczema in general. Some authorities, as in the Vienna school, insist upon the disease being local in all cases ; some of the French school are equally positive that eczema is an expression of a diathesis; while some, and perhaps these now are in the majority, take the safer middle course and teach that some cases are due to local causes, and some to constitutional conditions. By the use of croton oil an artificial eczema can be produced on any skin; but that does not prove that eczema is always due to a local cause. Water, with or without strong alkalies, may be used by a person for years with impunity; then there may come a time when it will cause an intense eczema of the hands. This seems to show clearly that there has arisen in the person some constitutional condition which renders the skin obnoxious to the use of water. Eczema, I believe, may be in some few cases a purely local disease; then it has rather the nature of a dermatitis and is of short duration. In most cases, however, there exists a predisposition to the disease, and when the predisposing causes are very pronounced and not easily removable, the disease will be very chronic and obstinate to treatment. This predisposition may be so pronounced that we may speak of the person having it as being eczematous, just as we would speak of another being rheumatic or gouty.

The causes of eczema of the scalp are predisposing and exciting. Here I do not intend to enter upon a discussion of all the causes of eczema as given by different authors, but shall content myself by giving such as have special relation to eczema of the scalp.

The predisposing causes of eczema capitis are infancy and a debilitated condition of the system arising from any cause.

Age is an important element in etiology. Children under five years of age furnish nearly one-quarter of all cases of eczema,[*] and in them the scalp is more fre-

---

[*] Bulkley's Eczema and its Treatment, New York, 1884.

quently attacked both independently and in connection with the disease elsewhere, than is the case in adults. In forty-nine personal cases of eczema of the scalp, thirty-five occurred in children under five years of age, and of these twenty-seven were in connection with eczema on other parts of the body, mostly of the ears and face. In the remaining fourteen cases which occurred in adults six were on the head alone. That children should be more predisposed to eczema than are adults is due to their skin being more delicate and hence more vulnerable. In them, too, we more often see evidences of struma, and vices of constitution either inherited or acquired.

A debilitated condition of the system predisposes to eczema in general, and hence to eczema of the head. Fat anæmic babies are prone to eczema. Over-fed or under-fed children, who live in bad hygienic surroundings and are uncared for, furnish some of our most obstinate cases. Disturbances of the digestion, expressed either as nausea, vomiting, diarrhœa, constipation, or simply as malassimilation without any tangible symptom, by their effect upon the constitution of the blood, and hence on the nutrition of the skin, predispose to eczema. Gout and rheumatism, nervous exhaustion, the chlorotic state, predispose to the disease. In almost every case there will be found some deviation from the standard of health.

The exciting causes of eczema capitis are all injuries to or irritations of the scalp. In children a very frequent exciting cause is well meant, but badly directed efforts at cleanliness. The vernix caseosa is attacked too vigorously with soap and water before it is properly soaked with oil; the fine-toothed comb is employed for the removal of dandruff and the scalp is scratched by it; or a too stiff brush is used for the tender scalp. In both children and adults all cutaneous irritants may give rise to eczema, such as mercurial ointments applied for the destruction of lice, and the like; tincture of arnica for bruises; water fomentations; too strong applications for the cure of seborrhœa or loss of hair, and the like. Pediculi are very frequently the cause of eczema, especially of the occipital region. The too vigorous use of the comb and brush may act like any other irritant in causing the disease. It is also possible that in an infant the growing hair may irritate the scalp.

DIAGNOSIS.—A moist eczema of the scalp is easy of diagnosis if we remember that in it the scalp is reddened; more or less thickened; exuding in some places and crusted in others; itchy; and the hair is stuck together as if by mucilage. Only one other disease simulates it and that is pediculosis. On the other hand the erythematous and squamous forms are at times difficult of diagnosis from pityriasis, ringworm, erysipelas, lupus erythematosus, or a simple dermatitis; and pustular eczema of the scalp in its more chronic form when it occurs in patches may be mistaken for psoriasis, seborrhœa sicca, favus, or syphilis.

A moist, exuding eczema is to be differentiated from *pediculosis* by its occurrence generally all over the scalp, while pediculosis is more limited to the occipital and temporal regions. In eczema there are no pediculi or ova, while in pediculosis they are readily to be found. When pediculosis is well marked there is always a pustular eczema, and in a pustular eczema of any continuance there are liable to be pediculi. But then the diagnosis is not essential, the indication for parasiticides being so plain, that in treatment they are to be used first, and afterwards remedies for the eczema.

In *pityriasis* the scalp is but slightly reddened, not at all thickened, the scales are abundant, readily detached and furfuraceous, the hairs are not stuck together, but frequently are found piercing a small scale. In eczema the scalp is decidedly red and more or less thickened; the scales are moderate in amount, often quite adherent, and come off in the form of plates; and the hairs here and there may be found stuck together, if there are any moist spots on the head.

Ringworm occurs in one or more circumscribed, round, scaly patches. The patches of eczema are not sharply circumscribed. In ringworm the hairs are brittle, broken off, twisted, and come out readily. In eczema the hairs are merely stuck together. In ringworm we have stumps of broken hair, which are pathognomonic of the disease. The crusts of ringworm are grayish in color, those of eczema are yellowish or greenish. Ringworm is never moist, has a well-marked history of contagion, the patches spread from one point peripherally, and it is not very itchy. Eczema is generally moist at some time, is not contagious, does not develop from one focus, and is very itchy. In ringworm the trichophyton fungus is readily demonstrable by the microscope. In eczema there is no fungus.

Erysipelas creeps over the scalp with a sharply cut advancing outline; there is a considerable swelling of the integument, and marked fever and constitutional disturbance. It runs its course rapidly. Eczema has an ill-defined border, less swelling of the scalp, and slight if any constitutional disturbance.

Lupus erythematosus develops slowly in the form of a dry, sharply defined, scaly, red patch. In a short time there will be loss of hair, and marked cicatrisation in the old parts of the patch. The scales are adherent and in close relation to the sebaceous glands. In eczema the patches are not sharply defined, there is no cicatrisation nor baldness, and the scales bear no relation to the sebaceous glands. Lupus is far more chronic than is eczema, and less amenable to treatment.

A simple *dermatitis* is an acute redness and swelling of the skin, arising from some readily ascertainable cause, and soon passes away. Should it continue for a few days it will pass over into an eczema with the characteristics of the latter disease.

Impetigo is regarded by many authorities as merely a pustular eczema, but there is at times an eruption of discrete pustules upon the head, which show no tendency to form patches, being isolated throughout their course. To this eruption the name of impetigo is given. It is less itchy than is eczema and the pustules are larger.

Psoriasis never occurs upon the scalp alone; characteristic patches will always be found elsewhere upon the trunk or extremities, and often there is a line of scaling papules on the forehead along the edge of the hair. Eczema is often limited to the scalp, with at most a patch of disease behind the ear. Psoriasis is dry and occurs as circumscribed, rounded, discrete, small patches, or larger ones evidently made up of separate smaller ones, and covered with thick whitish or dirty grayish crusts composed of heaped-up epidermic scales. When the crusts are removed the under-lying skin is dry or slightly moist, with a few minute bleeding points. The patches of eczema are larger, and less well defined; its crusts are made up of scales, dried pus and sebaceous matter, matting the hairs together, and when they are removed a moist exuding surface is exposed. Psoriasis is less itchy, and there is in it a well-marked history of relapses.

Seborrhœa sicca affects by preference the vertex; eczema has no sites of preference. In seborrhœa the scalp is normal in color or pale, in eczema it is always red. The scales and crusts of seborrhœa are greasy to the touch, and gray or yellow in color; those of eczema are harder, and yellowish green or even black. Seborrhœa is a dry disease, eczema is a moist one. Seborrhœa is less itchy than is eczema, and in it the hair is dry and, may be, powdery, while in eczema it is stuck together. Seborrhœa is followed frequently by permanent baldness. Alopecia may follow eczema but in most all cases it is transient.

Favus has yellow, cupped crusts which are never met with in eczema. Favus has a history of contagion, and of spreading from one or two foci; causes bald atrophic patches of a peculiar red color; is dry throughout its course; only slightly itchy; and has a stale straw or mousey odor. Eczema has no history of contagion; forms rapidly into moist patches; is exceedingly itchy; does not cause atrophy of the scalp nor baldness; and has a sickening odor.

Syphilis differs from eczema in its whole history and course. Its pustules and papules are discrete and unattended by itching. Its pustules tend to break down and ulcerate, and then are covered with striated crusts, which being removed expose circular deep ulcers, which heal and leave scars. Nothing like this is met with in eczema.

TREATMENT. — CONSTITUTIONAL TREATMENT. — The treatment of eczema capitis is both constitutional and local. Excepting where there is some evident local cause, such as the presence of pediculi or the use of some irritating application to the scalp, it is necessary to enquire very carefully as to the action of the various organs of the patient, and to use our best endeavors to aid them in properly performing their several functions. We must regard the patient as a sick man quite apart from his sick skin. There is no specific for eczema; each case must be treated on its own merits according to the principles of general medicine. Arsenic is one drug that is commonly administered in a routine way. In most cases it will do no good; in some cases it will do harm; and in a very few cases it will render excellent service. It should be kept as a last resort. The acetate of potash is another drug that is used in a routine way. It may do good; it probably will do no harm. I have seen but little if any benefit from its use in this affection, excepting where there is a rheumatic or gouty condition. Tonics, such as iron, strychnine and cod-liver oil, are of great service in debilitated, neurotic, and anæmic subjects, and will often aid us materially, especially in infants and children.

The care of the digestive and allied processes is all important. If we question mothers with eczematous babies we will find in most cases that the children are nursed every time they cry, and consequently at most irregular hours; or they are improperly fed either as to quantity, quality, or frequency, being allowed to eat "anything that is on the table," besides numerous apples, bananas and cakes between meals, and to drink beer, coffee, tea and the like. In adults we will find quite as frequently great indiscretions in eating and drinking. Now this is all wrong. We must insist upon an infant being fed at regular intervals and with the food proper to its age. We should inquire as to the quantity and quality of the breast milk, and the health of the mother. Children should not be allowed to eat "what is going" for the first few years of their life, and cakes, confectionery, pastry, beer, coffee, and tea should be rigorously excluded from their dietary. In children, as in adults, it is often well to forbid meat for a time, especially in the Summer. Children as well as adults must not drink beer, coffee or tea; it is best to limit their fluids to milk and water. It is not possible here and now to lay down any hard and fast rules as to diet; my object at present is to insist upon the importance of regulating the diet of an eczematous patient.

The action of the bowels is to be regulated, preferably by diet and exercise; by drugs if we must. In an acute case, a sharp purgative will be found useful, and my preference is for the old-fashioned remedy, calomel. It is best given in small, repeated doses, say to an adult one or two grains repeated every two hours till two or three doses are given. It unloads the bowels, and stimulates the liver. BYFORD * lays special stress upon this latter action in explanation of its good effect in eczema of children, and believes that as in them the liver is larger in proportion than in the adult, liver indigestion is probably an important etiological factor in the disease. He gives a quarter to an eighth of a grain of calomel twice a day to an infant until slight purgation is caused, and afterwards as often as the bowels need it. In children over two and one half years of age he gives a single purgative dose every six to eight days. The administration of calomel is also of service in chronic cases in adults, and I often give it in the form of tablet triturates in doses of from a tenth to a fifth of a grain three or four times a day for three or four days and then stop for a few days. Podophyllin may be used in proportionate doses and for the same purpose, and also with good results. Acids and alkalies are useful in appropriate cases for the regulation of the digestion.

LOCAL TREATMENT.—The indications to be met by the local treatment of eczema are to soothe an acutely inflamed skin; to stimulate it when in a state of subacute or chronic inflammation; and to protect it when it is endeavoring to reassume its normal condition. To know what we want to accomplish is of more importance in this disease than to know what is the lat-

---

*Jour. Amer. Med. Assoc., 1885, V. 317.

est drug that is "good for eczema." Before using any remedy all crusts and scales must be removed by soap and water, and after that is accomplished no water should be used upon the scalp excepting at intervals. If our remedies are properly used, crusts will not reform. Stiff ointments should not be used on hairy parts where the hair is thick, as it mats the hair and makes the scalp more unclean than before. Alcohol, water, oil and vaseline are our best excipients. The hair need never be cut, and should never be sacrificed in women. In young children and in boys our treatment is rendered easier by cutting the hair, but it is not necessary. In applying remedies to the scalp they should be worked in, and not merely smeared over it.

In an acute moist eczema of the scalp, the head should be saturated with olive or sweet almond oil, and covered with a linen cap. The oil should be reapplied until the acuteness of the inflammation has subsided. Sometimes the addition of an alkali to the oil will be useful, such as the bicarbonate of soda or borax (3 iv to. viii. ad. Oj); or lime-water in equal parts with oil. A weak carbolized oil, say one per cent. of carbolic acid, or glycerine diluted with water, answers admirably in some cases. In a few days the acuteness of the disease will subside under this soothing plan of treatment. Then the scalp is to be thoroughly cleaned with soap and water, and the treatment proper to the condition of the scalp used. The scalp will bear stimulating applications much better than the rest of the integument; so as soon as the acuteness of the disease is past we can use a weak tar lotion, say half a drachm of oil of cade to the ounce of olive oil.

In acute erythematous and papular eczema, besides carbolized oil or vaseline we may use black wash, or a weak white precipitate ointment, say ten to twenty grains to the ounce. The squamous stage will soon be reached and more stimulating measures called **for**. The treatment for acute moist eczema will be appropriate for an acute pustular eczema when it occurs in a violent outbreak of pustules forming patches and discharging freely. As generally met with, this form of eczema occurs in patches covered with crusts. The crusts must be first removed by soaking them in oil for twelve or twenty-four hours, and then washing the head with soap and water. After the scalp is clean it is to be freely anointed with oil either carbolized, or, better, with oil of cade, half a drachm to one drachm to the ounce. Sometimes sulphur one drachm in vaseline one ounce will do good service in this form of eczema; but it is uncertain, and apt to prove irritating. Mercurial ointments, say of calomel twenty to thirty grains to vaseline one ounce, or that of the white precipitate ointment already spoken of will be beneficial.

When the squamous stage is reached we need stimulation, and for this the best remedies are frictions with soap and water, and preparations of tar. Once or twice a week the scalp is to be washed with the tincture of green soap, or equal parts of soft soap and alcohol, or if the scalp is very tender and irritable a milder soap such as Sarg's liquid glycerine soap, or pure castile soap may be used. The washings should be repeated at intervals of a few days to a week or more according to their effect on the scalp. If they prove very irritating they are to be less frequently used. After washing, the scalp is to be carefully dried and anointed with some oil or soft ointment containing tar such as;

      Ol. Cade, . . ℨ ss—ij = 2 to 16.
      Ol. Olivæ ad, . . ℨ j = 30.
      M.

and this is to be applied every morning and evening.

Instead of the olive oil as a vehicle we may use vaseline, oil of sweet almonds, or agnine or lanoline diluted, as with castor oil, sufficiently to make them supple. This has proved itself to be in my hands the most efficient mode of treating subacute and chronic eczema of the scalp. Its only objectionable feature is its odor, and this is so pungent and persistent that many patients will not use it. Unfortunately we have no means of successfully disguising the odor. Instead of the oil of cade we may use pix liquida or the oleum rusci, but there is no particular advantage in these other forms of tar, and the odor is the same or worse. The oil of cajeput may be substituted for the tar in five to ten per cent. strength, but it is not so good. Carbolic acid five to fifteen grains to the ounce may be used, and salicylic acid in three to five per cent. strengths.

Instead of using oily applications we may sometimes do better with oil of cade or pix liquida exhibited in alcohol, one to two drachms to the ounce. BULKLEY (5) recommends a lotion composed as follows:

Acetate of lead,	.	gr. viij.	.5
Oil of bergamot,	.	ℨ ss	2.
Castor oil,	. .	ℨ iv	16.
Alcohol ad,	. .	℥ iv	120.

M.

Napthol β, napthalin, pyrogallol, in the strength of five to ten per cent. may be tried if other things fail. Resorcin has of late been highly commended by IHLE[*] and others; and icthyol by UNNA.[†] In many cases of eczema in which the scalp is much thickened, the treatment is best begun by having the patient wear a close-fitting cap made of sheet rubber, and its use con-

---

[*] Monatshft f. prakt. Derm. 1885, IV. 430.
[†] Monatshft f. prakt. Derm. 1886, V. Erganzungsheft No. 2, in Mai.

tinued till the scalp becomes moist and less thick. Then tar may be employed.

After the eczema has been cured the scalp may be left dry and scaly, and then it should be treated according to the principles given in Chapter XIX. on Dandruff.

PROGNOSIS.—Though eczema capitis is a very obstinate disease at times, still it is perfectly curable. Some cases yield readily to treatment, while others will tax our patience and resources to the utmost.

ECZEMA BARBÆ.

SYNONYMS:—Impetigo sycosiforme; Achor barbatus; Barber's itch.

SYMPTOMS:—Eczema of the beard, which has been often erroneously called "barber's itch," has nearly the same symptoms as eczema of the scalp. It may be of the erythematous, vesicular, papular, pustular or squamous variety, though by far the great majority of cases are of the last two forms. The erythematous, papular, and vesicular varieties have precisely the same clinical picture as the corresponding affections of the scalp. Most all of the cases are pustular in character and we shall speak only of that form.

Pustular eczema of the beard may be either acute or chronic; may affect only one limited area, or the whole beard; may be either symmetrical or asymmetrical; and it may be confined to the bearded portion of the face or pass over to the unbearded portions and to the neck. Not infrequently the eyebrows and eyelashes will be affected at the same time and in the same way. When the disease is acute the affected part or parts are swollen, tender and red, and feel stretched and burning. Upon the red skin there will be an eruption of small pin-head sized pustules which have no special relation to the hairs, though many of them do occur about the months of the hair-follicles. These pustules break of themselves in a very short time and discharge their purulent contents, which glues the hairs together. If the disease is very acute and violent the beard will look as if some mucilaginous fluid had been poured over it. Soon the exudation will dry into greenish or yellowish-green thick crusts, in which the hairs will be entangled and matted together. When the crusts are removed a moist exuding surface will be exposed. The hair will be unaffected excepting that it is stuck together; and attempts at epilation will be painful.

In subacute and chronic cases the hair will be somewhat thin, leaving the under-lying skin exposed. This will be red and scaling in some places, covered with greenish crusts in other places, and studded here and there with small pustules, either about the mouths of the hair-follicles or independently of them. It will be seen that the skin between the hairs is diseased as well as about the hairs. To these subacute and chronic cases sycosis is apt to join itself, or the eczema may pass over into a sycosis.

Squamous eczema of the beard is the final stage of the other varieties, through which they pass on their way towards recovery. In it the skin is reddened and thickened and covered with scales. Very often the progress of the disease is stayed at this point and recovery does not set in. On the contrary, relapses are liable to occur even after the squamous stage has lasted for some time.

Itching accompanies all varieties of eczema of the beard, and is, excepting the deformity, the most annoying feature of the disease.

ETIOLOGY.—This form of eczema is comparatively rare. In 1800 consecutive skin cases occurring in DR. GEO. HENRY FOX'S service at the New York Skin and Cancer Hospital, it was seen only sixteen times. Of course it occurs only in men. It is predisposed to by the same constitutional disturbances as we noted under eczema of the scalp. Its exciting causes are also in large part the same as noted in the previous section of this chapter. Exposure to dust and wind and weather; the irritations from tobacco smoke, or the secretions from nasal catarrh; the application of poultices for the relief of neuralgia; scratching of the beard either from habit or on account of the itching due to seborrhœa, and shaving, may be noted as special exciting causes.

DIAGNOSIS.—Eczema of the beard is to be diagnosed from sycosis, trichophytosis barbæ, acne, a pustular or tubercular syphilide, and an epithelioma.

Eczema is a more superficial process than *sycosis*; its pustules have no close relation to the hairs, it is very itchy, its crusts are thick and seated upon a moist and oozing surface, the hair is unaffected, and it is often associated with the same disease on other parts of the face or general integument. Sycosis is a deep perifollicular inflammation, its lesions are seated about the hair-follicles, the skin between the hairs being unaffected except in old and severe cases, its crusts are not so thick as those of eczema and do not mat the hairs together, and it occurs only upon places supplied with hair. It burns rather than itches, and the hairs are readily pulled out.

Trichophytosis barbæ has a history of spreading from one or two points, and often can be traced to a source of contagion. It is generally confined to the chin and throat, and its lesions are tubercles and nodules. The hair is early affected, is broken off and is readily pulled out, or falls out of itself. The crusts are small or wanting. It often extends beyond the beard as a typical ringworm, or ringworms may be found on other parts of the body. These symptoms are so characteristic that the two diseases should not be confounded.

Acne occurs as isolated and discrete pustules or small cutaneous abscesses on all parts of the face, and comedones are generally to be found. The skin between the pustules is unaffected; there is no exudation, and no itching.

In *syphilis* there will be other lesions elsewhere on the body, and a history of an initial lesion or of some other manifestation of the disease. The lesions are discrete, and if there are crusts, well-marked specific ulcers will be found beneath them when they are removed.

It would hardly seem possible to confound an eczema with an *epithelioma*, but it has been done. An epithelioma occurs as a single lesion, rounded or irregular in outline, of varying size from that of a one-cent piece to that of the palm, and forms a well-marked ulcer with raised hard waxy borders over which run delicate branching blood-vessels. The ulceration may be only superficial and covered with a delicate crust, but it is always sharply defined and bleeds easily. An eczema has no well-defined shape or border, and no raised edge. An epithelioma is painful, the pain being lancinating at times, while an eczema is itchy.

PROGNOSIS.—The obstinacy of the disease, and it is often exceedingly rebellious to treatment, is due to the irritation of the skin by the stiff hairs of the beard. Though the disease is obstinate it is perfectly curable; but it is liable to relapses, and it is good policy to let our patient know this before beginning treatment.

TREATMENT.—The treatment of eczema of the beard is conducted on the principles we have learned in studying the same disease of the scalp, but modified in some particulars to suit the changed conditions. When we can prevail upon the patient to cut his beard close, we have taken a long step towards curing the disease. Many patients prefer to allow the beard to grow, as it serves to disguise the disfigurement to a certain extent: but we can apply our dressings much better if we get rid of the hair. Shaving may be practised and is recommended in many text-books: but it is very painful in the acute stage of the disease and unnecessary if the hair is clipped close. In the squamous stage it should be advised because of its stimulating effect. The first step in treatment is to remove from the patient any irritant that has been acting upon his skin. If the case is a bad one we should have the patient stay at home until the violence of the disease is abated. Even then a cure can be secured more speedily if the patient will consent to stay in the house during its whole course.

In acute eczema of the beard, after the hair is cut close, a soothing application should be used. To this end we may employ a carbolized oil, simple oil, black wash, lime-water and oil, zinc oxide ointment, simple ointment, or Lassar's paste, the latter composed as follows:

Starch,	
Zinc oxide,	āā ʒ ij = 8.
Vaseline ad,	ʒ j = 32.
M.	

answers admirably. Diachylon ointment is also excellent. In this, as well as the more chronic form, tumenol oil will often render good service if the patient does not object to the black color. Whatever dressing is selected, it should be used freely, the ointments being spread thickly upon cloths, and bound down to the part with a roller bandage. In Vienna even the acute cases are treated by scrubbing the skin with green soap or its tincture, and following this with the diachylon ointment.

As soon as the acuteness of the disease has somewhat subsided we should proceed to more active measures. All crusts must be gotten rid of by soaking with oil and subsequent washing with soap and water. All hairs protruding from pustules should be plucked from their follicles. Now the tar preparations spoken of in the section on eczema capitis may be used, or mild sulphur ointment. Lassar's paste with the addition of about fifteen grains of salicylic acid to the ounce is soothing and efficient at this time. When the squamous stage is reached the beard should be shaved off every day or so, and then the stimulating remedies applied. As the disease becomes reduced to a slight redness and the patient is going about his usual avocations, the skin should be protected by a slight film of vaseline or ointment, or with a dusting powder. In fact throughout the treatment, if the patient is obliged to go out he must protect the skin in this way, and employ the more radical dressings at night.

ECZEMA PALPEBRARUM.

This form of eczema, which is also called *Eczema marginis ciliaris palpebrarum*, and *Blepharitis ciliaris*, is seen by the opthalmologist more often than by the dermatologist. Eczema of the eyelids usually accompanies the same disease on other parts of the head, though it may occur on them alone. It is always pustular in character at first, and afterwards squamous. The edges of the lids are swollen and rounded, and more or less thickly strewn with pustules that quite commonly are situated about the cilia and take on a sycosiform character. The lids after sleep are stuck together so as to be opened with difficulty. The pustules rupture, and then the edges of the lids become cov ered with yellowish or greenish crusts. When the crusts are removed excoriations are left, and sometimes even small ulcers. The skin itches, and the rubbing by the patient for its relief aggravates the disease. When the process passes over into the squamous form the edges of the lids are merely red and scaly. If the inflammation about the cilia has been intense and ulcers have formed, cicatrisation will follow and loss of the eyelashes. Both lids are usually affected, and the disease is commonly symmetrical.

With the blepharitis there will be more or less conjunctivitis, with its characteristic symptoms. The eyebrow, also, is often eczematous. Sometimes the disease will remain confined to the corners of the eye, especially the outer canthus, for a long time.

ETIOLOGY.—This disease is seen most commonly in strumous subjects, and is a frequent complication of phlyctenular ophthalmia and other strumous eye troubles in children. When occurring with eczema elsewhere it is an expression of the eczematous condition. When occurring alone it is most frequently caused either by some trouble with the eye itself, or by applications made to the eye for the relief of such troubles. Sometimes it is difficult to determine whether a conjunctivitis has caused the eczema, or an eczema has provoked a conjunctivitis.

TREATMENT.—The eyes should be bathed with warm water to remove the crusts, and then anointed to prevent their sticking together. This should be done especially before the patient goes to sleep. What ointment is to be used will depend upon the condition. In mild cases simple ointment will suffice. Generally some form of mercurial ointment will do better, such as that of the red or yellow oxide diluted eight or ten times with vaseline or simple ointment. A. R. ROBINSON recommends an ointment composed of one grain of the biniodide of mercury in one ounce of vaseline. DELIGNY has found the best application to be: Plumbi acetat. 0.25, Adeps. 25. LAWSON regards ointments as too irritating, and advises the following wash:

℞ Boracis vel zinc oxid, . ℈ij, say 1.5
   Glycerin, . . . ℈iv " 3.
   Aquæ sambuci, . . ℥ij " 12.
   Aquæ destil. ad . . ℥viij " 100.
M.

Another good remedy I owe to PROF. D. WEBSTER, and is as follows:

℞ Ac. salicylici, . . . gr. x, say .8
   Ungt. hydrarg. oxid. rubri, ℈i " 5.
   Ungt. aquæ rosæ, . . ℈vi " 30.
M.

Whether we use ointments or lotions we must be sure that any powders that enter into them are of the finest, so that matters are not made worse by the irritation of coarse grains left by the applications.

When the process becomes sycosiform the cilia must be plucked from the lids. If the disease prove very obstinate we may evert the lids and brush over them a solution of caustic potash, ten grains to the ounce, absorbing the superfluous fluid with blotting paper and then washing with water. This should be done every day until there is an amelioration of the symptoms, when one of the mercurial ointments may be used. In all cases constitutional treatment must be prescribed, as well as special treatment for the eyes, where these are diseased.

ECZEMA OF OTHER HAIRY PARTS.

The hair on any part of the body may be attacked by eczema and give rise to symptoms corresponding to those just detailed. Thus we have eczema of the axilla and of the pubes; a very annoying eczema of the hairs in the nostrils; and a peculiar form of follicular eczema which affects the hairs of the extremities or trunk.

The treatment of eczema of the axilla and pubes is the same as for the same disease of the beard.

Eczema of the hairs *of the nares* is often obstinate. We must endeavor to cure the nasal catarrh that is usually present, and to apply ointments to the part on pledgets of lint. Sometimes we may have to epilate and apply a solution of nitrate of silver. HARDAWAY* believes this condition is associated in many cases with a broken-down state of health, and recommends active internal treatment. For the local treatment he has found Squire's glycerole of the subacetate of lead of special service. This is made by mixing together acetate of lead 5 parts, litharge $3\frac{1}{3}$ parts, and glycerine 20 parts, by weight, heating to 350° F. and filtering through a hot-water funnel. He opens the little abscesses with a knife and when the disease is abated applies

Squire's glycerole plumb. subacetat, ℨ ss say 2.
Glycerine, ... ℨ iss " 6.
Ungt. aq. rosæ ad, . . . ℥ j " 32.
Ceræ albæ, ... qs. " qs.
M.

In very obstinate and relapsing cases he epilates the offending hairs by electrolysis. Eczema seated about the hair-follicles of the trunk and extremities is, happily, rare, and is sycosic in its nature. I have seen it only in a very few cases and then upon the legs of strumous subjects. It occurs as disseminated and discrete papules and pustules pierced by hairs, and is very itchy, so that excoriations are frequently seen. It

---

* Jour. Cut. and Ven. Dis., 1886, iV. 360.

has always proved obstinate to treatment. A cure can be effected only by attention to the general physical state. Locally the disease is to be treated on the principle of sycosis.

CHAPTER XXII.

PLICA POLONICA.

SYNONYMS:—Trichosis plica; Trichoma; Koltun (Polish); Weichselzopf (Ger.); Plique polonaise (Fr.); Polish ringworm (Eng.).

SYMPTOMS:—The term Plica Polonica is used rather to designate a condition than a disease, the condition arising from various causes and producing the clinical picture of the long hair of the head or other parts matted together into various shaped masses, in and on which rest all sorts of extraneous matters deposited from the floating dust of the atmosphere, and very often vast hordes of pediculi with their ova. At times these masses of matted hair will be close to the skin or scalp; at times removed to a considerable distance from it. If close to the scalp we will find beneath them a moist and oozing condition of the under-lying part; if at a distance from the scalp, the latter may be of normal appearance, or scaly. This is due to the age of the plica, the freshly formed ones are near the scalp, and when the disease of the scalp is recovered from, the growing hairs will push the mass further and further from the scalp until it either falls off of itself or is cut off. Various names have been used to designate the different shapes these masses may take; thus when the hair is short and there are many locks matted separately it is called plica multiformis, or caput Medusa; when but a single coil, and this long and in the shape of a tail, it is called plica caudiformis, and such tails may be so long as to reach below the knees. The odor of the head in plica polonica is pene-

A CAUDIFORM PLICA POLONICA.

trating and disgusting; somewhat like that of rancid fat. There are no special constitutional symptoms; the patient may be cachectic, strumous, phthisical or what not; or he may be perfectly sound and healthy. The only subjective symptom is itching.

Such is the disease as viewed in the light of modern medicine since the time when HEBRA demonstrated its dependence upon lack of care of the hair. Previous to that it was described as a disease with four stages namely: 1. Prodromal stage; 2. Stage of exudation; 3. Stage of efflorescence; 4. Stage of convalescence. It was regarded as a safety-valve in severe illnesses, and it was thought that if the patient could raise a plica he would recover from his disease. It was on that account held sacred from injury, and cases were reported in which a man or woman died in consequence of cutting off his or her plica. Plicas have been described as having taken on a fleshy condition, and reported as having bled from the stump when cut off, or at least to have exuded a reddish fluid. These were without doubt errors in observation.

ETIOLOGY.—The cause of this condition of the hair is want of cleanliness combined with the presence of lice, giving rise to pediculosis; or of any disease of the scalp which is accompanied by moisture, principally eczema. But this was not always thought to be the case, and the literary war was long and stubborn before this theory gained the victory over the old theories that regarded the condition as a specific disease. It used to be regarded as a dyscrasia occurring endemically in certain countries, especially in certain provinces of Russia and Poland. It was thought to occur at the crisis of many diseases, and was regarded as of favorable prognosis. The common people had such great faith in the salutary influence of a plica that they often produced one by the use of wax, pitch and the like.

Such an one was called a false plica. It was and even now is endemic in certain parts of Poland, and especially in the low-lying provinces, because the people are wonderfully careless about personal cleanliness, and are in the habit of wearing fur caps constantly upon their heads, inducing sweating and favoring an eczematous condition. It is seen both in men and women, and while most common amongst the lower ranks of society, it is met with also in the nobility of Poland and Russia. We encounter the condition now and again in this country, but of mild grade, and as an unmistakable result of want of care of the scalp.

TREATMENT.—The condition is removed very readily by soaking the hair with oil, washing with soap and water, and carefully combing; all combined with the exercise of great patience and perseverance. If the mass has grown away from the scalp the easiest way of treating it is by cutting it off. After the plica is disposed of, we must apply ourselves to the cure of the disease of the scalp that is at the bottom of the trouble. The older writers endeavored to treat the disease by internal medication. ROSENBERG (76) gives seventy remedies recommended at different times for the cure of the disease.

Besides this true or plica polonica, there is a form of felting of the hair which is apparently due to nervous influences, probably of an hysterical nature. Cases of this kind have been reported from time to time under the name of neuropathic or nervous plica. Recently LE PAGE (620) has reported a case of spontaneous and rapid felting of the hair in a girl seventeen years of age; and PESTONJI (623) has related another occurring in a woman twenty years old. JAROCHEVSKI (427 ap.) reported another case in an hysterical young woman. Microscopical examination showed the cuticle of the hair to be entirely separated from the cortex, rough, and its fibres split up. Its cortex showed irregular disposition of pigment, and its medulla was broken up. DE AMICIS (426 ap.) has also reported a case. In all cases the hair had been wet before it began to twist up, and in both there was headache and an entire absence of disease of the scalp.

CHAPTER XXIII.

DERMATITIS PAPILLARIS CAPILLITII.

SYNONYMS:— Dermatitis papillomatosa capillitii; Framboesia; Sycosis framboesia (HEBRA); Sycosis capillitii (RAYER); Mycosis framboesioides or acne keloidique or Pian ruboide (ALIBERT); Acne keloid.

SYMPTOMS:—This exceedingly rare disease of the

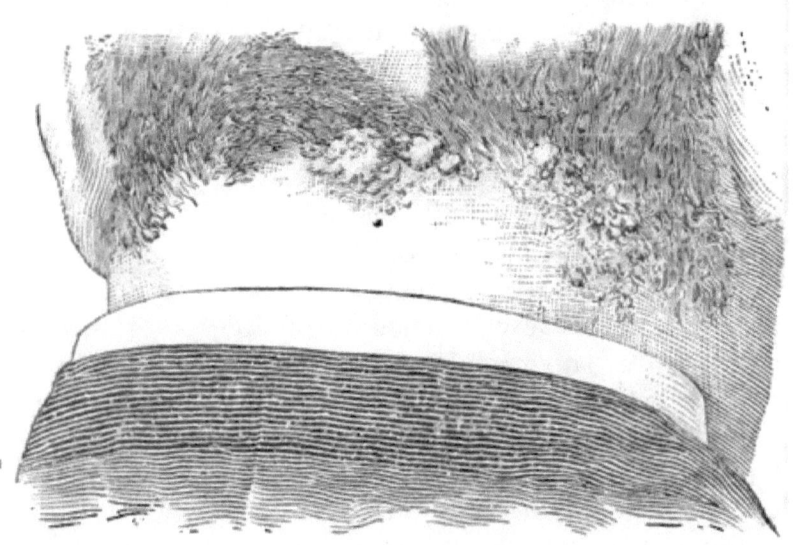

Dermatitis Papillaris Capillitii.

skin was first described as a separate entity by KAPOSI (629) in 1869. The same group of symptoms had already been described by ALIBERT in 1814, but regarded by him as a manifestation of syphilis. From time to time the disease had been observed and mentioned by the older authorities, and to it the name of Framboesia had been applied on account of the resemblance of its peculiar lesions to a raspberry, that being the meaning of the word "framboesia." Both in HEBRA and KAPOSI'S *Lehrbuch der Hautkrankheiten* 1872, and KAPOSI'S *Pathologie und Therapie der Hautkrankheiten*," 1880, the disease is placed in the chapter or section on sycosis. KAPOSI, in the article referred to and in his book, described the disease as follows: "The primary manifestations consist of an eruption of large and small papules which are very hard, of uniform consistence, and contain no pus. These lesions are elevated many lines above the skin, and form variously sized, deeply furrowed, uneven, lobulated tumors. They generally are covered with a thick epidermis, bleed only when cut or deeply wounded, are rarely moist, and then but in a few places; and the few vesicles or pustules that may be present are superficial and purely accidental. The papules are discrete at first, but soon run together to form the tumors, which at last flatten down into cicatricial tissue. The hair is deeply seated in the furrows between the lobules of the tumors; is formed into little bundles or wisps; and resists attempts at epilation, often breaking instead of coming out. Sometimes, on the other hand, it seems to be retained only mechanically in the furrows, and comes out easily when pulled upon. It often appears atrophied, and sometimes is entirely wanting in large areas, but never is connected with a pustule. The disease begins in most cases upon the occiput low down at the edge of the hair, and from there spreads upwards; sometimes involving the whole posterior part of the head. When so extensive, the lesions form a papillomatous vegetation which exudes a foul-smelling secretion, bleeds easily and is covered with crusts; sometimes abscesses form."

HANS HEBRA, JR. (626) in 1874 reported a case of "sycosis framboesioides" which occurred upon the side of the head in the form of raised and sclerosed patches sown with millet-seed sized papules and somewhat larger pustules, out of which protruded little bundles of hair. In 1881 SANGSTER (630) reported a case of papillary tumor of the scalp which was about the shape and size of a pigeon's egg; situated low down on the occiput; about half an inch in height; studded with hempseed seized papules; brown colored at the circumference, violaceous toward the centre; slightly ulcerated here and there, with thin crusts. It had begun at birth, and was increasing gradually in size. The hair was plentiful at the margin, but sparse and devoid of pigment towards the centre, and grew up between the papillæ without piercing them. There was a second tumor over the left ear. KAPOSI pronounced this to be a case of *dermatitis papillaris capillitii*. HYDE (628) has reported two cases, in each of which puncture caused a discharge of mucoid fluid followed by an exudation of thin venous blood and serum. He describes the first stage as an eruption of pin-head sized papules, vesicles, or vesico-pustules. BAKER (625) and WILLIAMS (632) have reported cases under the name of "*Acne Keloid.*" HERVOUET (627) reported a case of papilliiform hypertrophy of the scalp, which corresponds in many particulars to the disease as described by KAPOSI. VERITÉE (631) has also reported a case under the name of *Acné Keloidique*. Another case is by EVE.[*]

Through the kindness of Dr. Geo. H. Fox, of New York city, I am able to add the following case which I had the opportunity of studying with him: J. R., æt. 30. Large, well-developed and healthy man. One year ago (1884) there appeared on his neck an eruption of hard nodules, which have been steadily increasing in number and size. The eruption is situated upon the occiput amongst the hair, and upon the contiguous

---

[*] Path. Trans., 1884, xxxv., 397.

portions of the neck. It consists of numerous small, reddish, hemispherical tumors, each one of which is perforated by hairs. A few of these tumors showed a tendency to suppurate about the mouths of the follicles. Some of the tumors tend to become aggregated, and to form mulberry-like masses of dull red color. The smaller nodules are very firm, and hard and shotty to the touch. They do not bleed readily when punctured. They are not tender nor sore, and he is troubled only by the deformity caused by their presence.

From a study of these cases we learn that the disease begins as an eruption of papules of small size, which are usually few in number, and situated upon the back of the neck at the margin of the hair. These papules are of the color of the skin, or slightly more reddish, and sometimes have an inflammatory halo. They are exceedingly hard and firm, and do not contain pus. If pricked they give vent to a little bloody serous fluid. Slowly they increase in number, and form larger raspberry-like elevations which have uneven, lobulated surfaces. New lesions appear from time to time upon the neck following the line of the hairs, or the disease encroaches upon the hairy scalp, and in the course of months or years it may reach even up to the vertex. When these lobulated masses attain a certain size they often become softened a little, and when cut may be found to contain pus. At times they secrete a foul-smelling fluid, and become crusted. When the papules run together the inflammatory halo disappears. Gradually the growths become sclerosed and assume a keloidal appearance. When the hairy scalp is attacked, pustules may form out of which little bundles of hair protrude; but the hair is not loosened. The keloidal masses are generally bald; but some have a few bundles or tufts of hair in them, which usually is firmly seated and resistant to attempts at epilation. Sometimes these tufts are apparently only mechanically held in the furrows of the growths and are readily pulled out. Sometimes the hair when pulled appears healthy; sometimes atrophied; sometimes it breaks upon slight traction.

The disease may give rise to some tenderness or may be absolutely painful. Sometimes there are no subjective symptoms, and the patient will complain only of the deformity and the inconvenience. The course of the malady is exceedingly slow, but steadily progressive.

ETIOLOGY.—The etiology of the disease is very obscure. From its location at the back of the neck, at about the place where the band of the shirt or other clothing rubs, it has been suggested that the rubbing of the shirt-collar may be an etiological factor. It occurs both in women and men, and may begin at any age.

PATHOLOGY.—The disease process is described by KAPOSI as a chronic inflammation of the chorion principally, attended by a great increase in the number of the blood-vessels and in the size of the papillæ. This is followed by the formation of new connective tissue, which by pressure destroys the sweat and sebaceous glands, and finally compromises the existence of the hair-follicles, and the hair itself.

DIAGNOSIS.—The peculiar location of this disease upon the back of the neck at the margin of the hair, the arrangement of the hair in little tufts in the keloidal masses, and the great hardness of the lobulated or papillomatous tumors, distinguish this malady from all others. In *sycosis* we have no hard tumors, and the hairs are surrounded by pustules; in our present disease the pustules that may be present either have no relation to the hairs, or else surround a number of them. *Warts* or *papillomas* of the scalp lack the hardness of dermatitis papillaris capillitii, do not occupy the same region, do not tend to increase in size, and do not take on a keloidal condition. The large uneven tumors of *mykosis fungoïde* resemble the growths of dermatitis papillaris capillitii, but they are not quite so soft; they are more generally distributed over the whole body, come and go in a capricious manner, often break down and ulcerate, and do not slowly change into keloidal tumors or streaks.

PROGNOSIS.— So far as reported the growths are benign and are compatible with a good state of health. But they do not yield readily to treatment, and left to themselves are progressive, and show no tendency to spontaneous recovery.

TREATMENT.—The best treatment is the most radical, that is to scrape out the small growths with the sharp spoon, to cut off the longer small ones with the scissors, and to excise the large ones. In any case, care must be used to get down to the true skin and to remove the whole growth. After removal by any means the base should be cauterised, and for this purpose we may use the nitrate of silver stick, which both cauterises and stops the sometimes not insignificant bleeding following the operation. In some cases the stronger caustics may be employed with benefit without operation. HEBRA, JR., (16) speaks highly of the galvano-cautery in this disease, and has found that it was as effectual in removing the growths, and more active in preventing their return, than any other operative procedure.

CHAPTER XXIV.

NÆVUS PILOSUS.

Hair-moles or birth-marks are usually congenital, and even in cases in which they have developed within a few years after birth, the hair grows upon a congenitally hyperpigmented surface, a pigmentary

mole. They are of all sizes from that of a split pea up to that of a huge patch big enough to involve the surface of the body from a line between the angles of the scapulæ to half way down the thigh. They may be unilateral or bilateral, and sometimes symmetrical. There may be only one of them, or there may be scores of them; and they may be located on any region of the body. It is when they are located on the face or arms that they come most often under our observation.

Wherever they are located, and whatever their size may be, they possess the same characteristics, namely: upon a thickened and pigmented patch of skin, usually of dark brown or black color, there is a more or less luxuriant growth of dark stiff hair, and the whole patch is slightly raised above the level of the skin, and, if large, its surface is more or less uneven. The color of the patch varies with the complexion of the individual, being light brown in blondes, and dark brown or black in brunettes. The hair is nearly always darker upon these moles than it is on the head of the same person, excepting when the latter is black. The amount of hair present varies exceedingly; sometimes it is very luxuriant and grows close like the fur of an animal; sometimes there are only a few stiff hairs in the mole. The thick growth is seen more often on the large moles; while the sparse growth is more common on the small moles, as on the face. These hairs are coarser than those on the head; even when a hairy nævus occurs upon the scalp, the hairs covering it will be coarser than those about it. Exceptionally the hairs are fine. Pigmentary moles may or may not coexist with these hairy moles; molluscum fibrosum has been met with in several cases.

Most of the cases reported as "circumscribed hypertrichosis" are examples of hairy nævi. Some of these have been given in our chapter on Hypertrophia Pilorum; sometimes these moles undergo a change into epithelioma.

The *etiology* of nævus pilosus is obscure; our only

supposition is that they are due to some nervous influence. At times they occur along the course of nerves, and cases of extensive hairy moles of the back and thighs are not infrequently associated with spina bifida. They are sometimes hereditary, and I have seen them on the face in a number of cases of facial hirsuties. The popular idea that they are due to maternal impressions received during pregnancy is supported by not a few instances in ancient and modern times.

BULL (635) in 1882 reported the case of a child with an extensive hair-mole whose mother was frightened by a dog in the third month of pregnancy; and SOMMER (640), in 1885, described another case with the history of the mother having been frightened by a bear. How far maternal impressions really influenced these and other similar cases is not for me to determine.

Histologically these growths consist of a slight hypertrophy of the papillary layer of the skin, with a deposit of brown and black pigment granules in the rete mucosum. MORRIS (24.) The hairs are hypertrophied; the large hair-follicles are close together and possess small accessory hair-follicles in which are developed hairs; and there is a more rapid fall and new growth in them than in the skin of normal hairy parts. (MICHELSON (40), in Ziemssen.) CHIARI (428 ap.) found in one extensive case a heaping-up of cells in the corium. They had spindle-shaped nuclei and but little protoplasm. The papillæ were broadened by the cells. The rete appeared thinned, and the corneous layer thickened. The hairs were thickened, contained little medulla, and stood in more or less marked clumps.

The *diagnosis* of hairy moles from circumscribed hirsuties is determined by the presence of pigmentation and thickening of the skin in the former, and their absence in the latter.

TREATMENT.—The best treatment of hairy moles is

electrolysis, practised as taught in our chapter on Hypertrophia Pilorum. This is an exceedingly brilliant operation in the small nævi of the face, not only destroying the hair but also removing the discoloration. In large nævi it will certainly destroy the hair, and if any discoloration remains, it may be readily gotten rid of by the careful application of nitric or acetic acid, or the like. Of course it is possible to remove the growths by the knife or by powerful caustics, but electrolysis is a less painful method, and if done with sufficient care it will leave scarcely any scar.

CHAPTER XXV.

SYPHILIS—LUPUS—VITILIGO.

These three diseases affect the scalp, either as part of a general eruption, or as limited to that region, alone. When occurring upon the scalp, their etiology and pathology are the same as when upon other regions. Their symptoms differ slightly from those seen on non-hairy parts, and in this chapter will be noted only such variations.

Syphilis.

The erythematous, papular, pustular, tubercular and gummatous forms of syphilis are met with on the scalp; while the squamous, bullous, vesicular, and pigmentary forms are not met with there. When syphilis attacks hairy parts it tends to assume the pustular form, the pustules forming about the hair-follicles; and this holds true during the whole of the active stage of the disease, or what has been named the secondary period, whether the eruption upon the trunk and extremities be erythematous, papular, or pustular. Thus Bassereau (1), found, in 153 cases of erythematous syphilis, the scalp affected with a

Pustular eruption,	106 times
Pityriasis,	13 "
Papular eruption,	4 "
Macular "	2 "
No eruption,	28 "

Erythematous syphilis of the scalp does occur, however, with macules alone, which here as elsewhere are round or oval, rosy or red spots, out of which the red color may be driven by pressure. They cause the patient no annoyance, and he would be ignorant of their presence were it not that at this time there is often some seborrhœa, and the accumulation of sebaceous matter about the mouths of the hair-follicles forms little crusts upon which the comb catches in combing the hair. Pustules located about the hair-follicles are associated very often with the macules, and the little scales they form in drying aid in producing that symptom of "catching of the comb" which is one of the diagnostic marks of syphilis. The macular syphilide of the scalp is rare, and is met with most often along the margin of the hair upon the forehead and occiput. If it is accompanied by a good deal of seborrhœa we may have a marked fall of the hair and alopecia.

The PAPULAR SYPHILIDE is seen upon the scalp more commonly than the macular syphilide. It is often accompanied by pustular syphilides, and by seborrhœa, and is sometimes itchy. Its most frequent site is along the margin of the hair, and the papules may be either small or large. The *small papular syphilide* is round and slightly elevated above the surface; red in color at first, afterwards becoming coppery or raw ham colored; in close relation to the hairs; and sometimes a little scaly. It becomes absorbed after some weeks' duration, and as it disappears the hair falls out. The alopecia is only partial and transient, and the hair soon grows again. The *large papular syphilide* is of greater diameter and more elevated than the preceding variety. Like it, it is scaly, and when it becomes absorbed the hair falls. This syphilide sometimes ulcerates and heals with a cicatrix. Sometimes the papules become greatly hypertrophied, and run together to form raspberry-formed masses to which the name of "framboesoid," vegetative, or verrucous syphilide has been applied. Each mass is formed of a number of hypertrophied papules and resembles the papillomatous formation of a wart. It is round, of varying size, and gives vent to a foul-smelling secretion; its surface is sometimes moist and sometimes crusted, and when the crust is removed a shallow ulcer is uncovered. These masses may be present in great numbers so as to involve a large part of the scalp. They at last become absorbed or break down, and always leave permanent baldness.

The PUSTULAR SYPHILIDE is very common upon the scalp, and probably occurs in every case of syphilis with cutaneous manifestations. It is seen very early in the disease with the erythematous eruption; occurs also with the papular eruption; it may be part of a general pustular eruption, or may occur as a localized and relapsing syphilide. The pustules are either scattered, or grouped in circles or segments of circles, and usually occur about the hair-follicles. They may be small and superficial, or large, deep and ulcerating; and are surmounted by a greenish or blackish crust varying in size, thickness and color. The small superficial and pustular syphilide of the scalp, the *acne-form syphilide*, is seen usually within the first six months of the disease, but it may relapse and appear later. It is a papulo-pustular lesion, the papule appearing first and the pustule slowly forming on top of it. The pustule is conical or slightly rounded in form, and of pin-head size, or slightly larger. The pustule soon opens and discharges its contents, which dries into the characteristic greenish crust of syphilis. This syphilide is often very slow in its course, especially when it occurs as a part of a general specific pustular eruption, and on healing leaves a small cicatrix. The hair falls out of the follicles in relation with the pustules, but new hair grows again, excepting in cases in which the disease has been more severe than usual, so that the deep parts of the hair-follicles have been destroyed. The pustular syphilide may be in the form of lesions which are pustules from the beginning and tend to run together and form patches. This is known as the *impetigo-form syphilide*. The patches are covered with greenish or blackish thick crusts; are evidently made up by the coalescence of several pustules; and when the crusts are removed an ulcerating surface is exposed. The ulceration may be superficial or deep, and in debilitated subjects it may become serpiginous. This form of pustular syphilide occurs later than the acne-form syphilide, usually after the first six months; it may be met with in the second or third year of the disease. Sometimes it is apparently the form in which the acne form of syphilide relapses. Its course is slow; it always heals by cicatrization; and is always followed by permanent baldness.

The third and last form assumed by the pustular syphilide on the scalp is called the *ecthyma-form syphilide*. This syphilide may occur during the second half of the first year, or be one of the late or tertiary manifestations of the disease. Occurring early in the disease the pustules may be very numerous, and tend to group; occurring late in the disease there are but few of them, and these are grouped in circles or segments of circles. The pustules are larger than the other pustular syphilides, rapidly become ulcers; and are covered soon with a thick greenish or brownish crust. These lesions are usually superficial in the early period of the disease; deep and often serpiginous when occurring as a late lesion. The ulcers heal by cicatrisation and leave a bald spot.

The TUBERCULAR SYPHILIDE is one of the late forms of syphilis, and occurs sometimes upon the scalp, either alone or in connection with the same lesion elsewhere upon the body. They begin as deep red spots, which increase in size and become elevated. They are from a half to one inch in diameter; tend to group in circles or segments of circles; are sometimes surmounted with a scale; sometimes ulcerate and become serpiginous; and sometimes become verrucous, and assume that frambœsoid character described under the papular syphilide. Whether ulcerating or not they leave a cicatricial spot behind them, and this is absolutely bald. This lesion is prone to relapse, so that at times a large part of the scalp becomes bald.

The GUMMATOUS SYPHILIDE is not very common on the scalp. There may be only one gumma, or there may be a number of them. Small ulcerations frequently take place about the hair-follicles, so that the whole gummatous mass is covered with ulcers. The gumma is either absorbed or breaks down and ulcerates, and then the bones of the skull may be more or less damaged. It occurs most frequently upon the frontal and parietal region, and causes permanent baldness. Though these syphilides have been described here as affecting the scalp, they occur quite commonly on all the other hairy regions and then present similar symptoms. Erysipelas may complicate syphilis of the scalp.

The DIAGNOSIS of the early syphilides is unattended, as a rule, with difficulty, other unmistakable symptoms of constitutional syphilis being present upon the general integument. It is probable that unless they cause baldness they are frequently overlooked. When questioning a patient, with some doubtful skin lesion, as to his having had syphilis, the two most important facts to ascertain in regard to the scalp are the occurrence of baldness in patches coming on suddenly; and the catching of the comb upon the little scales on the scalp. Sometimes the later lesions offer considerable difficulty in diagnosis, and we are called upon to determine whether a pustular eruption is an eczema, simple impetigo, or non-specific ecthyma, or a syphilide; or we may have to decide whether a given lesion is lupus or an ulcerating tubercular syphilide. The gumma resembles a kerion; and a papular syphilide may be mistaken for psoriasis.

From *eczema* or *simple impetigo* a pustular syphilide is differentiated by the history of the initial lesion and preceding specific eruptions; by the greater slowness of the development of the pustules, and their not breaking down readily; by the absence of itching and burning; by the greenish or blackish crusts; by the little cicatrices left by the pustules; and by the baldness it causes.

From *non-specific ecthyma*, the specific form may be known by the history of other specific lesions; by the grouping of its pustules and their slower course, by the ulcers they form, which are often serpiginous, always have abrupt edges, and are deep, with their floors covered with a thick puriform fluid; by the crusts being thicker and more heaped up; and by the smooth, white, bald cicatrices they leave.

From *lupus*, the ulcerating tubercular-syphilide differs in occurring upon the scalp alone at times, while lupus vulgaris never occurs there without being found elsewhere upon the face or extremities; in syphilis there is an entire absence of the characteristic brownish nodules or papules of lupus. Syphilitic ulcers are rounded, often serpiginous, always punched out and deep, are covered with thick, heaped-up greenish or blackish crusts, and heal by a smooth, white, non-deforming cicatrix. Lupus ulcers are more irregular in shape, their crusts are thinner, and they heal, if at all, with more or less puckered and unsightly cicatrices. Syphilis is a disease of adult or advanced age, while lupus is a disease that begins most often in childhood.

The history and course of *kerion* is entirely different from that of a gummatous syphilide, as it occurs in childhood, is usually a single lesion, forms rapidly, is painful and tender, and most often stands in some relation to trichophytosis capitis. A gumma has a marked tendency to break down and ulcerate, whereas kerion has no such disposition.

Psoriasis may readily be distinguished from a papular syphilide in groups by the fact that it never occurs on the scalp alone. It is, moreover, very decidedly scaly when upon the scalp, while the syphilide is not scaly though it may be crusted. Psoriasis causes neither cicatrices nor baldness; syphilis gives rise to both.

It is possible that an *epithelial cancer* may be mistaken for an ulcerating syphilide; but its hard, waxy and raised edge, with delicate blood-vessels running over it; its much slower course; the shooting pains that accompany it; the great proneness to bleeding that it evinces; and the fact that it is uninfluenced by antisyphilitic remedies, sufficiently establish its diagnosis.

The TREATMENT of the syphilides of hairy parts is the same as that of the same lesions located elsewhere. Internally mercury is indicated for the early lesions; mercury with or without the iodide of potassium for the intermediary or late lesions; the iodide of potassium in increasing doses and pushed rapidly until the nose runs and the eyes water, in the ulcerating lesions. These drugs, combined with tonics as needed, and the enforcement of the laws of hygiene, will enable us to effect a cure of the disease in most cases even without local treatment. Local treatment is demanded to combat certain symptoms and to hasten the disappearance of the lesions. The early syphilides usually do not require treatment. If the scalp is covered with scabs and crusts they should be removed with soap and water, and an ointment of the white precipitate of mercury with or without vaseline, or of the nitrate of mercury in the strength of one or two drachms to the ounce of vaseline, may be applied. If there are superficial ulcerations, the same ointments may be used; and if the ulcers are deep, iodoform in powder will form a good dressing. The treatment of alopecia resulting from syphilis has been given already in our chapter on Alopecia.

Lupus Vulgaris.

This never occurs primarily upon the scalp. When seen in this region it is usually an extension from the forehead, and presents but a single patch. It then, according to Hans von Hebra (16), takes the form of a flat, little elevated, even infiltration of the skin which slowly proceeds to ulcerate. The hairs may grow with scarcely impaired vigor for some time after the disease has invaded the scalp. The disease is steadily progressive, heals with cicatricial tissue in one place while spreading in another, and in the course of years may involve the whole scalp, converting it into a mass of puckered cicatrix, which of course is without hair.

The DIAGNOSIS and TREATMENT of the disease as it affects the hairy parts is the same as when it attacks other parts, and for this the reader is referred to the text-books of dermatology.

Lupus Erythematosus.

This disease which originates in the hair-follicles and the sebaceous and sweat-glands of the skin is not infrequently met with on the scalp and hairy parts of the body. The scalp may be invaded from a patch upon the face, or occur coincidentally with the disease on the nose, cheeks or other parts. It may occur upon the scalp alone. It begins in one or a number of round red spots situated about the hair-follicles. These increase in size, and, new spots appearing, they at last coalesce to form a patch, which is irregular in shape and of various sizes up to one large enough to involve the greater part of the scalp. A fully formed patch is of red or violaceous color, covered with closely adherent, thin, parchment-like scales, which are attached to the follicles of the scalp; sharp in outline, always dry, and having its centre formed of delicate cicatricial tissue. Eventually cicatrization will take place in the whole patch, and the part will become absolutely and permanently bald. The disease is slow and chronic in its course, and at times is attended with burning or itching.

The DIAGNOSIS of the disease is easy. It is most apt to be confounded with ringworm and psoriasis. *Ringworm* differs from it in having a history of contagion, in its rounded shape, in its abundant scales and crusts, in its broken off and diseased hairs, and in its spontaneous recovery without leaving a cicatrix or baldness. *Psoriasis* is distinguished by having characteristic patches upon other parts of the body, by its more abundant scaling, by its not affecting the hair, by its history of recovery and relapses, and by its leaving no cicatrix.

The TREATMENT of lupus erythematosus of the scalp is the same as that of the same disease on other parts. I would here only lay special stress upon the great value of the local application of pure carbolic acid, a plan of treatment proposed by DR. GEO. H. FOX, of New York, and one that is attended with wonderful results in many cases. The acid is to be carefully applied by means of a little cotton on a bit of wood, and the application repeated every week, or more often, the frequency depending upon the fall of the crusts left by the previous burning. Cutler's fluid of equal parts of carbolic acid, tincture of iodine, and chloral hydrate, painted on every four or five days, is also good. Phosphorus in the dose of $\frac{1}{1}$ to $\frac{1}{16}$ of a grain in pill form may be administered by the mouth at the same time, care being had to intermit its use at intervals.

VITILIGO.

Leucoderma, acquired albinism, or vitiligo often befalls the hair, and causes white patches or tufts of hair to appear among the darker hair of the part. The skin of the scalp beneath the tuft is perfectly healthy, and the only change either it or the hair suffers is the loss of color on account of a loss of pigment. There may be only a single patch of white hair, or there may be so many that the whole hair of the head is nearly white. The hair may be affected on any part of the body. The disease is unattended with symptoms, and is chronic. Electricity in the form of galvanism or in the static form offers the only chance for improving the condition, but that chance is very small.

Besides the diseases already described many other of the cutaneous diseases may occur upon hairy parts, but the situation in no wise affects the symptoms of such diseases, and therefore their consideration has not been included in this book.

BIBLIOGRAPHY.

NOTE.—References in the text (1, 2, 3, etc.), are to be found in the following list of bibliography and journal literature, under the corresponding numbers.

A. TREATISES ON THE SKIN AND SYPHILIS.

1. Bassereau.—Traité des Affections de la Peau, Paris, 1852.
2. **Bateman**, Thos.—Synopsis of Cutaneous Diseases, London, 1829.
3. **Behrend, G.**—**Lehrbuch der Hautkrankheiten**, 2d Ed., Berlin, 1883.
4. Bulkley, L. D.—Manual of the Diseases of the Skin, 2d Ed., New York, 1882.
5. Bulkley, L. D.—**Eczema and its** Management, 2d Ed., New York, [1884].
6. Bumstead **and Taylor.—Venereal Diseases**, 2d Ed., Phila., 1883.
7. **Coats**, J.—Manual of Pathology, Phila., 1883.
8. **Deligny**, L.—L'eczéma, Paris, 1885.
9. **Diday** and Doyon—Therap. des Mal. **Cutan. et** des Mal Ven., Paris, 1876.
10. Duhring, L. A.—**Diseases of the Skin**, 2d Ed., Phila., 1881.
11. Finger, E.—Die **Syphilis und die venerischen Krankheiten,** Wien, 1886.
11a. Fournier.—Leçons sur la Syphilis, Paris.
12. Fox, T.—Skin Diseases, 3d Ed., New York, 1877.
13. Green, J.—Diseases of the Skin, 2d Ed., Phila., 1859.
14. Guibout, E.—Traité des Maladies de la Peau, Paris, 1885.
15. von Hebra and Kaposi—Lehrbuch der Hautkrankheiten, 2d Ed., Stuttgart, 1876.
16. von Hebra, H.—**Die krankhaften** Veränderungen der Haut, Braunschweig, 1884.
17. Hogg.—Parasitic Origin of Skin Diseases, London, 1873.
18. Hyde, J. N.—Diseases of the Skin, Phila., 1883. 2d Ed., 1888.
19. Kaposi, M.—Pathologie und Therapie der Hautkrankheiten, Wien, 1880.
20. Leloir, **H.—Recherches sur les** Affections Cutanées, Paris, 1882.
21. Lesser, E.—Lehrbuch der Haut und Geschlechtskrankheiten, Leipzig, 1885.
22. Liveing, R.—Treatment **of Skin Diseases**, 4th Ed, New York, 1878.
23. **Liveing**, R.—Diagnosis of Skin Diseases, New York, 1879.
24. **Morris, M.**—Manual of Skin Diseases, London, 1879.

25. **Morrow, P. A.**—Venereal Memoranda, New York, 1885.
26. **Neligan, J. M.**—Diseases of the Skin, Philadelphia, 1852.
27. **Neumann, I.**—Hautkrankheiten, 5th Ed., Wien 1880.
28. **Otis, F. N.**—Syphilis and Genito-urinary Diseases, New York, 1886.
29. **Piffard, H. G.**—Elementary Treatise on the **Diseases of the Skin**, New York, 1876.
29a. **Piffard, H. G.**—**Materia Medica and** Therapeutics **of the Skin**, New York, 1881.
30. **Purdon**—Cutaneous Medicine.
31. **Rayer**—Maladies de la Peau, Paris.
32. **Ricord.**—Lettres sur la Syphilis, Paris.
33. **Robinson, A. R.**—Manual of Dermatology, New York, 1884.
34. **Schwimmer, E.**—Die neuropathischen Dermatonosen, Wien, 1883.
34a. **Squire, B.**—In Reynold's System of Medicine, Vol. V., London, 1879.
34b. **Squire, B.**—A Manual of the Diseases of the Skin.
35. **Van Buren and Keyes—Genito-urinary** Diseases and Syphilis, New York, 1877.
36. **Van Harlingen, A.**—Hand-book of Skin Diseases, Phila, 1884.
37. **Wilson, E.**—Diseases of the Skin, 4th Ed., London, 1857.
38. **Wilson, E.**—Lectures on Eczema, London, 1870.
39. **von Zeissl**—Pathology and Treatment of Syphilis, 2d Ed., New York, 1886.
40. **von Ziemssen, H.**—Handbuch der Hautkrankheiten, Leipzig, 1884.

B. TREATISES ON THE HAIR.

41. **Bartels, M.**—Ueber abnorme **Behaarung beim** Menschen, Berlin, 1876.
41a. **Bazin, E.**—Leçons sur les Affections Parasitaire, Paris, 1862.
42. **Behrend, G.**—Beitrag zur Pathogenese und Behandlung der Acme disseminata und der Sycosis, Berlin, 1881.
43. **Beigel, H.**—Ueber Auftreibung und Bersten **der Haare**, Wien, 1855.
44. **Beigel, H.—The** Human Hair, London, 1869.
45. **Bergeron, E. J.**—Étude sur la Geographie et la Prophylaxie des Teignes, Paris, 1865.
46. **Besnier.**—Parasitic Diseases of the Skin, Paris, 1884.
47. **Boullaud, C. H.**—De la Trichophytie. Thése de Paris, 1865. No. 232.
48. **Cazenave, A.**—Traité des Maladies du **Cuir Chevelu, Paris**, 1850.
49. **Chausit—Sycosis ou Mentagre**, Paris, 1859.
50. **Chevrier—Du Sycosis, Thése** de Montpellier, 1871.
51. **Clasen, F. E.—Die Haut und** das Haar, Stuttgart, 1886.
52. **Cottle, E. W.—The Hair in** Health and Disease, **London**, 1877.
53. **Courrèges, A.—Étude sur la** Pelade, Paris, 1874.
53a. **Davey, W.—Treatise upon the** Human Hair, London, 1871.
54. **Eble, B.**—Die Lehre von den Haaren, Wien, 1831.
55. **Ebner, V. v.—Mikroskopische Studien** über Wachsthum und Wechsel der Haare, 1876.
56. **Feiertag—Ueber die Bildung der Haare,** Dorpat, 1875.

57. Feulard, H.—Teignes et teigneux, Paris, 1886.
58. Fox, Geo. H.—The use of Electricity in the Removal of Superfluous Hair, etc., Detroit, 1886.
59. Gamberini—Le Malattie dei Peli e delle Unghie, Bologna, 1882.
60. Godfrey, B.—Diseases of the Hair, London, 1872.
61. Grecesco—De l'Achorion Schoenleinii, Thése de Paris, 1868. No. 137.
62. Hildebrandt, H.—Ueber abnorme Haarbildung beim Menschen, Königsberg, 1878.
63. Hildesheim, W.—Das Haar und seine Krankheiten, Berlin, 1846.
64. Leonard, C. H.—The Hair, Detroit, 1881.
65. Leturc, A.—Le Nature et le Traitement de la Pelade, Paris, 1878.
66. Michelson, P.—Ueber Herpes tonsurans und Area Celsi, Leipzig, 1877.
67. Morris, M.—The Management of the Skin and Hair, London, 1886.
68. Oesterlen, O.—Das menschliche Haar und seine gerichtsärztliche Bedeutung, Tübingen, 1874.
69. Perry, B. C.—The Human Hair and the Cutaneous Diseases which affect it, 2d. Ed., New York, 1866.
70. Pfaff, E. R.—Das menschliche Haar, 2d Ed., Leipzig, 1869.
71. Pincus, J.—Haarkrankheiten und Haarpflege, 2d Ed., Leipzig.
72.—Pincus, J.—Der Einfluss des Haarpigments und des Markcanals auf die Färbung des Haares, 1872.
73. Pincus, J. P.—Das polarisirte Licht als Erkennungs-Mittel für die Erregungs-Zustände der Nerven der Kopfhaut, Berlin, 1886.
74. Ranke, J.—Haarmenschen, in allgemeine Naturkunde, Leipzig, 1886.
75. Robinson, T.—On Baldness and Grayness, 2d Ed., London, 1883.
76. Rosenberg, H.—Der Weichselzopf, München, 1839.
77. Rouquayrol, E.—Prophylaxie et Traitement de la Teigne tondante, Paris, 1879.
78. Schultz, H.—Haut, Haare und Nägel, 3d Ed., Leipzig, 1885.
78a. Sexton, Geo.—The Hair and Beard, and Diseases of the Skin, London, 1858.
79. Smith, Alder.—Ringworm, its Diagnosis and Treatment, 3d Ed., London, 1885.
80. von Steinkühl, W.—Der Weichselzopf in Deutschland, Hadamar, 1817.
81. Tobolewski, F. R.—Kurze Uebersicht über Bau, Zweck, und Krankheiten der Haare, Leipzig, 1884.
82. Truefit, H. P.—New Views on Baldness, London, 1863.
82a. Unna, P. G.—Anatomie und Physiologie de Haut, in Ziemssen's Handbuch de Hautkrankheiten, 1884.
83. Waldeyer, W.—Atlas der Menschlichen und Tierischen Haare, Lahr, 184.
84. Wilson, E.—Healthy Skin, 8th Ed., London, 1876.
    Andoque, A.—La Physiologie des Cheveux, etc., Paris, 1876.
    Anonymous—The Art of Preserving the Hair, London, 1825.
    Boeck, F.—Area Celsi, Greifswald, 1867.
    Braunstein, H.—Alopecia Areata als Trophoneurosen, Freiburg, 1873.

Chevalier, Sarah A.—A Treatise on the Structure of the Human Hair, New York, 1868.
Couillebault, L.—Quelques Considerations sur l'Herpes Parasitaire dans les Pays Chauds et Traitement par le Cassia Aiata, Paris, 1886.
Debay—Hygiene compléte des Cheveux et de la Barbe, Paris, 1851.
Deville, F.—Theoretical and Practical Exposition of the Diseases of the Hair, Baltimore, 1849.
Ecker, A.—Ueber abnorme Behaarung des Menschen, 1878.
Jahn, G. W.—Der Haararzt, Prag, 1828.
Kneiphof, J. G.—Abhandlung von Haare, 1777.
Ledeganck—Pathologie des Maladies des Follicules Pileux et Sebacées, Bruxelles, 1872.
Lovet, H. T.—Treatise on the Human Hair, New York, 1851.
Merkel, J. F.—Der erfahrene Haararzt, Leipzig, 1840.
Obert—Traité complete des Maladies des Cheveux, etc., Paris, 1848.
Reissner—Beiträge zur Kenntnis der Haare des Menchen und der Tiere, Breslau, 1854.
Rowland—The Human Hair, 1858.
Schweninger, E.—Ueber Transplantation und Implantation von Haaren, München, 1875.
Voigt, C. A.—Abhandlung über der Richtung der Haare, 1859.
Wilson, E.—Ringworm, London, 1847.

C. JOURNAL LITERATURE.

85. Arnstein—Die Nerven der behaarten Haut, Wien. Akadem. Sitzungsbr. 1876. lxxiv. 1. (Abst. Vrtljschr f. Dermat. und Syph. 1878. v. 283.)

85a. Cottle, A.—Practical Remarks on some Points of Tricopathy, etc.. Lancet, 1846. ii. 9.

86. v. Ebner.—Mikroskopische Studien über das Wachsthum und den Wechsel der Haare, Wien. Akad. Sitzungsbr. 1876. lxxiv. 339.

87. Esoff, Johannes—Beitrag zur Lehre von der Icthyosis und von den Epithelwucherungen bei derselben nebst Bemerkungen über den Haarwechsel, Virchow's Archiv, 1877. lxix. 417.

88. Fleming W.—Ein Drillingshaar mit gemeinsamer innerer Wurzelscheide, Montshaft f. prkt. Derm. 1883. ii 163.

89. Flesch, M.—Locken von gekräuselten Haar inmitten der sonst schlichten Kopfhaares. Vrhndl. der Berliner Anthrop. Gslshft. April, 1886, Abst. Mntshft. f. Prakt. Derm. 1886, v. 522.

90. Jobert—Des Poils considerés comme Agents tactiles chez l'Homme, Gaz. Med. de Paris, 1875. iii. 74.

91. Lewis, W. J.—Hair microscopically examined and medicolegally considered, Proc. Amer. Soc. of Microscopists, 1884.

92. Lowther, T. D.—Does the Hair grow after Death? Louisville Med. News, 1877, iv. 186.

93. Pincus, J.—Der Einfluss des Haarpigments und des Markcanals auf die Färbung des Haares, Archv f. Derm. u. Syph. 1872, iv. 1.

94. Senulin—Beiträge zur Histologie der Haare, Zeitschr. f. Anat. und Entwcklngs. gsch. 1877, ii. 654. Abst. Vrtljschr. f. Derm. und Syph. 1877, iv. 574.

95. Wertheim, G.—Ueber den Bau des Haarbalges beim Menschen, Wien. Akad. Sitzungsbr. Math. naturw kl. 1864. i. 302. Abst. Schmidt's Jahrbuch, 1865, cxvii. 286.

Champuis & Moleschott—Untrsuch. z. Naturl. d. Mensch. u. d'Thiere, 1860, vii. 325.
Götte—Cntrlbl. f. d. Med. Wissensch. 1867, v. 769; also, Archv. f. Mikro. Anat. 1868, iv. 273.
Hodgkinson & Sorby—J. Chem. Soc. Lond. 1877, xxxi. 427.
Pincus, J.—Archv. f. Anat. Phys. u. Wissenschft. Med. 1871, pg. 55.
Renaut, J.—Compt. rend. Acad. d. Sc. Par. 1880, xci. 1084.
Schweninger, E.—Ztschr. f. biol. München, 1875, xi. 341.
Stieda, L.—Archv. f. Anat. Phys. u. Wissensch. Med. 1867, pg. 511.

CANITIES.

96. Anonymous—Periodic Change of Color of Hair, Lancet, 1884, ii. 603.
97. Berger, O.—Zwei Fälle von **Canities praematura**, Virchow's Archiv. 1871, liii. 533.
98. Brown-Sequard—Expériences démontrant que les Poils peuvent passer rapidement du Noir au Blanc. Archv. d. Phys. Norm. et Path. 1869, ii. 442.
99. Charcot, J. M.—Apropos d'un Cas de Canitie survenue très rapidement, Gaz. Hebdom. de Paris, 1861, viii. 445.
100. Ehrmann, S.—Ueber das Ergrauen der Haare und verwandte Processe, Allg. Wien. Med. Zeit. 1884, xxix. 331.
101. Ehrmann, S.—Untersuchungen über die Physiologie und Pathologie des Hautpigments, Vrtljschr f. Derm. und Syph., 1885, xii. 507, and 1886, xiii. 57.
102. Ferguson, J.—Sudden **Canities**, Canad. Jour. Med. Sc. 1882, vii. 113.
103. Godlee—**Hereditary White Patch of** Hair, Med. Times & Gaz. 1884, i. 180.
104. Isdell, V. C.—**Case of the Restoration of the Natural** Color **of Human Hair after having been Gray for several Years**, Med. Times & Gaz. 1884, ii. 680.
105. Jackson, Geo. T.—**Canities, Jour.** Cutan. & Ven. Dis. 1885, iii. 38.
106. Jeffries—**Case of sudden** Canities, Boston Med. & Surg. Jour. 1871, lxxxiv. 45.
107. Landesberg, M.—**Jaborandi and Pilocarpine**, Med. Bul. Philadelphia, 1882, iv. 43.
108. Landois, L.—Das **plötzliche** Ergrauen der Haupthaare, Virchow's Archiv. 1866, xxxv. 575.
109. Landois, L.—E. Wilson's Fall von Intermittenden Ergrauen der Haupthaare, Virchow's Archiv. 1869, xlv. 113.
110. Lesser—Ueber Ringelhaare, Allg. Wien. Med. Zeit. 1885, **xxx.** 441; and Montshft. f. Prakt. Dermat. 1885, iv. 371.
111. Miner, J. F.—**Change of the Color of the** Hair in a Night.—Case, Buffalo Med. & Surg. Jour. 1864-'65, iv. 93.
112. Murray—Hair **Bleaching from Neuralgia**, Lancet, 1869, i. 324.
113. Pincus, J.—**Ueber Canities senilis und** prematura, Virchow's Archiv. 1869, xlv. 129.
114. Pincus, J.—Der Einfluss des Haarpigments und des Markcanals auf die Färbung des Haares, Archiv Derm. und Syph. 1872, ii. 1.
115. Pohlmann, J—An experimental study on the action of Pilocarpine, Buffalo Med. & Surg. Jour 1882-'83, xxii. 441.
116. Raymond—Un Cas de Décoloration rapide de la Chevelure, Rev. de Med. 1882, ii. 770.
117. Shaw, H.—On Degradation of **Type in the Insane**, St. Barth. Hosp. Rep. 1884, xx. 169.
118. Smythe, A. G.—Changing of the **Color of the** Hair **without known** Cause, Archiv. Derm. 1880, vi. 246.
119. Wallenberg—Ein Fall von bleibender Veränderung des **Haar** und Hautfarbe nach Scharlachfieber, Vrtljschr. f. Derm. **und Syph.** 1876, i. 63.

120. Welch, F. H.—On Change of Color in Hair, Lancet, 1873, i. 754.
121. Wertheim, G.—Ueber das Ergrauen, Weisswerden und Ausfallen der Haare beim Menschen, Wien. Med. Wochschr. 1878, xxviii. 181 et seq.
122. Wilson, E.—Ringed Hair, Trans. Roy. Soc. Lond. 1867.

DISCOLORATION OF THE HAIR.

123. Beigel, H.—Blaue Haare, Virchow's Archv. 1867, xxxviii. 324.
124. Billi—Un Caso di Tricolorosi, Gior. Ital. d. mal d. Pelle, 1872, xiii. 243. (abst.) Ann. d. Derm. et Syph. 1872-'73, iv. 138.
125. Cattell, Thos.—Practical Remarks on some Points of Trichopathy and the Chemical Pathology of the Human Hair, Lancet, 1846, ii. 9.
126. Cattell, Thos.—Practical Remarks on Diseases manifested in the Hair, Lancet, 1846, 316.
127. Hauptmann—Rothwerden dunkler Haare einer Leiche bei der Verwesung, Virchow's Archiv, 1869, xlvi. 502.
128. Jackson, Geo. T.—Discoloration of the Hair, Jour. Cutan. & Ven. Dis. 1884, ii. 173.
129. Leonard, C. H.—The Hair, Detroit, 1881.
130. Oesterlen, O.—Das menschliche Haar, Tubingen, 1874.
131. Orsi—Grüne Haare (abst.) Virchow's Jahresbericht, 1871, ii. 523.
132. Petri—Ueber die grüne Färbung der Haare bei alteren Kupferarbeitern, Brl. Klin Wchnschft. 1881, 18, 762.
133. Pfaff—Das menschliche Haar, Leipzig, 1869.
134. Prentiss, D. W.—Remarkable Change of Color of the Hair, while under Treatment by Pilocarpine, Phila. Med. Times, 1881, xi. 609.
135. Reinhard, C.—Ein Fall von periodischen Wechsel der Haarfarbe, Virchow's Archiv. 1884, xcv. 337.
136. Smyly, W. J.—Sudden Change in Color of the Hair and Skin, Med. Press. & Circ. 1883, xxxv. 184.
137. Squire, B.—An extremely rare Condition of the Hair, Lancet, 1881, ii. 74.

ALOPECIA.

139. Anonymous—The Prevention of Baldness, Med. Record, N. Y., 1886, xxix. 101.
140. Barlow, Thos.—Alopecia in Congenital Syphilis. Lancet, 1877, ii. 276.
141. Bierbaum, J.—Alopecia Partialis, Jour. f. Kinderheil, 1863, xli. 167.
142. Chinchole, F.—De la Nature parasitaire du Pityriasis Capitis et de l'Alopecia Consécutive, Paris, 1874.
143. Cattle, Thos.—Practical Remarks on Diseases manifested in the Hair, Lancet, 1846, ii. 316.
144. Cattle, Thos.—Practical Remarks on some Points of Tricopathy and the Chemical Pathology of the Human Hair Lancet, 1846, ii. 9.
145. Cockburn, J. B.—Syphilitic Alopecia treated with Lee's Mercurial Vapor Bath, Lancet, 1867, i. 763.

146. Crisp, E.—General Alopecia with Microscopic Specimens of the Hair and Nails, Trans. Path. Soc. Lond., 1871, xxii. 305.
147. Douet—Syphilis constitutionelle—Alopécie, Gaz. de Hôp. 1864, xxxvii. 259.
148. Dulaurier, A. B.—Syphilis constitutionelle—Alopécie, Guerison, Gaz. de Hôp, 1864, xxxvii. 310.
149. Eaton, V. G.—A Bald and Toothless Future, Popular Science Month., N. Y. 1886, xxix. 803.
150. Ellinger, L.—Zur Aetiologie und Prophylaxie der Alopecia Prematura, Virchow's Archiv. 1879, lxxvii. 549.
151. Finch, H.—The Treatment of Alopecia, Lancet, 1873, ii. 101.
152. Fournier, A.—Des Alopécies, Gaz. des Hôp, Paris, 1879, lii. 1067, et seq.
153. Fournier, A.—De l'Alopécie, de l'Onyxis, etc., comme Accidents de la Période secondaire de la Syphilis, Ann. de Derm. et de Syph. 1870-71, iii. 12.
154. Gaskoin, G.—Alopecia Vitiligo, Brit. Med. Jour. 1873, i. 642.
155. Gowers—Cases of Universal Alopecia and Epilepsy, Med. Times and Gaz. 1878, ii. 379.
156. Heitzmann, C.—Remarks on Electrolysis and other Practical Topics, Trans. Amr. Derm. Asso., 1885, pg. 32; also, Jour. Cutan. & Ven. Dis., 1885, iii. 339.
157. Hill, J. H.—Hairless Australian Aborigines, Brit. Med. Jour., 1881, i. 177.
158. Hutchinson, J.—Congenital Absence of Hair with Atrophic Condition of the Skin, and its Appendages, Brit. Med. Jour., 1886, i. 929 ; also, Proc. Roy. Micro. Soc., Lond. 1885-6, ii. 116; also, Lancet, 1886, i. 923.
159. Ihle, M.—Beitrage zur Behandlung der Haut-Krankheiten mit Resorcin, Montshft. f. prakt. Dermat. 1885, iv. 429.
160. Jones and Atkins—Microscopical Appearances in a Case of Congenital Alopecia, Dublin Jour. Med. Sc. 1875, lx. 200.
161. King, F. A.—On the Causes of Alopecia and its greater Frequency in Males than in Females, Amr. Jour. Med. Sc. 1868, April, 416.
162. Kinney, Thos. H.—Alopecia from Nervous Shock, Virg. Med. Month. 1881, March, 937.
163. Lassar and Bishop.—Die Uebertragbarkeit der Alopecia Prematura, Montshft. f. prakt. Dermat. 1882, i. 131.
164. Lassar, O.—Ueber Alopecia Prematura, Berl. Klin. Wochnshrft. 1883, xvi. 233.
165. Loisch, T.—Zur Kasuistik der Alopecia (A. generalis), Wien. Med. Wochnshrft. 1879, xxix. 888.
166. Luce, J. B.—Recherches sur un Cas curieux d'Alopécie, Thèse de Paris, 1879, No. 579.
167. McGuire, J.C.—Alopecia; its Etiology, Diagnosis and Treatment, Am. Pract. & News. 1886, i. 359.
168. Mackey.—Alopecia, Brit. Med. Jour. 1885, ii. 797.
169. Malassez, L.—Note sur la Champignon du Pityriasis simple —Archv. de Physiologie Normale et Patholog. 1874, pg. 451.
170. Malassez, L.—Note sur l'Anat. Patholog. de l'Alopécie Pityriasique, Archv. de Physiologie Normale et Patholog. 1874, pg. 465.
171. Michelson—Ueber die malignen Formen der Alop. Pityrodes und der Alop. Areata, Verein f. Wissnschft. Heilkund zu Königsberg, Abst. Monatshft. f. prakt. Dermat. 1882, iv. 124.

172. Pincus, J.—**Zur** Diagnosis des ersten Stadium der Alopecia, Virchow's Archiv. 1866, xxxvii. 18.
173. Pincus, J.—**Das zweite** Stadium **der Alopecia Pityrodes,** Virchow's Archiv. 1867, xli. 322.
174. Pincus, J.—**Ueber den** Krankheits-Charakter **der** chronischen Alopecia und ihre gewöhnliche Beschränkung **auf** der Vorder—und Mittelkopf, Berl. Klin. Wochnschrft. 1875, pg. 42, 59.
175. Pincus, J.—Zur **Therapie der Alopecia** Pityrodes, Virchow's Archiv. 1868, xliii. 305.
176. Pincus, J.—**Ein Fall von** Alopecia Pityrodes **vor Eintritt der Pubertat, Berl. Klin.** Wochnshrft. 1869, vi. 341.
177. Pincus, J.—**Uber die** Alopecia und den Indurativen Krankheits **Process** überhaupt, Berl. Klin. Wochnshrft. 1883, xx. 645.
178. **Rohé, Geo.** H.—The Causes of Premature Baldness, Atalanta **Med. &** Surg. Jour. 1878-79, **xvi. 391.**
179. **Schede,** M.—Ein Fall von **totaler** angeborener Alopecia, **Archv. f. Klin.** Chirurg. 1872, xiv. **158.**
180. Schmitz, Geo.—Ueber eine **noch** nicht bekannt gewordene **Wirkung des** Pilocarpinum Muriaticum, Berl. Klin. Wochnshrft. 1879, xvi. 48.
181. Sedgwick, **Wm.—On the Influence of** Sex in Hereditary Disease, Brit. & For. **Med. Chir. Rev.** 1863, i. 452.
182. Shoemaker, **J. V.—Loss of Hair,** Med. Bulletin, Phila. 1879, March, pg. 19.
183. Shoemaker, **J. V.—Weitere** Untersuchungen über die Natur und Wirkung **der Oleate,** Montshft. f. prakt. Dermat. 1884, iii. 358.
184. Startin—**Treatment of Alopecia, Brit.** Med. Jour. 1880, **ii.** 157.
185. **Todd, R. Cooper—Case of Total Loss of Hair,** Lancet, 1869, ii. 41.
186. Unna—**Aphorismen über Schwefel**-therapie und Schwefelpraparaten, **Monatshft. f. prakt. Dermat.** 1882, Bd. i. No. 10.

Bärensprung—Annal. de Charitee, Berl. Bd. viii. fo. 8 et seq.
Clemens, A.—Würz. Med. Ztschr. 1865, vi. 365.
Fomin—Archv. Vet. Nauk. St. Petersburg, 1886, xvi. 105.
Frodsham, J. M.—Med. Mirror, Lond. 1864, i. 408.
Kane, H. H.—Pub. Health, N. Y. 1879-80, i. 82.
Miklucho—Vrhndlung. d. Berl., Gslschft. f. Anthrop. 1881, pg. 143.
Nayler, G.—Proc. Roy. Med. & Chir. Soc. 1864, iv. 289.
Pincus, J.—Deutsche Klinik, 1871, xxiii. 3 et seq., and 1872, xxiv. 114 et seq.
Waldenström—Deutsche Klinik. 1873.
Wicherkiewicz, B.—Klin. Montsbl. f. Augenheil, 1886, xxiv. 139.
Wyss—Archiv der Heilkund, 1870, xi. 395.

Alopecia Areata.

187. **Anderson,** M'C.—**Parasitic Affections of the Skin,** Med. Times & Gaz. 1861, i. 298.
188. **Anderson,** M'C.—**Relationship between** Addison's Disease, Vitiligo, **and Alopecia Areata, Glasgow Med. Jour.** 1879, xi. 14.
189. Balman, T.—**Alopecia Areata followed by** Universal Loss of **Hair,** Brit. Med. Jour. 1865, i. 204.
190. Bazin—**Microsporon, Article in Dict. Encycloped.** des Sciénces Méd. Paris, 1873.
191. **Bender,** M.—Ueber **de Ætiologie der** Alopecia Areata, Deutsch Med. Wochnshrft. 1886, xii. 817.
192. **Boeck,** F.—Beobachtung über Area Celsi, Virchow's Archv. 1868, **xliii.** 336.

193. Bordoni-Uffreduzzi—The Microphyten of the Normal Skin, Fortschr. d. Med. 1886, No. 5; Abst. Vrtljschrft. f. Derm. u Syph. 1886, xiii. 257.

194. Bristowe, J. S.—Observations on the Diseases of the Skin which are generally supposed to be due to the Growth of Vegetable Parasites, St. Thos. Hospital Rep. 1870, i (n. s.) 157.

195. Buchner, H.—Kritisches Bemerkungen zur Ætiologie der Area Celsi, Virchow's Archv. 1878, lxxiv. 527

196. Collier, Jno.—On the Causation of Alop. Areata, and its accidental co-existence with Tinea Tonsurans, Lancet, 1881, i. 951.

197. Cumming, J.—Alopecia Areata, The Practitioner, 1878, i. 110.

198. Cummisky, J.—Notes on T. Decalvans, Phila. Med. Times, 1882, xiii. 41.

199. De Young, A. H.—Oleate of Mercury, Phila. Med. & Surg. Reprtr. 1881, xliv. 315.

200. Dyce-Duckworth—On the Nature and Treatment of Porrigo Decalvans, St. Bartholomew's Hosp. Rep. 1872, viii. 144.

201. Dyce-Duckworth—Case of Area Celsi (Porrigo Decalvans), in which the Parts were Examined after Death, Trans. Path. Soc. (Lond.) 1882, xxxiii. 386.

202. Duhring—Pathology of Alopecia Areata, Amr. Jour. Med. Sc. 1870, July, 122.

203. Duhring—Alopecia Areata, Med. Times, Phila. 1872, ii. 469

204. Ebstein, Wm.—Zur Ætiologie der Alop. Areata, Deutsche Med. Wochnshrft. 1882, liii. 724.

205. Eichhorst, H.—Beobachtungen über Alopecia Areata, Virchow's Archiv. 1879. lxxviii. 197.

206. Finch, H.—The Treatment of Alopecia, Lancet. 1873, ii. 101.

207. Fournier, A.—Des Alopécies, Gaz. des Hôp. 1879, lii. 1067 et seq.

208. Fox, T.—Practical Notes on Cutaneous Subjects, Lancet, 1874, ii. 510.

209. Fox, T.—On Alopecia Areata and Tinea Tonsurans, Med. Times & Gaz. 1874, ii. 630.

210. Fox, T.—Tinea Decalvans, Lancet, 1874, ii. 510.

211. Gambourg—Drei Fälle von Alopecia Areata in derselben Familie, Tidsskr. f. prakt. Med. 1882, No. 22 (Abst.) Vrtljschr. f. Derm. und Syph. 1883, x. 654.

212. Gibney, V. P.—A Case of Scleroderma with Hemiatrophia Facialis, Alopecia Areata and Canities, Archv. Derm. 1879, v. 155.

213. Graham, J. E.—A Case of Alopecia Areata, Canad. Jour. Med. Sc. 1880, v. 138.

214. Gruby—Recherches sur la Nature, le Siege, et le Développement du Porrigo Decalvans ou Phytoalopécie, Compt. rend. de l'Acad. des Sc. 1843, xvii. 301.

215. Guibout—Teigne Pelade, Gaz. des Hôp, 1886, lix. 76.

216. Hardaway, W. A.—Two Cases of Recurrent Alopecia Areata, Jour. Cutan. & Ven. Dis. 1884, ii. 260.

216 a. Hilbert, R.—Partielle Hypertrichosis, Virchow's Archiv. 1885, xcix. 569.

217. Hillier, Thos.—Is Alopecia Areata or Tinea Decalvans Contagious? Lancet, 1864, ii. 374.

218. Horand—Considerations sur la Natur et le Traitement de la Pelade, Ann. de Derm. et de Syph. 1874-5, vi. 408 et seq.

219. **Hutchinson. J**—Alopecia **Areata, Lancet, 1882, i. 395;** also Med, **Times & Gaz.** 1858, i 165 et seq.
220. **Hutchinson,** J.—Cases of Alopecia Circumscripta, in which Contagion occurred, Trans. Path. Soc., Lond. 1862, xiii. 265.
221. Jackson, Geo. T.—Alopecia Areata; its Ætiology, Diagnosis and Treatment. N. Y. Med. Jour. 1886, xliii. 210.
222. Jenner, Wm.—Clinical Lectures on the Diseases of the Skin, Med. Times & Gaz. 1857, ii. 650.
223. Joseph, Max—Experimentelle Untersuchungen über die Ætiologie der Alopecia Areata, Mntshft. f. prakt. Derm. 1886, v. 483.
224. **Lailler, M.**—La Pelade, Gaz. des Hôp. 1875, cv. 858.
225. **Lassar, O.**—**Alopecia Areata,** Deutsche Med. Wochnshrft. 1885, xi. 531.
226. **Liveing. R.**—Remarks on Alopecia Areata and Tinea Tonsurans, Med. Times & Gaz. 1874, ii. 601.
227. **Liveing. R.**—**On the Causes** of Alopecia Areata, Tr. Intrnat. Med. Cong., Lond. 1881,iii. 158 ; also Brit. Med. Jour. Oct. 1st, 1881.
228. **Malassez, L.**—Note sur le Champignon de la Pelade, Archv. de Phys. Norm. et Path. 1874, i. 203.
229. Michelson, P.—Zur Discussion **über die** Ætiologie der Area Celsi, Virchow's **Archiv. 1880, lxxx. 296.**
230. Michelson, P.—Ueber die malignen Formen der Alopecia Pityrodes und der Alopecia Areata, Verein f. Wissnschft. Heilkund. zu Königsberg, April 3d, 1882.
231. Michelson, P.—Ueber Herpes Tonsurans und **Area** Celsi, Volkmann's Klinische Vorträge, No 120.
232. Michelson, P.—Bemerkung zu der Arbeiten des Herrn Dr. v. Sehlen, **über die** Ætiologie der Alopecia **Areata,** Virchow's Archiv. 1885. xcix. 572, and 1885, c. 576.
232a. Michelson, **P.**—Zum Capitel der Hypertrichose, Virchow's Archiv. 1885, c. 6.
233. Michelson, P —Zur Symptomatologie der Alopecia Areata, Verein f. Wissnschft. Heil. zur Königsberg, Nov. 1885 Abst. Mntshft. f. prakt. Derm. 1886, v. 77.
234. Michelson, P.—Ueber die sogenannten Area Coccen, **Frtschrt.** d. Med. 1886, iv. 230.
235. **Nayler,** Geo.—Alopecia Areata, Med. **Times &** Gaz. 1864, i. 266.
236. Nieden, A.—Vier Fälle von Alopecia **Totalis Persistens,** Centrbl. f. prakt. Augenheilk. 1886, x. 133.
237. Nystrom, A.—Note sur la Nature de la Pelade ou Alopecia Areata, Annal. de Dermat. et de Syph. 1875, vii. 440.
238. Piffard, H. **G.**—Alopecia Areata, Med. **Rec.,** N. Y. 1876, xi. 77.
239 Pincus, **J.**—Ueber die Aufeinanderfolge von Alopecia Areata und Vitiligo, Virchow's Archiv. 1872, liv. 433.
240. Reid, Thos.—**Alopecia** Areata, Glasgow Med. **Jour. May,** 1869.
241. **Rindfleisch—Area Celsi, Archv. f. Derm.** und Syph. 1869, i. 483.
242. Robinson—A **Case of** Alopecia Areata, Lancet. 1882, i. 395.
243. Robinson, T.—**Case** of Symmetrical loss of Hair with Absorption of Pigment in the invaded Patches, Tr. Path. Soc., Lond. 1882, xxxiii. 391.
244. Robinson, T —Alopecia Areata, **Parasitic Sycosis and Ring-**

worm of the Body in the same Individual, Brit. Med. Jour. 1885, i. 1040.
245.—Robinson, T.—Alopecia Areata and Ringworm, Brit. Med. Jour. 1885, ii. 47.
246. Scherenberg, H.—Beobachtung über Area Celsi, Virchow's Archiv. 1869, xlix. 493.
247. Schulthess, W.—Klinische Beobachtungen über Alopecia Areata, Corrspdzbl. Schweiz. Arzte, 1885; Abst. Vrtljschr. f. Derm. und Syph. 1886, xiii. 287.
248. Schultze, H.—Die Theorien über Area Celsi, Virchow's Archiv. 1880, lxxx. 193.
249. Pye Smith—Alopecia Areata, Guy's Hosp. Rep. 1880-81, xxv. 139.
250. Smith, Alder—Alopecia Areata and Ringworm, Brit. Med. Jour. 1885, i. 1297.
251. Smith, T. C.—A case of Alopecia Calva, Detroit Lancet, 1884, viii. 111.
252. Startin, Jas.—Alopecia Areata and Ringworm, Brit. Med. Jour. 1885, ii. 375.
253. Stowers, J. H.—Tinea Tonsurans, accompanied by Alopecia Areata, Lancet, 1881, i. 326.
254. Stowers, J. H.—On the Nature and Treatment of Alopecia Areata, Brit. Med. Jour. 1875, ii. 226.
255. Tay, Warren.—Cases of Alopecia Areata and Tinea Tonsurans, Lancet, 1874, ii. 659.
256. Thin, Geo.—Alopecia Areata, Brit. Med. Jour. 1882, ii. 783.
257. Thin, Geo.—Four Cases of Alopecia Areata occurring in one Family, Tr. Path. Soc., Lond. 1882, xxxiii 390.
258. Thin, Geo.—A Further Contribution to the Treatment of Alopecia Areata, Brit. Med. Jour. 1882, ii. 828.
259. Thin, Geo.—On Bacterium Decalvans; an Organism associated with the Destruction of the Hair in Alopecia Areata, Proc. Roy. Soc. 1881-82, xxxiii. 247
260. Thin, Geo.—A Case of Alopecia Areata showing Contagion, Lancet, 1882, i. 395.
261. Thin, Geo.—Alopecia Areata und Bacterium Decalvans, Mntshft. f. prakt. Dermat. 1885, iv. 241.
262. Tyson, W. J.—Three Cases of Universal Alopecia, with Remarks, Brit. Med. Jour. 1886, i. 345. Lancet, 1886, i. 351
263. Uchermann—Ein Fall von Alopecia Completa Traumatica, Tidsskr. f. prakt. Med. 1883; Abst. Vrtljschr. f. Derm. und Syph. 1883, x. 626.
264. Unna—Ein Fall von hochgradiger Alopecia Acuta, Deutsch. Med. Wochnschrft. 1881, vii. 565.
265. Veiel—Specieller Bericht über die Resultate der Heilanstalt für Flechtenkrankheit in Canstadt, 1855-1861; Schmidt's Jahrbch. 1863, pg. 117 et seq.
266. Vidal, M.—Des Pelades, Gaz. des Hôp. 1879, lii. 459.
267. Von Sehlen, H.—Zur Ætiologie der Alopecia Areata, Virchow's Archiv. 1885, xcix. 327, and c. 362.
268. Von Sehlen, H.—Alopecia Areata, Mntshft. f. prakt. Dermat. 1885, iv. 380.
269. Waldenström—Behandlung der Alopecia Areata, Deutsch. Klinik. 1873, xxv. 273.
270. Watson, Jas.—Notes of a Case of Alopecia Areata, treated by Carbolic Acid, Edin. Med. Jour. 1864-5, x. 234.

271. Wilson, E.—On the Phytopathology of the Skin, Brit. & For. Med. Chir. Rev., Jany. 1864, pg. 199.
272. Wyss, O.—Case of Alopecia Areata following the Use of Arsenic, Amr. Jour., Syph. & Derm. 1870, pg. 349.
273. Ziemssen—Area Celsi seu Alopecia Circumscripta, Schmidt's Jahrbch. 1864, cxxiii. 299.

>Besnier E.—Moniteur J. de Méd., Par. 1877, ii. 305
>Chaboux—Union Med. de la Seine, Inf. Rouen. 18⁹   116.
>Drysdale, C.—Jour. Cutan. Med., Lond. 1868-69, i 3{ .
>Jacobi, A.—Amr. M. Month. & N. Y. Rev. 1861. xvi. 9.
>Nachtigal—Mitth. an der Med. Klin. zu Würzburg, 1886, ii. 405.
>Pincus, J.—Deutsche Klinik, 1869. xxi. 1 et seq.
>Pirocchi, P.—Giorn. Ital. delle Mal. Ven. e della Pelle, Dec. 1880.
>Wyss, O.—Archv. d Heilkund. 1870, xi. 395.
>Ziemssen, H.—Greifswald. Med. Beitr. 1864, ii. 111.

Atrophia Pilorum Propria.

274. Anderson, M'C.—Unique Case of Hereditary Trichorrhexis Nodosa, Lancet, 1883, ii. 140.
275. Behrend, G.—Ueber Knottenbildung an Haarshaft, Virchow's Archiv. 1886, ciii. 437; Berl. Klin., Woch. 1885, xxii. 270.
276. Beigel—Ueber Auftreibung und Bersten der Haare, Sitzungsbr. der Mathem. Naturw. Klasse der Wien. 1855, xvii. 612.
277. Bulkley, L. D.—Curious Knotting of the Hair, producing an Appearance similar to the Nits of Pediculi Pubis, Arch. of Dermat., N. Y. 1881, vii. 403.
278. Cheadle & Morris—Piedra, Trichorrhexis Nodosa and Tinea Nodosa, Lancet, 1879, i. 190.
279. Devergie—Note sur la Tricoptilose, Ann. de Derm. et de Syph. 1870-71, iii. 5.
280. Duhring, L. A.—Case of an Undescribed Form of Atrophy of the Beard, Amr. Jour. M. Sc. 1878, July, 88.
281. Eichhorst, H.—Beobachtungen über Trichorrhexis Nodosa, Zeitschrft. f. Klin. Med. 1884, vii, supplement Hft. 58.
282. Ferber—Eigenthumliches Verhalten des Haupthaares, Virchow's Archiv., 1866, xxxvi. 598.
283. Fox, T. C.—On the Atrophic Changes in the Hair known as Trichorrhexis Nodosa, Lancet, 1878, ii. 803.
284. Hoggan, Geo.—Piedra, Lancet, 1878, ii. 347.
285. Jackson, Geo. T.—Atrophia Pilorum Propria, Jour. Cutan. & Ven. Dis. 1884, ii. 261.
286. Kohn, S.—Ueber Trichorrhexis Nodosa, Vrtljschrft. für Derm. und Syph. 1881, viii. 581.
287. Landois—Ringelhaar, Virchow's Archiv. vol. xxxv. pg. 575, and 1869, xlv. 113.
288. Lesser, E.—Ueber Ringelhaare, Algemeine Wien. Med. Zeit. 1885, lxxx. 441, 452; Montshft. f. p. Derm. 1885, iv. 371; Deutsche Medicinalzeit, 1885, No. 22, pg. 249.
289. Leonard, C. H.—The Hair, etc., Detroit, 1881.
290. Luce—Sur un Cas curieux d'Alopécie, Thése de Paris, 1879, No. 578.
291. Malley, A. C.—Piedra, Lancet, 1878, ii. 276.
292. Morris, M.—Piedra, Lancet, 1879, i. 407; also Med. Times & Gaz. 1879, i. 409.
293. Morris, M.—Piedra, a new Disease of the Hair, Trans. Path. Soc., Lond. 1879, xxx. pg. 441.
294. Oesterlen—Das Menschliche Haar, Tübingen, 1874.

295. Paxton, F. V.—On a Diseased Condition of the Hair of the Axillæ, Jour. Cutan. Med., Lond. 1869, iii. 133.
296. Pfaff—Das Menschliche Haar, Leipzig. 1869.
296a. Pincus, J.—Haarkrankheiten. Leipzig.
297. Purdon, H. S.—Note on Fragilitas Crinium, Jour. Cutan. Med., Lond. 1871, iv. 253.
298. Pye Smith, P. H.—Specimen of the Affection of the Hair which has been described as Piedra, etc., Trans. Path. Soc., Lond. 1879, xxx. 439.
299. Robinson, Tom—Baldness and Grayness, Lond. 1883.
300. Roeser—De la Tricoptilose, Ann. de Derm. et de Syph. 1877-78, ix. 185.
301. Schwimmer, E. und Eberth—Ueber Trichorrhexis Nodosa Barbae, Vrtljschr. f. Dermat. u. Syph. 1878, v. 585.
302. Sherwell, S.—Case of Trichorrhexis Nodosa, or Beigel's Disease, Arch. Dermat., N. Y. 1879, v. 240.
303. Smith, W. G.—On a rare Nodose Condition of the Hair, Brit. Med. Jour. 1879, ii. 291 ; 1880, i. 654.
304. Spiess, A.—Das Verhalten der Centraltheile der Haare, etc., Zeitsch. f. Rat. Med. 1859, Reihe, iii. Bd. v. fo. 1.
305. Startin, Jas.—Piedra, Lancet, 1878, ii. 866.
306. Thin, Geo.—Case of Congenital Abnormality in the Hair Production on the Scalp, Arch. Dermat., N. Y. 1882, viii. 237.
307. Waldeyer—Atlas der Mensch. und Tierische Haare, Lahr, 1884.
308. Whitla, W.—Trichorexis Nodosa, Dublin Jour. Med. Soc. 1879, lxvii. 104.
309. Wilks—Lect. Path. Anat. 1857.
310. Wilks, Sam.—Piedra, Lancet, 1878, ii. 347.
311. Wilson—Syphilitic Disorganization of the Hair, Jour. Cutan. Med. 1869, iii. 309.
312. Wolfberg—Zur Ætiologie der Trichorrhexis Nodosa, Deutsche Med. Wochnschrft. 1884, x. 490.

    Breda—Gaz. Med. Ital. prov. Venete, 1882, xxv. 303.
    Desenne—Compt. rendue de l'Acad. des Sc. Par. 1878, lxxxvii. 34.
    Engel—Wien. Akad. Sitzungsbrt. 1856.
    Küchenmeister—Oestr. Zeitschr. f. prkt. Heilk. 1867, xiii. 218.

Hypertrophia Pilorum.

313. Adams, A. L.—Redundancy of Hair on the Body, Lancet, 1874, i. 688.
314. Anderson, McC.—Analysis of 11,000 consecutive Cases of Skin Disease, Lancet, 1871, ii. 708.
315. Anonymous—A Freak of Nature, Lancet, 1873, ii. 612.
316. Anonymous—The Kostroma People, Med. Times & Gaz. 1874, i. 245.
317. Ayer—Excessive Growth of Hair following Applications to the Skin, Bost. Med. & Surg. Jour. 1863, lxviii. 162.
318. Bartels, M.—Ueber abnorme Behaarung beim Menschen., Gslshft. f. Natur. u Heil., Berlin, 1876, pg. 110.
319. Bartoux—L'epilation par la Galvanocaustique Chimique, Rev. Med. Fr. et Etrang. 1886, March 13th.
320. Behrend, G.—Ueber dauernde Beseitigung krankhaften Haarwuchses, Berl. Klin. Wochnschrft. 1886, xxiii. 170.
321. Beigel, H.—Ueber abnorme Haarentwickelung beim Menschen, Virchow's Archiv. 1868, xliv. 418.

322. Boudet, M.—Sulphuret of Soda as a Depilatory, Jour. de Pharmacien, vol. xviii., pg. 119; Abst. Brit. & For. Med. Chir. Rev. 1851. i. 279.
323. Brocq, L.—De la Destruction des Poils par l'Electrolyse, Gaz. Hebdom de Med. Paris, 1886, xxiii. 391; also Bull. et Mém. Soc. Méd. de Hôp. de Par., iii. 1886, 248; also Therap. Contemp. 1886, vi. 375.
324. Bulkley, L. D.—A new Method of permanently removing Superfluous Hair, Arch. Derm. 1878, iv. 287.
325. Butler, Jno.—The permanent Removal of Superfluous Hair by Electrolysis, Med. Chir. Quart. N. Y. 1880, i. 43.
326. Carpenter. Julia W.—The Removal from the Skin of Papillary Growths, Pigmentary Moles and Superfluous Hair, Cincin. Lancet-Clinic, 1886, xvii. 515.
327. Chisolm, J. J.—Treatment of Wild Hairs by Electrolysis, Maryland Med. Jour. 1880-81, vii. 553.
328. Chowne, W. D.—Remarkable Case of Hirsutic Growth in a Female, Lancet, 1852, i. 421 et seq; 1852, ii. 51.
329. Cummins—Impotence in the Male, Lond. Med. Gaz. 1836-37, xix. 263.
330. Duhring, L. A.—Case of a Bearded Woman, Arch. Derm. 1877, iii. 193.
331. Duhring, L. A.—An Instrument for the Removal of Superfluous Hair, Amr. Jour. Med. Sc. 1881, lxxxii. 142.
332. Elliot, Geo. T.—Keratosis Sebacea; a Case associated with Hypertrichosis, Med. Rec. N. Y. 1886, xxix. 64.
333. Eschricht—Ueber die Richtung der Haare am menschlichen Körper, Müller's Archiv. f. Anat. u. Phys. 1837.
334. Finger—Hypercrinosis mit Amenorrhœa, Allg. Wien. Med. Zeit. 1873, xviii. 604.
335. Fischer—Ueber Trophische Störungen nach Verletzungen an den Extremitäten, Berl. Klin. Wochnschrft. 1871, viii. 145.
336. Fox, Geo. H.—On the Permanent Removal of Hair by Electrolysis, Med. Rec., N. Y. 1879, xv. 270.
337. Fox, Geo. H.—The Permanent Removal of Hair by Electrolysis. Med. Rec. N. Y. 1882, xxi. 253.
338. Fürst, L.—Hypertrichosis Universalis mit Hypertrophie der Kiefer-alveolarränder, Virchow's Archiv. 1884, xcvi. 357.
339. Hamilton, A. McC.—Upon the Significance of Facial Hair growths among Insane Women, Med. Rec., N. Y. 1881, xix. 281.
340. Hardaway, W. A.—A General Account of 110 Cases of Skin diseases, St. Louis Med. & Surg. Jour. 1877, xiv. 291.
341. Hardaway, W. A.—The Treatment of Hirsuties, Trans. Amr. Derm. Asso 1878; Archv. Derm., 1878, iv. 337.
342. Hardaway, W. A.—The Permanent Removal of Superfluous Hair by Electrolysis, Phila. Med. Times, 1879-80, x. 247.
343. Hayes, P. S.—The Removal of Hair by Electrolysis, St. Louis Med. & Surg. Jour. Nov., 1881.
344. Heitzmann, C.—Experiments on Epilation, Arch. Derm., 1881, vii. 130.
345. Heitzmann, C.—Remarks on Akido-galvano-cautery for Epilation, St. Louis Cour. Med. 1882, vii. 16.
346. Heitzmann, C.—Die dauernde Entfernung von Haaren mittelst Elektrolyse, N. Y. Medicinisch. Presse, 1885-86, i. 2.
347. Hilbert, R.—Partielle Hypertrichosis neben angeborner Icthyosis circumscripta, Virchow's Arch. 1885, xcix. 569.

348. Jackson, Geo. T.—Superfluous Hair; the Russian Dog-faced Boy, and Facial Hirsuties in Women, Med. Rec., N. Y. 1885, xxvii. 568.
349. Jelly, W.—Traumatic Paraplegia: Remarkable Growth of Hair, Brit. Med. Jour. 1873, i. 671.
350. Karewski—Zur Therapie der Hypertrichosis, Deutsche Med. Wochschr. 1886, xii. 587.
351. Krebs, C.—Case of Hypertrichosis, Hosp. Tidende, 1878, v. 609; Arch. Derm. 1879, v. 161.
352. Lustgarten, D.—Bemerkungen über radikal Epilation mittelst Elektrolyse, Wien. Med. Wochnschrft. 1886, xxxvi. 1226.
353. Michelson, P.—Ueber abnorme Haarentwickelung beim Menschen, Schrift. der Physikökonom. Gslschft. z. Königsberg, Bd. xxv.
354. Michelson, P.—Zur Capitel der Hypertrichose, Virchow's Archiv. 1885, c. 66.
355. Michelson, P.—Die Elektrolyse als Mittel zur radicalen Beseitigung an abnormer Stelle gewachsener Haare, Berl. Klin. Wochnschrft. 1885, xxii. 674.
356. Michelson, P.—Ueber die galvano-chirurgischen Methoden zur Beseitigung an abnormer Stelle gewachsener Haare, Bericht der 59th Naturforscherversammlung in Berlin, Montshft. f. prakt. Derm. 1886, v. 502.
357. Michelson, P.—Neuere Arbeiten über Elektrolytische Radikaldepilation, Montshft. f. prakt. Derm. 1886, v. 167.
358. Möller, Max—Ueber radikal Epilation mittelst galvanischen Stromes, Wien. Med. Presse, 1885, xxvi. 1415.
359. Piffard, H. G.—An improved Instrument for the Removal of Superfluous Hair, Jour. Cutan. & Ven. Dis. 1882-83, i. 183.
360. Prince, M.—On the exact Measurement of the Electric Current and other Practical Points in the Destruction of the Hair by Electrolysis, Bost. Med. & Surg. Jour. 1886, cxv. 429.
361. Rohé, Geo. H.—Experiences with Electrolysis in Dermatological Practice, Med. Times, Phila. 1884-85, xv. 832.
362. Rohé, Geo. H.—Electrolysis and some of its Applications in Medicine and Surgery, Maryland Med. Jour. Nov. 20th, 1886.
363. Schiefferdecker, P.—Trophischer Störungen nach peripheren Verletzungen, Berl. Klin. Wochnshrft. 1871, viii. 160.
364. Shaw, H.—On the Degeneration of Type in the Insane, St. Barth. Hosp. Rept. 1884, xx. 169.
365. Slocum, C. E.—Case of Hirsuties Gestationis, Med. Rec., N. Y. 1875, x. 470.
366. Smith, G.—The Removal of Superfluous Hair by Electrolysis, Birmingham Med. Rev. Dec. 1885; and Brit. Med. Jour. 1886, i. 151.
367. South, Jno. F.—A Case of Premature Puberty, Med. Chir. Trans. 1823, xii. 76.
368. Startin, Jas.—Removal of Superfluous Hair by Electrolysis, Lancet, 1886, ii. 969.
369. Stowers, J. H.—Depilatories, Brit. Med. Jour. 1879, ii. 117.
370. Stricker, Wm.—Zwei ältere Fälle von Hypertrichose, Virchow's Archiv. 1877, lxxi. 111.
371. Stricker, Wm.—Noch eine Familie von Haarmenschen, etc., Virchow's Archiv. 1878, lxxiii. 622.
372. Stricker, Wm.—Weitere Mittheilungen über Hypertrichosis, Virchow's Archiv. 1880, lxxxi. 567.

373. Turner—Case of Hirsuties, Med. Times & Gaz. 1865, ii. 507.
374. Unna, P. G.—Criticism on Waldeyer & Grimm's Atlas, Montshft. f. prakt. Derm. 1885, iv. 169.
375. Virchow—Die russischen Haarmenschen, Berl. Klin. Wochnschrft. 1873, x. 337.
376. White, J. C.—The Use of Electrolysis in the Treatment of Hirsuties, Bost. Med. & Surg. Jour. 1881, civ. 412.
377. Wicks, W. C.—Depilatories, Brit. Med. Jour. 1879, ii. 159.
378. Wilson, E.—Bearded Women, Lancet, 1873, ii. 756.

>Bartels, M.—Zeitschr. f. Ethn. Berl. 1881, xiii. 255.
>Brown, A. M.—Cincin. Med. & Dent. Jour. Dec. 1885.
>Flesch, M.—Archv. f. Anthrop. Brnschwg. 1880–81, xiii. 125.
>Hildebrandt—Schrift. d. physik. okon. Gslshft. z. Königsberg, vol. xix.
>Hovelacque—Bull. Soc. d'Anthrop. de Par. 1878, i. 272.
>Keane, A. H.—Nature, Lond. 1882–83, xxvii. 245.
>Leonard, C. H.—Med. Advance, Detroit, 1879, iii. 58.
>McDowall, T. W.—J. Ment. Sc. Lond. 1877, xxiii. 86.
>Ornstein, B.—Vrhandl. d. Berl. Gesellsch. f. Anthrop. 1875, pg. 91 and 279; 1876, pg. 287; 1877, pg. 485; 1880, pg. 172; 1881, pg. 740.
>Perrin, E. R.—Bull. Soc. d'Anthrop. de Par. 1873, viii. 741.
>Ranke, H.—Archiv. fur Anthrop. Brnschwg. 1883, xiv. 339.
>Rattone, G.—Giorn. d. v. akad. di Med. di Torino, 1885, xxxiii. 534; also, Gazz. d. Clin. Torino, 1886, xxiii. 314.
>Royer, C.—Bull. Soc. d'Anthrop. de Par. 1873, viii. 718.
>Schulenberg—Verhandl. d. Berl. Gslschft. f. Anthrop. 1880, pg. 295.
>Von Siebold—Archv. f. Anthrop. Brnschweig. 1877–78, ix. 253.
>Weir, J. J.—Nature, Lond. 1886, xxxiv. 223.
>Yemans—Tr. Detroit M. & Libr. Ass. 1879, pg. 7.

Trichiasis and Distichiasis.

379. Benson, A.—On the Treatment of Partial Trichiasis by Electrolysis, Brit. Med. Jour. 1882, ii. 1203.
380. Michel, C. E.—Trichiasis and Distichiasis, with an Improved Method for their Radical Treatment, St. Louis Clin. Rec. 1875, ii. 145.
381. Michel, C. E.—Trichiasis and Distichiasis. Reflections upon their Nature and Pathology, with a Radical Method of Treatment, St. Louis Cour. of Med. 1879, i. 121.

Sycosis.

382. Barthélemy—Case, Annal. de Derm. et Syph. 1881, ii. 523.
383. Behrend, G.—Beitrag zur Pathogenese und Behandlung der Acne desseminata und der Sycosis, Deutsche Med. Wochnschrft. 1881, vii. 283.
384. Chausit, M.—Sycosis ou Mentagre, Paris, 1859.
385. Chevrier—Du Sycosis, Thése de Montpellier, 1871, No. 47.
386. Cooke, W.—Sycosis, Med. Times & Gaz. 1858, i. 306.
387. Devergie—On the Treatment of Mentagra, Jour. d. Med. et d. Chir. prat. July, 1861; Abst. Rankin's Abst. 1861, xxxiv. 117.
388. Duhring, L. A.—Sycosis non Parasitica, Phila. Med. Times, 1874–75, v. 390.
389. Duprez—Sycosis, Gaz. de Hôp. 1859, pg. 232.
390. Fox, T.—Treatment of non-Parasitic Sycosis, Lancet, 1873, ii. 902.
391. Gaskoin, G.—A Case of True Sycosis, Med. Times & Gaz. 1873, ii. 89.
392. Hardy—Sycosis, Nouv. Dict. d. Med. e. d. Chir. prat. 1883.
393. Hebra—Acne Mentagra, Allgm. Wien. Med. Zeit. 1860, v 221.

394. Hebra—Sycosis, Wien. Med. Blätr. 1884, vii. 511.
395. Hunt, T.—On the Modern Treatment of Chronic Diseases of the Skin, Med. Times & Gaz. 1855, ii. 389.
396. Köbner, H.—Ueber Sycosis und ihre Beziehungen zur Mykosis Tonsurans, Virchow's Archiv. 1861, xxii. 372.
397. Laycock—Sycosis Menti vel Mentagra of Two Years' Standing, Med. Times & Gaz. 1864, ii. 650.
398. Mombert—Zur Behandlung der Acne Mentagra; Sykosis, Med. Centrl. Zeit. 1860, xxix. 85; Abst. Schmidt's Jahrb. 1861, cx. 305.
399. Piffard, H. G.—Calx Sulphurata and its Uses, Jour. Cutan. & Ven. Dis. 1883, i. 105.
400. Robinson, A. R.—Sycosis, N. Y. Med. Jour. Aug. & Sept. 1877.
401. Rohé, G. H.—Non-Parasitic Sycosis, Med. Chron. Balt. 1883-84, ii. 157.
402. Shoemaker, J. V.—Inflammation of the Hair Follicles of the Beard, Trans. Am. Med. Asso. 1879, xxx. 195.
403. Stark, G. A.—A Case of Sycosis Treated by Carbolic Acid and Canada Balsam, Canad. Med. & Surg. Jour. Sept. 1876, pg. 97.
404. Veiel—Treatment of Sycosis in the Cannstadt Hospital, Blätr. f. Heilwisnshft. 1873, iv. No. 11; Abst. Vrtljschr. f Derm. u. Syph. 1874, iv. 430.
405. Wood, C. A.—Sycosis and allied Affections, Canad. Med. Rec. 1885, xiii. 100.
406. Yemans, C. C.—Sycosis, Mich. Med. News, 1880, iii. 124.
407. Younkin, E.—The Etiology, Diagnosis and Treatment of Sycosis, Amr. Med. Jour. 1885, xiii. 56.

Trichophytosis Capitis.

408. Anderson, T. McC.—Parasitic Affections of the Skin, Med. Times & Gaz. 1861, i. 7, et seq.
409. Anonymous—Histological Researches in Favus and Ringworm, Med. Times & Gaz. 1881, ii. 582.
410. Atkinson, I. E.—The Botanical Relations of the Trichophyton Tonsurans, N. Y. Med. Jour. 1878, xxviii. 561.
411. Beck, J. T.—Notes on a Case of Tinea Tonsurans, Lancet, 1875, ii. 554.
412. Besnier—Observations upon Parasitic Diseases of the Skin, Paris, 1884; Abst. Vrtljschr. f. Derm. u. Syph. 1885, xii. 612.
413. Boullaud, C. H.—De la Trichophytie, Thése de Paris, 1865, No. 232.
414. Brown, E. A.—A Method of Treating Tinea Tonsurans, Pract. 1874, xii. 327.
415. Cane, L.—Cases of Ringworm Treated by Ol. Hydrarg., Lancet, 1873, ii. 227.
416. Cottle, W.—Notes on the Treatment of Ringworm, Brit. Med. Jour. 1880, i. 806.
417. Cottle, W.—The Treatment of Ringworm, Lancet, 1880. i. 482 et seq.
418. Crocker, H. R.—Goa or Araroba Powder in Ringworm, Lancet, 1877, i. 124.
419. Curtis, F. C.—Ringworm of the Scalp: Suggestions on its Treatment and the Recognition of its Cure, N. Y. Med. Jour. 1886, xliii. 214.

420. Denslow, L. G. N.—The Parasitic Diseases of the Skin, &c., Northwest. Lancet, 1884, iv. 37 et seq.
421. Duckworth, D.—Clinical Observations upon Certain Skin Diseases, St. Barth. Hosp. Rept. 1873, ix. 100.
422. Duckworth, D.—On an Improved Forceps for Depilation. Lancet, 1878, i. 490.
423. Duhring, L. A.—Tinea Tonsurans of Unusual Form, with Tinea Circinata. Med. & Surg. Reptr. 1878, xxxix. 96.
424. Elliot, Geo. T.—Pyrogallic Acid, N. Y. Med. Jour. 1885, xlii. 257.
425. Elliot, Geo. T.—The Diseases of the Skin caused by the Vegetable Parasites; their Symptoms and Treatment, N. O. Med. & Surg. Jour. 1886, xiv. 116 et seq.
426. Fayrer, J.—Indian Ringworm and its Treatment by Goa Powder, Med. Times & Gaz. 1874, ii. 470.
427. Finny, J. M.—The Croton Oil Treatment of Tinea Tonsurans, Brit. Med. Jour. 1881, i. 302.
428. Foulis, Jas.—The Treatment of Ringworm of the Scalp, Brit. Med. Jour. 1885, i. 536.
429. Fox, T.—Ringworm, Lancet, 1871, i. 412.
430. Fox, T.—Ringworm in Schools, Lancet, 1872, i. 5.
431. Fox, T.—Suspected Ringworm, Lancet, 1873, ii. 733.
432. Fox, T.—Alopecia Areata and Tinea Tonsurans, Med. Times & Gaz. 1874, ii. 630.
433. Fox, T.—On Ringworm of the Head and its Management, Lancet, 1877, ii. 602 et seq.
434. Gee, Saml.—The Treatment of Tinea Tonsurans, Lancet, 1874, i. 318.
435. Grawitz, P.—Ueber die Parasiten des Soors, des Favus und Herpes Tonsurans, Virchow's Archiv. 1877, lxx. 546; 1886, ciii. 393.
436. Harrison, A. J.—A new Method of Treating Tinea Tonsurans, Brit. Med. Jour. 1885, ii. 434.
437. Hillier—Notes on Skin Diseases, Med. Times & Gaz. 1867, i. 34.
438. Hoggan, G.—Comparative Growth of the Fungi of Favus and Ringworm, Lancet 1878, ii. 918; also, Trans. Path. Soc. Lond. 1879, xxx. 444.
439. Horand—L'herpes Tonsurant, Lyon Med. 1874, xvii. 34.
440. Hunt—On the Constitutional Treatment of Scald Head, Med. Times & Gaz. 1857, ii. 445.
441. Hutchinson, J.—Identity of the Fungus in Pityriasis Versicolor and Tinea Tonsurans, Med. Times & Gaz. 1859, i. 123.
442. Hutchinson, J.—A Clinical Report on True Ringworm, Med. Times & Gaz. 1861, i. 12 et seq.
442a. Jackson, Geo. T.—The Chronic Contagious Diseases of the Skin of the Head and Face, Mis. Val. Med. Month. July, 1886.
443. Jeffreys, R.—Treatment of Ringworm, Brit. Med. Jour. 1881, i. 76.
444. Jenner, Wm.—The Pathology and Treatment of the Diseases of the Scalp, popularly Known by the Name of Ringworm, Med. Times & Gaz. 1853, ii. 182.
445. Jenner, Wm.—Tinea Tonsurans, Med. Times & Gaz. 1857, ii. 545.
446. Ladreit de Lacharrière—Note sur la Traitement de la Teigne Tonsurante par l'Huile de Croton Teglium, Bul. gen. de Therap. 1876, xci. 97.

447. **Laillier—Maladies** Contagieuses du cuir chevelu chez les Enfants, **Annal. de Hygiéne** Publique, 1885, xiv. 377.
448. Lambert, **W. H.—Scurf** after Ringworm, Lancet, 1880, **ii.** 158.
449. Lancereaux—Note sur la Transmission de l'herpes Circiné du chat à l'Homme, L'Union Med. 1874, xvii. 969.
450. Leftwich, R. W.—A Hint on the **Treatment** of Ringworm, Lancet, 1886, i. 278.
451. Lesser, E.—Eine augenblicklich herrschende Epidemie von Herpes Tonsurans, **Deutsch** Med. Wochnschrft. Feby. 11th, 1886.
452. Liveing, R.—Remarks on Alopecia **Areata and Tinea** Tonsurans, Med. Times & Gaz. 1874, ii. 601.
453. Liveing, R.—Peculiarities of Ringworm and its Treatment, Lancet, 1879, ii. 642.
454. Liveing, R.—The Treatment of Ringworm by Croton Oil, **Brit. Med. Jour.** 1881, i. 227.
455. Liveing, R.—Remarks on Bald Tinea Tonsurans and **Vegetable** Parasites, **Brit.** Med. **Jour,** 1882. i. 496.
456. M'Guire, J. C.—Treatment of Trichophytosis, **The** Amr. **Pract.** & News, 1886, **ii. 201.**
457. Macleod—Ringworm **of the** Scalp, Lancet, 1880, i 995.
458. Majocchi—A New Form of Trichophytosis; Granuloma Trichophyticum, Bull. della R. Acad. Med. di Roma, Oct. 1883; Abst. Vrtljschr. f. Derm. u. Syph. 1884, xi. 177
459. Maynard, W. **J.—Tinea** Tonsurans, **St.** Louis Med. & Surg. **Jour.** 1881, xli. 291.
460. Morris, M.—**Diagnosis and** Treatment of Ringworm, **Lancet,** 1881, i. 164 et seq.
461. Morris, M.—Isolation in Ringworm Cases not **necessary,** Lancet, 1882, i. 291.
462. Neumann, I.—**Ueber** Behandlung der Psoriasis vulgaris, **des** Herpes Tonsurans, und der Pityriasis versicolor mit Chrysophansäure und Goa Pulver, Wien. Med. Presse, 1878, xix. 417 et seq.
463. Rabitsch, J—Die Salicylsäure bei der Berhandlung des "Ringworms," Wien. Med. Wochnschrft. 1882, xxxii. 394.
464. Railliet—De la Teigne Tonsurante **chez les** Animaux, Annal. d. Derm. et d. Syph. 1880, i. 232.
465. Richard, P.—Les **formes** Cliniques de l'herpes Tonsurans, La. Fr. Med. 1886, i. 697.
466. Richardson, **B. W.**—Ethylate **of** Sodium in **the** Treatment of Naevus and other **Forms of** Disease, Lancet, 1881, i. 242.
467. Robinson, A. R.—The Anatomical Seat of the Fungus in Tinea Tonsurans Capillitii, N. Y. Med. Jour. 1881, xxxiii. 289.
468. Robinson, A. R.—Mycological Studies in Ringworm and Favus, Trans. Amr. Derm. Soc. 1885, pg. 21.
469. Rouquayrol—Prophylaxie et Traitement de **la Teigne** Tondante, Thése de Paris, 1879, No. 451.
470. Rudkin, G. M. A.—Scurf **after** Ringworm, Lancet, 1880, ii. 158.
471. Saalfeld, **E.**—Eine langdauernde Epidemie **von** Dermatomykosis Tonsurans in Berlin, Berlin Klin. Wochnschrft. 1886, xxiii. 643.
472. Sangster, A.—**Tinea Tonsurans accompanied** by Alopecia **Areata,** Lancet, 1880, ii. 9.

473. Sangster, A.—Isolation in Ringworm Cases not Necessary, Lancet, 1882, i. 329.
474. da Silva Lima, J. T.—Goa Powder, Med. Times & Gaz. 1875, i. 249.
475. Smith, A.—Ringworm of the Head; Diagnosis and Treatment, Lancet, 1880, i. 52 et seq.
476. Smith, A.—The Croton Oil Treatment of Ringworm, Lancet, 1880, i. 581.
477. Smith, A.—Scurf after Ringworm, Lancet, 1880, ii. 158.
478. Smith, A.—On the Treatment of Chronic Ringworm, Brit. Med. Jour. 1882, ii. 682.
479. Smith, A.—Report on the Treatment of a very Extensive Outbreak of Ringworm of the Head in a School, Brit. Med. Jour. 1882, ii. 1195.
480. Smith, J. N.—Treatment of Ringworm of the Scalp, Brit. Med. Jour. 1879, ii. 641.
481. Startin, Jas.—Treatment of Ringworm of the Scalp, Brit. Med. Jour. 1879, ii. 641.
482. Startin, Jas.—The Croton Oil Treatment of Ringworm, Lancet, 1880, i. 696.
483. Startin, Jas.—Scurf after Ringworm, Lancet, 1880, ii. 158.
484. Stowers, J. H.—Tinea Tonsurans accompanied by Alopecia Areata, Lancet, 1881, i. 326.
485. Stretton, W. H.—Isolation in Ringworm Cases not Necessary, Lancet, 1882, i. 330.
486. Taylor, F.—On the Condition of the Skin in Tinea Tonsurans, Lancet, 1878, ii. 695; also, Med. Chir. Trans. 1879, lxii. 177.
487. Thin, Geo.—On the Condition of the Skin in Tinea Tonsurans, Lancet, 1878, i. 439; also, Med. Chir. Trans. 1878, lxi. 179.
488. Thin, Geo.—Isolation of Ringworm Cases not Necessary, Lancet, 1882, i. 250.
489. Thin, Geo.—Contribution to the Pathology of Parasitic Diseases of the Skin, Brit. Med. Jour. 1882, ii. 301.
490. Tipple, E.—Ringworm, Lancet, 1876, ii. 176.
491. Toulmin, F.—Treatment of Ringworm of the Scalp, Brit. Med. Jour. 1879, ii. 641.
492. Tuffen, W.—Parasitic Fungi affecting Plants, Med. Times & Gaz. 1868, ii. 233 et seq.
493. Unna, P. G.—Mykologische Beiträge, Vrtljschr. f. Derm. u. Syph. 1880, vii. 165.
493a. Unna, P. G.—Icthyol. Montshft. f. prakt. Derm. 1882, i. 333.
494. Van Harlingen, A.—Notes on the Management of Ringworm of the Scalp, Med. News, Phila. 1883, xlii. 297 et seq.
495. Watson—Boracic Acid in the Treatment of Ringworm, Lancet, 1875, ii. 750.
496. Weller, G.—Treatment of Ringworm of the Scalp, Brit. Med. Jour. 1879, ii. 641.
497. Welsh, Jas.—Treatment of Ringworm, Lancet, 1875, ii. 823 et seq.
498. Wilson, E.—On the Phytopathology of the Skin and Nosophytodermata, Brit. & For. Med. Chir. Rev. Jany. 1864, pg. 199.

    Barduzzi—Comment. Clin. de Pisa, 1877, pg. 210.
    Bärensprung—Charite Annal. vol. vi.
    Hog—Quart. Jour. Mic. Sc. Jany. 1866.
    Malmsten, P. H.—Arch. f. Anat. Physiol. u Wissensch. Med. Brl. 1848, pg. 1.
    Morris—Jour. Roy. Microscop. Soc. 1883, iii. 329.

Schilling—Compend. Clin. del Mal. Cutan. Roma, 1877.
Startin—Trans. Willan Sc. Lond. 1885, i. 92.
Ziemssen—Greifswalder Med. Beiträge, 1864, ii. 99.

KERION.

499. Atkinson, I. E.—On Kerion Celsi, a Variety of Tinea Tonsurans, Arch. Derm. 1881, vii. 47.
500. Auspitz—Ueber das Sogenannte Kerion Celsi, Wien. Med. Presse, 1878, xix. 853 et seq.
501. Dubini, A.—Vespajo del Capillizio, Giorn Ital. del Mal. Ven. e. del Mal. del Pelle, 1866, i. 17.
502. Fox, T.—The Kerion of Celsus; a Phase of Tinea Tonsurans. Lancet, 1868, i. 156.
503. Majocchi—Ten Cases of Kerion Celsi, Gaz. Med. di Roma, 1877; Abst. Vrtljschr. f. Derm. u. Syph. 1878, v. 477.
Andronico, C.--Bull. d. Sc. Med. di Bologna, 1886, xvii. 377.

TRICHOPHYTOSIS BARBAE.

504. Anderson, McC.—On the Pathology of the so-called Sycosis Menti, Edin. Med. Jour. June, 1868, pg. 1089.
505. Anderson, McC.—Tinea Barbae, Lancet, 1879, ii. 485.
506. Anonymous—Herpes Tonsurans Barbae, Allg. Wien. Med. Zeit. 1884, xxix. 63 et seq.
507. Bulkley, L. D.—Trichophytosis Barbae, Arch. Dermat. 1880, vi. 249.
508. Burton, J.—Treatment of Porrigo Decalvans, Med. Times & Gaz. 1856, i. 10.
509. Cane, L.—Cases of Ringworm Treated by Ol. Hydrarg. Lancet, 1873, ii. 227.
510. Chevrier—Du Sycosis, Thése de Montpelier, 1871.
511. Dunlop—Sycosis Menti Resembling a Malignant Tumor, Glasgow Med. Jour. 1881, xv. 56.
512. Fox, T.—Parasitic Sycosis, Lancet, 1873, ii. 141; also Med. Times & Gaz. 1873, ii. 477.
513. Gerlier—Sur l'Épidémie Trichophytique de Ferney-Voltaire, Lyon Méd. April 24th, 1881.
514. Hardy—Quelques Considerationes sur l'Etiologie, la Nature et la Traitement des Maladies Contagieuses du Systeme Pileux, Annal. de Derm. & Syph. 1876-77, viii. 401; Trans. Internat. Med. Cong. Germ., 1878.
515. Ihle, M.—Beiträge zur Behandlung der Hautkrankheiten mit Resorcin, Montshft. f. prakt. Derm. 1885, iv. 424.
516. Jamieson, W. A.—On Tinea Barbae, the so-called Parasitic Sycosis, Lancet, 1879, ii. 314.
517. Jenner, Wm.—Mentagra, Med. Times & Gaz. 1857, ii. 650.
518. Köbner, H.—Ueber Sykosis und ihre Beziehungen zur Mykosis Tonsurans, Virchow's Archiv. 1861, xxii. 372.
519. Köbner, H. and Michelson, P.—Ueber Parasitare Sycosis, Archiv. f. Derm. u. Syph. 1869, i. 7.
520. Lang, E.—Ueber eine Seltenere Form der Parasitären Sykosis und einige entzündliche Geschwulste, Vrtljschr. f. Derm. u. Syph. 1878, v. 393 et seq.
521. Lewin—Sykosis Parasitica, Charite Annal. 1874, i. 639; Abst. Vrtljschr. f. Derm. u. Syph. 1876, iii. 100.
522. Piffard, H. G.—Trichophytosis Barbae, Illus. Quart. Med. & Surg. 1883, ii. 137.

523. Ravogli, A.—Sykosis Parasitaria Barbae, Cincin. Lancet Clinic. 1881, vii. 206.
524. de Silva Lima, J. C.—Goa Powder, Med. Times & Gaz. 1875, i. 249.
525. Smith, A.—Tinea Sykosis, Brit. Med. Jour. 1880, ii. 536.
526. Tanturi—Phytosycosis and its Dependence upon Herpes Tonsurans, Giorn. Ital. d. Mal. Ven. et d. Mal. Pelle, 1870; Abst. Archiv. f. Derm. u. Syph. 1870, ii. 643.

FAVUS.

527. Anderson, T. McC.—Parasitic Affections of the Skin, Med. Times & Gaz. 1861, i. 170 et seq.
528. Anderson, T. McC.—Tinea Favosa Epidermidis, communicated from Mice, Glasgow Med. Jour. 1880, xiii. 244.
528a. Anderson, T. McC.—On the Non-identity of the Parasites met with in Favus, Tinea Tonsurans, etc. Brit. & For Med. Chir. Rev. 1866, ii. 225.
529. Anonymous—De la Teigne Faveuse et de son Traitement par l'Emploi Topique de l'Huile de Naphte, Gaz. des Hôp. 1857, pg. 323.
530. Anonymous—Remedy for Favus, Med. Times & Gaz. 1854, i. 594.
531. Anonymous—Histological Researches on Favus and Ringworm, Med. Times & Gaz. 1881, ii. 582.
532. Aubert, P.—Diagnostique de la Teigne Faveuse, Annal. Derm. et Syph. 1881, ii. 34.
533. Aubert, P.—Rôle du Traumatisme dans l'Etiologie de la Teigne Faveuse, Annal. de Derm. et Syph. 1881, ii. 289.
534. Balzer, F.—Recherches Histologiques sur la Favus et la Trichophytie, Arch. Gen. d. Med. 1881, ii. 385.
535. Boer—Favus, Tagbl. d. Naturversaml. in Berlin, 1886; Abst. Montshft. f. prakt. Derm. 1886, v. 517.
536. Bredin, J. N.—Treatment of Favus, Lancet, 1874, ii. 436.
537. Bulkley, L. D.—Favus, Med. Times, Phila. 1878-79, ix. 178.
538. Bulkley, L. D.—Favus and its Treatment by a New Method of Depilation, Trans. Med. Sc. N. Y. 1881, pg. 155.
539. Charpy, A.—Du Favus Miliaire, Annal. Derm. et Syph. 1874-75, vi. 328.
540. Denslow, L. G. N.—The Parasitic Diseases of the Skin, etc. Northwest. Lancet. 1884, iv. 37 et seq.
541. Descroizilles—Favus des Parties Glabres Coincidente avec un Favus du cuir Chevelu, Rev. Mens. d. Mal. de l'Enfance, 1884, ii. 234.
542. Duckworth, D.—Case of Favus of the Scalp and Body, Trans. Clin. Soc. Lond. 1875, viii. 107.
543. Elliot, Geo. T.—The Diseases of the Skin Caused by the Vegetable Parasites: their Symptoms and Treatment, N. O. Med. & Surg. Jour. 1886, xiv. 116 et seq.
544. Fagge, C. H.—Remarks on Certain Cutaneous Affections, Tinea Favosa, Guy's Hosp. Rept. 1870, xv. 351.
545. Fuller—Treatment of Favus, Med. Times & Gaz. 1857, i. 263.
546. Gigard, G.—Sur une Épidémie de Teigne Faveuse sévissant a nantoin chez les Bêtes a Cornes et chez les Enfants, Lyon Med. 1880, xxxiv. 547.

547. **Grecesco**—De l'Achorion Schoenleinii, Thése de Paris, 1868, No. 137.
548. **Guibout**—Teigne Faveuse, Favus, ou Porrigo Favosa, Gaz. des Hôp. 1886, lix. 75.
549. **Hillier**—Favus; Treatment by Epilation; Cure, Med. Times & Gaz. 1864, ii. 118.
550. **Hillier**—Notes on Skin Diseases, Med. Times & Gaz. 1867, i. 34.
551. **Hutchinson, J.**—Radical Treatment of Favus, Med. Times & Gaz. 1855, i. 8.
552. **Hutchinson, J.**—Clinical Reports on Favus, Med. Times & Gaz. 1859, ii. 553, et seq.
553. **Hutchinson, G. W.**—Treatment of Favus, Lancet, 1874, ii. 365.
554. **Ihle**—Beiträge zur Behandlung der Hautkrankheiten mit Resorcin, Montshft. f. prakt. Derm. 1885, iv. 429.
555. **Jenner, Wm.**—The Pathology and Treatment of the Disease of the Scalp Popularly Known as Ringworm, Med. Times & Gaz. 1853, ii. 182.
556. **Jenner, Wm.**—Tinea Favosa, Med. Times & Gaz. 1857, ii. 649.
557. **Kaposi**—Ueber einen Fall von Favus Universalis, Sitzungsbr. d. Wien. **Gslshft. d. Artz.** Oct. 17th, 1884, Abst. Vrtljschr. f. Derm. u. Syph. 1885, xii. 350.
558. **Knoche, J. P.**—Favus, The Kansas City Med Index, 1885, vi. 59.
559. **Koser, S. S.**—Some Observations upon Favus, Med. & Surg. Rept. Phila. 1873, xxix. 271.
560. **Laillier**—Maladies **Contagieuses du** cuir chevelu chez **les** Enfant, Annal. d. Hygiene **Publique,** 1885, xiv. 377
561. **Mégnin**—Teigne Faveuse chez les Souris, Contagion de la **Teigne Tonsurante du** Cheval à l'Homme, Prog. Méd Jany 1st, 1881.
562. **Morrow, P. A.**—Report on a Case of Favus with Remarks on the Treatment of the Tineas, Jour. Cutan. & Ven. Dis. 1886, iv. 321.
563. **Neumann, I.**—Zur Entwickelungsgeschichte des Achorion, Archv. f. Derm. u. Syph. 1871, iii. 20 et seq.
564. **Pirrie, Wm., Jr.**—Observations on Favus, Lancet, 1860, ii. 557 et seq.
565. **Prior, C. E.**—The Treatment of Porrigo Favosa by Carbolic Acid, Brit. Med. Jour. 1867, ii. 358.
566. **Purdon, H. S.**—The Treatment of Favus, Archiv. Derm. 1881, vii. 138.
567. **Purser, J. M.**—Observations Tending to Show the Identity of the Fungi of Favus and Tinea Circinata, Dub. Jour. Med. Sc. 1867, xliv. 66.
568. **Quincke, H.**—Ueber Favus, Montshft. f. prakt. Derm. 1885, iv. 433; 1886, v. 308.
569. **Reed, J. C.**—Treatment of Favus, Lancet, 1874, ii. 365.
570. **Remy, Chas.**—Recherches sur l'Anatomie Microscopique du Favus, La Progrès Med. 1875, iii. 687.
571. **Reynolds, H. J.**—Favus, Chicago Med. Jour. & Exam. 1886, liii. 344.
572. **Sawicki**—Treatment of Favus, Przeglod. lekarski Krakowski, 1876; Abst. Vrtljschr. f. Derm. u. Syph. 1877, iv. 286.

573. **Seymour, W. W.**—Kerosene as a Remedy for Favus, **Bost. Med. & Surg. Jour.** 1882, cvii. 453.
574. **Simon, Th.**—Dermatologische Mittheilungen, **Archiv. f. Derm. u. Syph.** 1870, ii. 541.
575. **Smith, W. G.**—Notes on some Diseases of the Skin, Favus, Dub. Jour. Med. Sc. May, 1875, pg. 391.
576. **Smith, W. G.**—Cases of Favus from the Cat, with Histories of Contagion, Dub. Jour. Med. Sc. 1879, lxviii. 450.
577. **Spillmann, E.**—Observation de Favus, Annal. de Derm. & Syph. 1870–71, iii. 347.
578. **Squire, B.**—Case of **Favus Cured** without Epilation, Tr. Path. Soc. Lond. 1865, xvi. 256.
579. **Unna, P. G.**—Mykologische Beiträge, Vrtljschr. f. **Derm. u. Syph.** 1880, vii. 165.
580. **Unna, P. G.**—Icthyol. Montshft. f. prakt. Derm. 1882, i. 333.
581. **Walker, J. S.**—Treatment of Favus, Lancet, 1874, ii. 436.
582. **Weber**—Ueber die Behandlung des Favus, Corrspndzbl. f. Schweiz. Arzt. 1880, No. 17.
583. **West, T.**—Parasitic Fungi Affecting Plants, Med. Times & Gaz. 1868, ii. 233 et seq.
583a. **Wilson, E.**—On the Phytopathology of the Skin, Brit. & For. Med. Chir. Rev. Jany. 1864, pg. 199.

    Hallier—Die Natur des Favuspilze und sein Verhältniss zu Penicillium Glaucum, Jena. Zeitschr. vol. ii.
    Lenzberg—Die prakt. Arzt. Feby. 1881.
    Neumann, S.—Compt. rend. Sc. de Biol. Par. 1886, iii. 173.
    Pick—Vrhndlung. der K. K. Zoolog.-botan. Gslschft. in Wien. 1865.
    Wagner—Archv. d. Heilk. Leipz. 1866, vii. 472.
    White, J. C.—Extr. Rec. Bost. Soc. M. Improv. 1862, iv. 177.

PEDICULOSIS PUBIS.

584. **Anderson, McC.**—On Parasitic Affections of the Skin, Med. Times & Gaz. 1861, i. 597.
585. **Bertulus, E.**—L'école Moderne et le Phthiriasis ou Maladie Pédiculaire Spontanée, Gaz. Med. Aug. 19th, 1871, pg. 352.
586. **Bulkley, L. D.**—Phthiriasis, Arch. Derm. New York, 1881, vii. 394.
587. **Crane, A.**—Chrisma as a Parasiticide, Lancet, 1881, ii. 76.
588. **Duguet**—Les taches **Bleues;** leur Production Artificielle, Gaz. des Hôp. 1880, liii. 362.
589. **Gibier**—Nouvelle Étude sur la Corrélation qui existe entre les taches ombrées et la Phthiriase du Pubis, Gaz. Med. de Par. 1881, iii. 131.
590. **Hamal**—Pediculosis Pubis, Med. Times & Gaz. 1857, ii. 482.
591. **Moursou, J.**—Nouvelles Recherches sur l'Origine des Taches Ombrées, Annal. de Derm. et Syph. 1877–78, ix. 198.
592. **Ryding, Geo.**—Pediculosis Pubis, Lancet, 1858, i. 621.

PEDICULOSIS PALPEBRARUM.

593. **Ring, F. W.**—Case of Phthiriasis Palpebrarum, Med. Rec. N. Y. 1885, xxviii. 647.
594. **Rosenmeyer, L.**—Ueber Pediculosis Palpebrarum, Münch. Med. Wochnschrft. Mch. 2d, 1886, pg. 145.
595. **Stelwagon, H. W.**—A Case of Phthiriasis Palpebrarum, Arch. Derm. N. Y. 1881, vii. 301.

BIBLIOGRAPHY.

Beigel's Disease, etc.

596. Behrend, G.—**Ueber Knotenbildung** am Haarschaft, Virchow's Archiv. 1886, ciii. 437.
597. Duhring, L. A.—**Unknown Ova** upon Human Hair, Archv. Derm. N. Y. 1876, ii. 216.
598. Eberth, C. J.—**Untersuchungen über** Bakterien, Virchow's Archiv. 1875, lxii. 504.
599. Martin, Aloys—Zeitschrft f. Rationelle Med. 1863, xiv. 357; Abst. B. & F. M. C. Rev. 1863, xxxi. 527.
600. Pick—Ueber Dermatomykosis Palmellina, **Allg. Wien.** Med. Zeit. 1875, xx. 370.
601. Thin, Geo.—**Case of** Parasitic Affection of Moustache, Lancet, Nov. 4th, 1882.

Küchenmeister—Oester. Ztschr. f. prakt. Heilk. 1867, xiii. 218.

Dandruff.

602. Boeck, C.—Abst. Montshft. f. prakt. Derm. 1886, v. 90.
603. Bizzozero—The Microphyten of the Normal Human Skin, Gaz. d. Hosp. 1884, No. 29; Abst. **Vrtljschr. f.** Derm. u. Syph. 1884, xi. 523.
604. Chincholle—De la Nature Parasitaire du Pityriasis Capitis et de l'Alopecie Consecutive, Paris, 1874.
605. Duhring, L. A.—Seborrhœa of the **Scalp and Face,** Med. Times, Phila. 1879-80, x. 34.
606. Ferrari—Etiology of **Pityriasis, Atti** Acad. Gioencia di Sc. Natur. i **Catania, vol. xviii.; Abst. Montshft. f.** prakt. Derm. 1886, v. 84.
607. **Fournier—Pityriasis Capitis, Jour. de Med.** et de Chir. 1886, lvii. 63.
608. Heitzmann. C.—On the **Treatment of** Seborrhœa, Trans. Int. Med. Cong. Phila. 1876, pg. 723.
609. Jackson. Geo. T.—Dandruff; **What it is and How** to Cure it, Med. Rec. New York, 1884, xxv. 428.
610. Manino Lorenzo—The **Microsporon** dispar of Vidal in Seborrhœa, Giorn. Ital. d. Mal. Ven. Mch. & April, 1886; Abst. **Vrtljschr. f. Derm. & Syph.** 1886, xiii. 457.
611. Morison, R. B.—A New Instrument for the Treatment of Seborrhœa and **Eczema Capitis,** Maryland Med. **Jour.** 1883, x. 446.
612. Oudemanns & Pekelharing—Saccharomyces Capillitii, ein Spaltpilz der Behaarten Kopfhaut, Tiydschrift voor Geneeskunde; also, Archv. Neerl. d. Sc. Exactes, 1886, xx. 404; Abst. Montshft. f. prakt. Derm. 1886, v. 323.
613. Payne, J. F.—Microsporon Furfur in Pityriasis of Scalp. Brit. Med. Jour 1886, ii. 922.
614. Pellizzari, C.—Microphytes of the Normal Human Skin and of Alopecia Areata, Bollettino del Soc. Tra. i. Cultori del Sc. Med. i. Siena; Abst. Vrtljschr. f. Derm. u. Syph. 1884, xi. 523.

Plica Polonica.

615. Beigel, H.—Specimen of **Plica** Polonica, **Tr.** Path. Soc. Lond. 1866, xvii. 418.
616. Beigel, H.—The **Second Case of Plica** Polonica observed in England, Med. Times & Gaz. 1867, i. 509.

617. Dietl—Zur Streitfrage des Weichselzopfes, Wien. Med. Wochnshrft. 1863, xxx. 737.
618. Hamburger—Ueber die Irrlehre von der Plica Polonica, Zeit. v. Klin. Med. 1861.
619. Lessing, F.—Plica Polonica, Med. Times, Phila. 1882-83, xiii. 82.
620. Le Page, J. T.—On Neuropathic Plica, Brit. Med. Jour. 1884, i. 160.
621. Le Viseur—Fragments zur Nosographie des Weichselzopfs, Deutsch. Klinik. 1859, xi. 373.
621a. Le Viseur—Zur Weichselzopffrage, Deutsche Klinik. 1861, xiii. 349.
622. Mettenheimer, C.—Zur Entstehungsgeschichte der Weichselzopfsartigen Bildungen, Jahrb. für Kinderheil. 1875, ix. 149.
623. Pestonji, D. B.—On a Case of Neuropathic Plica, Lancet, 1885, ii. 431.
624. Urbanowicz—De la Plique Polonaise, Archv. Gen. de Med. 1871, i. 215.
NOTE.—For literature prior to 1839, see Rosenberg, Der Weichselzopf. München, 1839, pg. 65.

DERMATITIS PAPILLARIS CAPILLITII.

625. Baker, M.—Acne Keloid, Trans. Path. Soc. Lond. 1882, xxxiii. 367.
626. Hebra, H., Jr.—Bericht von Hebra's Klinik. in Wien. für 1874, Vrtljshr. f. Derm. u. Syph. 1876, iii. 98.
627. Hervouet, H.—Note sur un Cas d'Hypertrophie Papilliforme du cuir Chevelu, Annal. Derm. et Syph. 1883, iv. 421.
628. Hyde, J. N.—A Clinical Study of Dermatitis Papillaris Capillitii, Jour Cutan. & Ven. Dis. 1882, i. 33 et seq.
629. Kaposi (Kohn)—Ueber Sogenannte Framboesie, Archiv. f. Derm. u. Syph. 1869, i. 382.
630. Sangster, A.—A Papillary Tumor of the Scalp, Trans. Inter. Med. Cong. Lond. 1881, iii. 143.
631. Veritá, A.—Acne Keloidique, Acad. Med. Sc. May 9th, 1882; Abst. Gaz. Med. 1882, iv. 245.
632. Williams, R.—Acne Keloid, Brit. Med. Jour. 1884, i. 668.

NAEVUS PILOSUS.

633. Anonymous—Remarkable Case of Hairy Naevus, Lancet, 1869, ii. 276.
634. Baker, W. M.—On the Removal, by Operation, of a Hairy Mole occupying one-half the Forehead, Lancet, 1877, ii. 803.
635. Bull, W. A.—Case of Diffused Superpigmented Mole of Abdomen, Brit. Med. Jour. 1882, i. 304.
636. Despres, A.—Hétérotopie Pileuse Cutanée Congenitale; Naevus Pilosus occupant presque tout le Corps, Gaz. Hebdom. Par. 1874, xi. 244.
637. Hildebrandt, H.—Ueber Abnorme Haarbildung beim Menschen, Schrft. d. Physik. ökon. Gslshft. 1878, vol. xix.
638. Lawson, Geo.—Epithelioma of Large Mole, Trans. Path. Soc. Lond 1873, xxiv. 256.
639. Murray, Jno.—Extensive and Increasing Hair Mole in a Child, Trans. Path. Soc. Lond. 1873, xxiv. 257.
640. Sommer, W.—Ein Neuer Fall von Hypertrichosis circumscripta, Virchow's Archiv. 1885, cii. 407.

APPENDIX.

Note.—This appendix is intended to fill out the bibliography and journal literature of the diseases of the hair and scalp down to 1892 inclusive.

Treatises on the Skin and Syphilis.

1. Anderson, T. McC.—Parasitic Affections of the Skin, London, 1861.
2. Anderson, T. McC.—A Treatise on the Diseases of the Skin, London and Philadelphia, 1887.
3. Brocq, L.—Traitement des Maladies de la Peau, 2d Ed., Paris, 1892.
4. Campbell, C. M.—Skin Diseases of Infancy and Early Life, London, 1889.
5. Crocker, H. R.—Diseases of the Skin, London and Philadelphia, 1888; 2d Ed., 1893.
6. Duhring, L. A.—Epitome of Diseases of the Skin, Philadelphia, 1886.
7. Eichhoff, P. J.—Die Hautkrankheiten, Leipzig, 1890.
8. Erichsen, J. E.—Practical Treatise on the Diseases of the Scalp, London, 1842.
9. Fox, W. T.—Skin Diseases of Parasitic Origin, London, 1863.
10. Fox, T. and T. C.—Epitome of Skin Diseases, London and Philadelphia, 1883.
11. Friese, C.—Haut und Haare, Berlin, 1891.
12. Hardaway, W. A.—Manual of Skin Diseases, Philadelphia, 1891.
13. Hardy, A —Traité des Maladies de la Peau, Paris, 1886.
14. Hunt, T.—Pathology and Treatment of Certain Diseases of the Skin, London, 1847.
15. Jackson, G. T.—Ready-reference Handbook of Diseases of the Skin, Philadelphia, 1892.
16. Jamieson, W. A.—Diseases of the Skin, 3d Ed., Philadelphia and Edinburgh, 1892.
17. Kaposi, M.—Pathologie et Traitement des Maladies de la Peau, ed. Besnier et Doyon, Paris, 1891.
18. Keyes, E. L.—Genito-Urinary Diseases with Syphilis, New York, 1888.
19. Kippax, J. R.—Handbook of Diseases of the Skin, Chicago, 1884.
20. Klencke, H.—Dietetische Kosmetik, Leipzig, 1888.
21. Kopp, C.—Die Trophoneurosen der Haut, Wien, 1886.
22. Leloir et Vidal.—Traité descriptif des Maladies de la Peau, Paris, 1889, et seq.
23. Liveing, R.—Handbook of Skin Diseases, London, 1887.
24. Morris, M.—Management of Skin and Hair, London.

25. Neumann, C. E. O.—Die Haut, Haare, Nägel und Zähne des Menschen, Leipzig.
26. Ohmann-Dumesnil, A. H.—Handbook of Dermatology, St. Louis, 1889.
27. Piffard, H. G.—Practical Treatise on Diseases of the Skin, New York, 1891.
28. Ravogli, A.—The Hygiene of the Skin, Cincinnati, 1888.
29. Rohé, G. H.—Practical Manual of Diseases of the Skin, Philadelphia, 1892.
30. Schultz, H.—Haut, Haare und Nägel, Leipzig. 1885.
31. Shoemaker. J. V.—Practical Treatise on Diseases of the Skin, 2d Ed., New York. 1892.
32. Startin, J.—Lectures on Parasitic Diseases of the Skin, London, 1851.
33. Stelwagon, H. W.—Essentials of Diseases of the Skin, Philadelphia, 1890.
34. Van Harlingen, A.—Handbook of Skin Diseases, 2d Ed., Philadelphia 1889.

Treatises on the Hair.

35. Audrain, I.—Contribution à l'Étude de la Trichophytie tonsurante, Thèse de Paris, 1892.
36. Barteau, P. A.—De la Teigne tonsurante, Thèse. Paris, 1856.
37. Behr, Th. and E.—Das neue Haarerzeugungsverfahren, Leipzig.
38. Bernhardt, H—De Sycosi, Berlin, 1862.
39. Beschorner, F.—Der Weichselzopf, Breslau, 1843.
40. Besnier, E.—Sur la Pélade, Paris, 1888.
41. Blumenthal, B.—Ueber Sycosis vulgaris et parasitaria, Darmstadt, 1886.
42. Boeck, F.—Ueber die Area Celsi, Greifswald, 1867.
43. Bondi, E.—Pathologie des Weichselzopfs, Berlin, 1828.
44. Braschoss, J.—Merkwürdige Fälle von Favuserkrankung, Bonn, 1887.
45. Braunstein, H.—Ueber Alopecia areata, Freiburg, 1873.
46. Buchin, M.—De la Pélade: Nature, Traitement, Prophylaxie, Paris, 1887.
47. Bulkley, L. D.—Acne and Alopecia, Detroit, 1892.
48. Burnett, J. C.—Ringworm: its Constitutional Nature and Cure, London, 1892.
49. Cantani, A.—Un Caso de Atrofia progressiva, Napoli, 1887.
50. Carrere, G. A.—Étude sur le Traitement de la Teigne tondante, Paris, 1890.
51. Churlet, M.—Dissertation sur la Teigne, Strasbourg, 1811.
52. Claudat, F. N.—De la Teigne et de son Traitement, Paris, 1879.
53. Clemenceau.—La Pélade, Nantes, 1891.
54. Cleven, K.—Die Haarkur, Berlin, 1891.
55. Feiertag, I.—Ueber die Bildung der Haare, Dorpat, 1875.
56. Geyl.—Beobachtungen und Ideen über Hypertrichose, Hamburg und Leipzig, 1890.
57. Goossens, L C. H.—Over Area Celsi, Rotterdam, 1885.
58. Grenier, J. N.—Essai sur la Teigne, Strasbourg, 1810.
59. Gurney, Thos.—Specific Disease a Cause of Baldness, London, 1888.
60. Hamburger, E.—Ueber die Irrlehre von der Plica polonica, Berlin, 1861.

61. Hayes, P. S.—Electricity and the Methods of its Employment in the Removal of Superfluous Hair, etc.. Chicago, 1889.
62. Hennocque, C.—Du favus de la Peau et des Muqueuses, Thèse, Paris, 1885.
63. Kaeseler, G.—**Ueber Area Celsi seu** Alopecia areata, Greifswald, 1886.
64. Kruska, E.—Ein Beitrag zu dem Kapitel : **Abnorme** Behaarung beim Menschen, Jena, 1890.
65. Landau, C.—**Ueber Sycosis parasitaria, Bonn,** 1885.
66. Lanoix.—**Beobachtungen über die mit dem** Abschneiden der Haare verbundene Gefahr, 179-.
67. Lettre, M. D.—**Ueber Plica** oder Zopfkrankheit, Berlin, 1870.
68. Loriot, G.—Contribution à l'Étude de la Pélade, Paris, 1887.
69. Marcus, M.—Ueber Alopecia areata, Bonn, 1886.
70. Marianelli, A.—Achorion Schoenleinii, Pisa, 1892.
71. Muret-Depéret.—**De la** Folliculite conglomérée trichophytique, Thèse de Paris, 1892.
72. Nachtigal.—Ueber Area Celsi, **Würzburg,** 1885.
73. Nollet, H. C.—Études sur la **Nature de la** Pélade, Bordeaux, 1888.
74. Reissner, E.—Beiträge **zur Kenntniss der Haare des Menschen** und der Säugethiere, Breslau, **1854.**
75. Richard, J.—De la **Teigne** faveuse, Neuchatel, **1859.**
76. Richter, W.—**Ueber Area** Celsi, Würzburg, 1884.
77. Römisch, W.—**Ueber Favus und** Favusbehandlung, Freiburg, 1891.
78. Rothenberg, J.—**Der Weichselzopf, Würzburg, 1841.**
79. Rowland, A.—**The Human Hair,** London, 1853.
80. Sommerfeld, T.—**De Ziekten van het Haar,** Amsterdam, **1891.**
81. Thin, Geo.—**Pathology and** Treatment of Ringworm, **London,** 1887.
82. Wollermann, T.—**Ueber Plica polonica, Berlin,** 1868.

JOURNAL LITERATURE.

Anatomy and Physiology.

83. **Duclert, L.**—Determinisme de la frisure des productions pileuses, Jour. d. l'Anat. et Phys., 1888, xxiv., 103.
84. **Foley, J. L.**—The hygiene of the **hair, N. Y. Med.** Jour., 1887, xlv., 406.
85. **Garcia, R.**—Beiträge zur Kenntniss der Haarwechsels bei menschlichen Embryonen und Neugeborenen, Abst. Montshft. f. prkt. Dermat., 1892, xiv., 242.
86. **Giovannini, S.**—Ueber die normale Entwicklung und über einige Veränderungen der menschlichen Haare, Vrtljhr. **f.** Derm. u. Syph., 1887, xiv., 1049.
87. **Giovannini, S.**—**De la** régénération **des poils après** l'épilation, Arch. **Mikros. Anat.,** 1890, xxvi., 528.
88. **Giovannini, S.**—**Delle** alterazione **dei** follicoli **nella** depilazione, etc., **Giorn. ital. d. mal. ven.,** etc., 1890, **xxv.,** 378.
89. **Giovannini, S.**—Sur la keratination du poil et les altérations des follicules causées par l'épilation, Arch. de Biologie, 1890, x., 609.
90. **Giovannini, S.**—Ueber ein Zwillingshaar mit einer einfachen Würzelscheide, Arch. Derm. u. Syph., 1893, xxv., 187.

91. **Mertsching, A.**—Beiträge zur Histologie des Haares und Haarbalges, Archiv. Mikros. Anat., 1887-8. xxxi., 32.
92. **Recker, H.**—Eine Nachlese zu Erdls und Waldeyers Untersuchungen über die Haare, Jahrsbrcht. d. Westfälischen Prov.-Vereins, Münster, 1891, xix.
93. **Schein, M.**—Ueber das Wachsthum der Haut und der Haare des Menschen, Wien. klin. Woch., 1892, v., 86; Arch. Derm. u. syph., 1892, xxiv., 429.
94. **Schweninger, E.**—Ueber Transplantation und Implantation von Haaren, (his works) 1886, i., 1.
95. **Sewell, H.**—The cleansing functions of hairs, Science, 1893, xxi., 117.
96. **Stieda, L.**—Ueber den Haarwechsel, Biol. Centrlbl., 1887-8, vii., 353. et seq.
97. **Veraglia ed Conti.**—Contributo allo Studio delle ghiandolo cutanee e dei follicoli pilliferi, Giorn. del. R. Acad. di Med. di Torino, 1885 (abst. Monatshft. f. prkt. Dermat., 1887, vi., 720).

> Bonnet.—Morpholog. Jahrb., 1885-6, xi., 220; Sitzungsb. d. phys. med. Gslschft. z. Würzb., 1889, 129.
> Hoffmann, L.—Deutsche Zeitschr. f. Thiermed., 1885-6, xii., 51.
> Maurer, F.—Morphol Jahrbuch., 1891-2, xvi.1., 717.
> Topinard, P.—Rev. d'Anthrop., 1887, ii., 1
> Voigt, C. A.—Denkschrift d. K. Akad. d. Wissenschaft. Bd. xiii.

Canities.

98. **Breda, A.**—(Ringed Hair) Rivista Veneta di Scienze mediche; abst. Montshft. f. prkt. Dermat., 1888, vii., 291.
99. **Falkenheim, H.**—Zur Lehre von den Anomalien der Haarfärbung, Vrtljhr. f. Dermat. u. Syph., 1888, xv., 33.
100. **Lesser, E.**—Ueber Ringelhaare, Tagebl. d. Versamml. deutsche Naturforsch., etc., 1855, lviii., 160; also Ann. Derm. et. Syph., 1886, vii., 36.
101. **Lesser, E.**—Ein Fall von Ringelhaaren, Vrtljhr. f. Derm. u. Syph., 1886, xiii., 51.
102. **Morgan, J. H.**—Hereditary tuft of white hair on the forehead. Brit. Med. Jour., 1890, ii., 85.
103. **Robinson, T.**—Baldness and grayness, Wood's Med. and Surg. Monog., 1891, ix., 725.

> Obolonski, N.—Sborn. rabat Charkoff., 1886-7, ii., 74.
> Ottolenghi.—Arch di prichiat. Torino, 1889, x., 194.
> Schutze, J. C.—Vrhndl. d. Berl. Geselishft. f. Anthrop., 1886, p. 559.

Discoloration of the Hair.

104. **Prentiss, D. W.**—Change of color in the hair from the internal use of pilocarpin, Trans. X. Int. Med. Cong., 1891, iv., 24.

Alopecia.

105. **Anonymous.**—Therapie der Alopecia pityrodes, Montshft. f. prkt. Dermat., 1888, vii., 295.
106. **Arnozan, H.**—Folliculites depilantes des parties glabres, Ann. Derm. et Syph., 1892, iii., 491.
107. **Besnier, E.**—Alopécie cicatricielle innominée, Ann. Derm. et Syph., 1888, x., 104.

108. Blaschko —**Alopecia mit Trichorrhexis,** Montshft. f. prkt. Derm., 1891, xiii., 105.
109. Brocq, L.—Des folliculites et périfolliculites décalvantes, Bul. et Mem. Soc. Med. et Mem. Hôp. de Par., 1888, v., 399.
110. Brocq. L.—Treatment of alopecia syphilitica, Jour. Cut. and Gen.-Urin. Dis., 1889, vii., 346.
111. Brocq, L.—Des **rapports** qui existent entre les alopécies de la keratose pilaire et les alopécies dites séborrhéique, Ann. Derm. et Syph., 1892, iii., 773.
112. Carrier, A. E.—**Bald heads, Trans.** Mich. Med. Soc., 1891, **xv., 106.**
113. Curtis, R. J.—Brains or hair, Med. Rec., 1886, xxx., 526.
114. Darier.—Sur l'examen microscopique des cheveux dans l'alopécie **syphilitique,** Ann. Derm. et Syph., 1889, x., 198.
115. De Molines, P.—**Sur un cas** d'alopécie congénitale, Ann. Derm. et Syph., 1890, i., **54.**
116. Elliot, Geo. T.—Alopecia præmatura : its most frequent cause, seborrhœal eczema, N. Y. Med. Jour., 1893, lvii., 130.
117. Ferros.—Beitrag **zum Studium der** Alopecie, Montshft. f. prkt. Dermat., 1892, xiv., 528.
118. Fournier, A —De alopécies, Med. Modern., 1889-90, i , 957.
119. Fournier, A.—L'alopécie syphilitique, Union med., 1890, l., 793; Gaz. med. de Par., 1888, v. 49.
120. Giovannini —(Anatomical **Changes in Alopecia areata** and Alopecia syphilitica) abst. Montshft. f. prkt. Dermat., 1888, vii , 28.
121. Giovannini.—(Alopecia **syphilitica)** Arch. Derm. u. Syph , 1892, **xxiv.,** 1022.
122. Gouinlock, W. C.—**Hats as a cause of baldness,** Pop. Sc. Month., 1887, xxxi., 97.
123. Graetzer, E.—Die **Lassar'sche Haarkur in der Privatpraxis,** Therap. Montshft., 1889, iii., 452.
124. Holder, A. B.—Diseases **among Indians,** Med. Rec., 1892, xlii., **357.**
125. **Hyde,** J. N.—**Congenital alopecia, Internat.** Clinics, 1891, i , 321.
126. Hlingworth.—Tinea decalvans, Brit. Med. Jour., 1891, ii , 457.
127. Jackson, G. T.—**A practical treatise** on baldness, Wood's Med. and Surg. Monographs, 1889, iv., 601.
128. Jackson, G. T.—**Baldness : what can we do for it ?** Med. Rec., 1887, xxxi., 509.
129. Jardit, P.—Observation d'alopécie du cuir chevelu et de la barbe, etc., Ann. Derm. et Syph., 1891, ii., 461.
130. Lassar, O.—Ueber **Haarkuren,** Therap. Montshft., 1888, ii., 543.
131. Mapother.—Alopecia symptomatica, Brit. Med. Jour , July 5th, 1891.
132. Mayerhausen, G.—Die franklinische Kopfdouche als mittels gegen das Ausfallen der Haare, Int. klin. Rundschau, 1890, iv., 1841.
133. O'Neil, W.—Rejuvenescence of hair of head and beard, 1889, ii., 113.
134. Paschkis, H.—Traitement de l'alopécie d'origine séborrhéique, Nouv. Montpelr. méd., October 8th, 1892, J. d. Mal. cut. et Syph., 1892, iv., 667.
135. Paschkis. H.—Die Therapie des Haarausfalles, Centrlbl. **f. d. ges. Therap.,** 1892, x., 321.
136. Quinquaud.—Folliculite épilante décalvante, Ann. Derm. et **syph., 1889, x., 99.**

137. Robinson, T.—Baldness and grayness, Wood's Med. and Surg. Monographs, 1891, ix , 725.
138. Schultz, F. J. -Atrichia adnata, Vrhndl. der Rigaer **Gesellsch. deutsch.** Aerzte, 1891 ; ref. Montshft. f. prkt. Derm , 1893, **xvi., 242.**
139. **Semeleder,** F.—Congenital baldness, Med. Rec., 1888, **xxxiii., 441.**
140. Tyson, **W. J.**—Some remarks on premature baldness, **Lancet,** 1891, ii., 173.
141. Unna, **P. G.**—(Successful **inoculation of** alopecia pityrodes) Montshft. f. **prkt. Derm.**, 1892, xiv., **413.**
142. Ward, **E. B.**—**Baldness: what shall we do with it ?** Med. **Age,** 1887, v., 391.
143. Wheeler. **G. H.**—**About** bald heads, **New** York Evening **Post,** August 6th, 1887.
144. White, A.—Alopecia adnata. Med. Age, 1887, v., 318.
145. Wickham, L.—An undescribed form of alopecia, Brit. Jour. **Dermat.**, 1888-9, i., 227.

Alopecia Areata.

146. Arnozan, X.—Pseudopélade **avec** plaques achromateuses **et** plaques hyperchromateuses, Bul. Soc. **Franç.** d. Dermat., 1891, ii., 352.
147. Askanazy, S.—Casuitisches **zur Frage** der Alopecia neurotica, Arch. Derm, u. Syph., 1890, **xxii., 523.**
148. Barthélemy.—Pélade **généralisée au** cuir chevelu et à la face, traitée par le procédé du Dr. **Moty,** Bul. **Soc.** Franç. d. Dermat., 1891, ii., 421, 443.
149. Barthélemy.—Pélade traitée par les injections intradermiques de sublimé. Ann. Derm. et Syph., 1892, iii., 1165.
150. Behrend, G.—Ueber die klinischen **Grenzen der** Alopecia **areata,** Berl. klin. Woch., **1887,** xxiv., **108.**
151. Behrend. **G.**—**Ueber** Alopecia areata **und über die** Veränderungen der Haare bei derselben, Arch. path. Anat., **1887, cix.,** 493.
152. **Behrend, G.**—**Zur Frage von der Alopecia areata,** Berl. **klin. Woch.,** 1888, **xxv., 148.**
153. Behrend, **G.**—**Ueber** Nervenläsion und Haarausfall **mit Bezug auf der** Alopecia areata. **Arch.** path. Anat., 1889, cxvi., 173.
154. Besnier, E —Instruction provisoire sur les mesures à prendre a l'égard des sujets atteints de pélade, Trib. med., 1889, xx., 410.
155. Blaschko, A.—Vorstellung **eines** Falles **von** Alopecia areata, **Berl. klin.** Woch., 1891, **xxviii.,** 1152.
156. Blaschko, A.—Alopecia areata **und Trichorrhexis,** Berl. klin. Woch., 1892, xxix., 71 ; Trans. Ann. Derm. et Syph., 1892, iii., 211.
157. Brocq, L.—Doit **on** considérer la pélade comme une affection contagieuse, Gaz. hebdom., **1887,** xxiv., 307.
158. Brousse.—Un cas de pélade totale du **cuir** chevelu, Bul. Soc. **Franç.** de Dermat., 1891, ii., 195.
159. Bulkley, L. D.—Clinical **study on alopecia areata** and its treatment, Trans. Med. Soc. N. Y., **1889, 184 ; Med. Rec.,** 1889, xxxv., 231.
160. **Bulkley.** L. D.—A therapeutic **note on** alopecia areata, Jour. Cut. and Gen.-Urin. Dis., 1892, x., 47.
161. Busquet, **G. P.**—Du traitement antiseptique des teignes et en particulier de la **pélade,** Ann. Derm. et Syph., 1892, iii., 269.
162. Chatelain, E.—Du traitement de la pélade par le collodion iodé, **J. d. Mal. cutan.** et syph., 1890-1, ii., 221.

163. Chatelain, E.—Nouvelles observations sur le traitement de la pélade par le collodion iodé, J. Mal. cutan. et syph., 1891, iii., 605.
164. Crocker, H. R.—Alopecia areata: its pathology and treatment, Lancet, 1891, i., 478.
165. Crocq, J.—Ueber Wechselseitige Beziehungen von Alopecia areata, Psychosen, und Spermatorrhœa, Presse med. Belge, 1892, No. xiv.
166. Cutler, C. W.—The use of iodine, carbolic acid, and chloral in dermatology, Jour. Cut. and Gen.-Urin. Dis., 1892, x., 380.
167. De Tullio.—(Progressive areaform atrophy of hair follicles) Rif. Med., 1887; Montshft. f. prkt. Dermat., 1887, vi., 769.
168. Eichhoff, P. J.—Zur Frage der Kontagiosität der Alopecia areata, Montshft. f. prkt. Dermat., 1888, vii., 1025.
169. Ferros.—Contribution au traitement de la pélade, Annal. Soc. d'Hydrol. med. Par., 1892, xxxvii., 249; Ann. Derm. et Syph., 1892, iii., 546.
170. Feulard, H.—La favus et la pélade en France, Ann. Derm. et Syph., 1892, iii., 1118.
171. Feulard, H.—Pélade décalvante et vitiligo, Ann. Derm. et Syph., 1892, iii., 842.
172. Fournier, H.—La recrudescence de la pélade, Jour. de Med. Par., 1888, xv., 35.
173 Froelich, L.—Pélade et lesions oculaires, Rev. med. de la Suisse Romande, 1890, x., 745.
174. Froelich, L.—Ueber Augenerkrankung bei Alopecia areata, Berl. klin. Woch., 1891, xxviii., 343.
175. Giovannini, S.—(Anatomical changes in alopecia areata, etc.) abst. Montshft. f. prkt. Derm., 1888, vii., 23.
176. Giovannini, S.—Recherches sur l'histologie pathologique de la pélade, Ann. Derm. et Syph., 1891, ii., 921.
177. Giovannini, S.—Ueber die histologischen Veränderungen der syphilitischen Alopecie und ihr Verhältniss zu den Veränderungen der Alopecia areata, Montshft. f. prkt. Dermat., 1893, xvi., 157.
178. Grindon, J.—Etiology of alopecia areata, Wkly. Med. Rev., 1889, xx., 381.
179. Hallopeau, H.—Sur une nouvelle variété d'angionevrose donnant lieu à la plaques d'alopécie pseudo-péladique, Bul. Soc. Franç. de Dermat., 1891, ii., 157.
180. Hallopeau, H.—De la nature de la pélade et des antiseptiques propres à son traitement, Union med., 1889, xlviii., 337.
181. Havenith, Dubois.—La policlinique, Brux., 1892, No. 10.
182. Hoffmann R.—Ein Fall von Alopecia areata nach Trauma, Allg. Med. Centrl.-Ztg., 1889, lviii., 1361; Maryland Med. Jour., 1889-90, xxii., 344.
183. Hutchinson, J.—Alopecia areata usually a sequel of ringworm, Arch. Surg., 1889-90, i., 162.
184. Hutchinson, J.—The permanency of cures in alopecia areata, Arch. Surg., 1889-90, i., 163.
185. Hutchinson, J.—Universal alopecia in middle age with history of severe ringworm in childhood, Arch Surg., 1889-90, i., 370.
186. Hutchinson, J.—On alopecia areata and its relation to ringworm, Arch. Surg., 1893, iv. 289.
187. Joseph, M.—Zur Aetiologie der Alopecia areata, Centrlbl. f. d. med. Wissenschft., 1886, xxiv., 178.
188. Joseph, M.—Experimentelle Untersuchungen über die Aetiologie der Alopecia areata, Wien. med. Woch., 1886, xxxvi., 1642.

189. Joseph, M.—Zur Aetiologie und Symptomatik der Alopecia areata, Berl. klin. Woch., 1888, xxv., 82, etc.

190. Joseph, M.—Erwiderung auf Herrn Dr. G. Behrend's Aufsatz, etc., Arch. path. Anat., 1889, cxvi., 333.

191. Kaposi, M.—Alopecia areata oder Area Celsi, Internat. klin. Rundschau, 1889, iii., 536.

192. Kinney, Thos. H.—Alopecia from nervous shock, Virg. Med. Month., 1880–81, vii., 937.

193. Lavallée, A. Morel.—Sur un mode de traitement rapide de la pélade, etc., Union med., 1892, liii., 889; Ann. Derm. et Syph., 1892, iii., 713; Bull. Soc. Fr. Derm., 1892, iii., 318.

194. Leloir, H.—De la pélade et des péladoïdes, Bull. Acad. Med. Par., 1888, xix., 936; Gaz. Hôp., 1888, lxi., 700.

195. Leo.—Ein Fall von Alopecia areata, Montshft. f. prkt. Derm., 1887, vi., 1119.

196. Lewinski.—Ein eigenartiger Fall von Verlust des Kopfhaares, Montshft. f. prkt. Dermat., 1887, vi., 141.

197. Merklen, P.—Etiologie et prophylaxie de la pélade, Ann. Derm. et Syph., 1888, ix., 813.

198. Mibelli, V.—Sulla patogenesi dell' Alopecia areata, Giorn. ital. d. Mal. ven., 1888, xxix., 416.

199. Mibelli, V.—(Alopecia areata) Boll. n. r. Accad. d. fis. di Sienna, 1887, v., 63 (abst. Montshft. f. prkt. Dermat., 1887, vi., 629).

200. Morrow, P. A.—The treatment of alopecia areata, Jour Cut. and Gen.-Urin. Dis., 1891, ix., 381.

201. Moty.—Nouveau traitement de la pélade, Ann. Derm. et Syph., 1891, ii., 406, 864.

202. Nachtigal.—Ueber das Verhalten der electrocutaneous Sensibilität bei Area Celsi, Mitth a. d. med. Klin. z. Würzb., 1886, ii., 405.

203. Nimier, H.—De la folliculite microbienne tonsurante du cuir chevelu, Gaz. hebdom., 1890, xxvii., 234.

204. Ohmann-Dumesnil, A. H.—A case of alopecia areata due to traumatism, Tr. Med. Assoc. Missouri, 1889, p. 144; St. Louis Clin. Phys. and Surg., 1890, iii., 104.

205. Ohmann-Dumesnil, A. H.—Some successful methods of treating alopecia and alopecia areata, N. O. Med. and Surg. Jour., 1892, xx., 1; Med. News, 1892, lxi., 146; Monatshft. f. prkt. Dermat., 1892, xv., 49.

206. Ollivier, A.—La pélade et l'école, Bul. Acad. Med. Par., 1887, xviii., 725; Rev. de Hygiène, 1887, ix., 195.

207. Overall, G. W.—Alopecia the result of lesions of trophic nerve centres, Alienist and Neurol., 1886, vii., 254.

208. Petrini.—Note sur un cas de calvitie et de pélade généralisée, Bul. Soc. Fr. Derm. et Syph., 1892, iii., 250; Ann. Derm. et Syph., 1892, iii., 554.

209. Pontoppidan, E.—Ein Fall von Alopecia areata nach Operation am Halsen, Montshft. f. prkt. Dermat., 1889, viii., 51.

210. Putnam, C. P.—An epidemic of baldness in spots (alopecia areata?) in an asylum for girls, Arch. Pediat., 1892, ix., 595.

211. Queely, E. St. G.—Alopecia areata, Lancet, 1887, ii., 1266.

212. Quinquaud.—De la pélade, Semaine med., 1890, x., 301.

213. Raymond, P.—Considérations sur le traitement de la pélade, Bul. Soc. Fr. Derm. et Syph., 1892, iii., 386; Ann. Derm. et Syph., 1892, iii., 794.

214. Raymond, P.—Les nouveaux traitements des péladiques, Gaz. d. Hôp., 1892, lxv., 893.

215. Robinson, A. R.—The pathology and treatment of alopecia areata, Trans. IX. Inter. Med. Cong., 1887, iv., 241; Montshft. f. prkt. Dermat., 1888, vii., 476.
216. Samuel, S.—Ueber Dr. M. Joseph's "Atrophischen Haarausfall," Arch. path. Anat. und Phys., 1888, cxiv., 378.
217. Schachmann.—Contribution au traitement de la pélade, Ann. Derm. et Syph., 1887, viii., 178.
218. Schütz, J.—Beitrage zur Aetiologie und Symptomatologie der Alopecia areata, Montshft. f. prkt. Dermat., 1887, vi., 97.
219. Schütz, J.—Ein Fall von Alopecia neurotica, Montshft. f. prkt. Dermat., 1887, vi., 296.
220. Schütz, J.—Sechs Fälle von Alopecia neurotica, Münch. med. Woch., 1889, xxxvi., 124.
221. Stelwagon, H. W.—Alopecia areata, Internat. Med. Magazine, 1892, i., 726.
222. Sympson, E. N.—Case of alopecia of entire scalp, Arch. Pediat., 1892, ix., 840.
223. Thibierge, G.—Sur la question de la contagion de la pélade, Ann. Derm. et Syph., 1887, viii., 503.
224. Tison, M. E.—Traitement de la pélade par le collodion iodé, Bul. soc. Med. Prat. Par., 1892, p. 255; J. d Mal. cut. et Syph., 1892, iv., 239.
225. Vaillard, L., et Vincent, H.—Sur une pseudo-pélade de nature microbienne, Ann. de l'Institut Pasteur, 1890, iv., 446.
226. Variot.—Observations et reflexions sur la pseudo-alopécie, etc., Bul. et Mem. Soc. med. d. Hôp., Paris, 1891, viii., 253; Gaz. med., 1891, viii., 313.
227. Vidal, E.—Pélade généralisée, Ann. Derm. et Syph., 1889, x., 575.
228. Wermann.—(Cases) Korrespndzbl. d. ärztl. Kreis. u. Bezirks-Vereine Sachsens, 1891, viii., 1; Monatshft. f. prkt. Dermat., 1892, xiv., 294.
229. Wickham, L.—On a case of pseudo-pelade of Brocq, Brit. Jour. Dermat., 1890, ii., 251.

Archambault, P.—Jour med. d. Bordeaux, 1889-90, xix., 424.
Basu, B. J.—Indian Med. Rec., 1892, iii., 8.
Bourguet, L.—Gaz. hebdom. d. Sc, med. d. Montpellier, 1887, ix , 145.
Brocq, L.—Rev. gén. de clin. et de thérap., 1889, iii , 399.
Brooke, E. M. W.—Proc. Alumnæ Assoc. Wom. Med. Col. Penn., 1890, xv., 102.
Butte, L —Annal. d. la Po iclin. Paris, 1892, ii., 361.
Cantani, A.—Gior. intern. d. Sc. med., Napoli, 1887, ix., 305.
Colin, L —Arch d. Med. et Pharm. mil., 1888, xii., 81.
Colquhon, D.—Tr. Intern. Med Cong., Melbourne, 1889, ii., 972.
Coustan.—Rev. d Hygiène, 1887, vi., 555.
Dubreuilh, W.—Mem. et Bul. Soc. Med. et Chir. Bordeaux, 1889, page 288.
Dubreuilh, W.—Jour. d. Méd d. Bordeaux, 1888-89, xviii., 522.
Giovannini.—Gior. d. r. Accad. med. d. Torino, 1892, xl., 65.
Kazanli, A. I.—Vrach St. Petersburg, 1888, ix., 763.
Krotkoff, M.—Med. Obozr., 1889, xxxii., 391.
Mansuroff, N.—Klin. Sborn p. dermat. i sif , 1889, page 15.
Ploquin, A —Annal. de la Policlin. Paris, 1890-91, i., 247.
Pontoppidan, E.—Hosp. Tid. Kopenh., 1889, vii., 221.
Robinson, T.—Illust. Med. News, 1884, v , 171.
Schindelka, H.—Oestr. Ztschr. f. wissen. Veterin. 1887, i., 247.
v. Sehlen.—Sitzungsbr. d. Gslschft. f. Morph, u. Phys. in München, 1885, i., 117.
Seleneff, I. T.—Med. Obozr., 1890, xxxiv., 773.
Tommasoli, P. L.—Boll. d. Soc tra i cult. d. sc. med. in Sienna, 1886, iv., 379.
Vaillard et Vincent.—Arch. Med. et Pharm. mil., 1891, xviii., 369.
Wermann.—Koz. Bl. d. artzl. Kreis. in Sachsen, 1891, ii., 38.

APPENDIX.

ATROPHIA PILORUM PROPRIA.

Trichorrhexis Nodosa.—Fragilitas.—Ringed **Hair.**

230. Abramovitch, A.—(Contribution to the study of trichorrhexis nodosa) Russk. Med., 1888, vi., 457; abst. **Arch.** f. Derm. u. Syph., 1889, xxi., 106.

231. Archambault, **P.—Note** sur un cas de cheveux moniliformes, Ann. Derm. et Syph., 1890, i., 392.

232. Abraham, P. S.—Moniliform hair, **Brit. Med. Jour., 1891, ii.,** 1148.

233. Abraham, **P. S.—A case of** monilethrix, **Brit. Jour.** Dermat., 1892, iv., 21.

234. Beatty and Scott.—Moniliform hairs (monilethrix), **Brit.** Jour. Dermat., 1892, **iv.,** 171; Monatshft. f. prkt. Dermat., 1892, xv., 207.

235. Behrend, G.—Ueber Trichomycosis nodosa (Juhel-Renoy), **Piedra (Osorio),** Berl klin. Woch., 1890, xxvii., 464.

236. Blaschko.—Alopecia mit Trichorrhexis, Monatshft. **f. prkt.** Dermat., 1891, xiii., 105.

237. Giovannini, P.—Ueber normale Entwicklung und über einige Veränderungen der menschlichen Haare, Vrtljhr. f. Derm. u. Syph., 1887, **xiv.,** 1149.

238. Hallopeau, **H.—Sur** un aplasie moniliforme des cheveux, Bul. Soc. Franç. de Derm. et Syph., 1890, i., 78, 117; Bul. Med., 1890, **iv., 501.**

239. Hudelo.—Aplasie moniliforme familiale des cheveux, **Ann.** Derm. et Syph., 1892, iii., 1144.

240. Jamieson, A.—Case of **nodose hairs, Med. Press and** Circ., 1888, xlvi., 35.

241. Lesser.—**Aplasia** pilorum intermittens, Monatshft. f. prkt. Dermat., 1887, vi., 1099; Archiv f. Dermat. u. Syph., 1892, Erganzungsheft i.; Verhandl. d. Deutsch. Dermat. Gslschft., 1892, p. 248.

242. McMurray, W.—Notes on **some** abnormal conditions of the **hair,** Australasian Med. Gaz., 1891-92, xi., 279.

243. Montgomery, D. W.—Trichorexis nodosa, **Pacif. M. and S. Jour., 1887, xxx.,** 640.

244. Newton, R. C.—Nosoditas crinium or trichorrhexis nodosa, Med. Rec., 1889, xxxv., 375.

245. Parker, R.—A novel hair disease: (?) **acne** mentagra, Brit. Med. Jour., 1888, ii., 1335.

246. Payne, J. F.—Hair showing remarkable nodose condition, or "beaded hairs" Trans. Path. Soc. Lond., 1886, xxxvii., 540.

247. Ravenel, M. P.—Trichorrhexis nodosa, Med. News, 1892, lxi., 489.

248. Raymond, B.—Recherches sur la trichorrexis nodosa, Bull. Soc. Franç. **de Derm. et Syph.,** 1891, ii, 339; Ann. Derm. et Syph., 1891, ii., **568.**

249. Sabouraud, R.—Dix-sept cas de cheveux moniliformes (moni'ethrix) dans une même famillé. Bul. Soc. Franç. de Derm., 1892-93, 362;

249a. Sabouraud, R.—Sur les **cheveux** moniliformes, Ann. Derm. et Syph., 1892, iii., 781.

250. Steven, **J.** L.—Cases **of** trichorexis **nodosa** with hereditary history, Glasgow Med. Jour., 1889, xxxi., 459.

251. Tenneson.—Kératose pilaire et aplasie moniliforme des cheveux, Ann. Derm. et Syph., 1892, iii., **1146.**

Breda, A.—Rev. veneta di St. med., 1887, vii., 457.
Sasaki.—(Trichor. nod.) Sei-I-Kwai M. Jour., 1890, ix., 249.

Hypertrophia Pilorum.

252. Aulas.—Note sur une cas d'hypertrichose de la main, Loire Med., 1888, vii., 113.
253. Bloom, I. N.—Permanent removal of superfluous hair, Amer. Pract. and News, 1887, iv., 9.
254. Brocq, L.—De la déstruction des poils par l'électrolyse, Ann. Derm. et Syph., 1887, viii., 460.
255. Brocq, L.—De la déstruction des poils par l'électrolyse, Bul. et Mem. Soc. Med. Hôp. Paris, 1888, v., 147 and 387.
256. Brocq, L.—Nouveaux détails sur la déstruction des poils par l'électrolyse, La Semaine med., 1891. xi., 127
257. Chiari, H.—Ueber Hypertrichosis des Menschen, Prag. med. Wochenschr., 1890, xv., 495.
258. Clasen, E.—Elektrolytische Operationen in der ärztlichen Praxis, Deutsch. med. Zeitg., 1892, No. 63.
259. Dodd, A. H —A case of lumbar hypertrichosis, Lancet, 1887, ii., 1063.
260. Dubreuilh.—De l'épilation par l'électrolyse, Jour. d. Med. d. Bordeaux, 1890-91, xx., 506.
261. Dubreuilh.—Épilation électrolytique, Bul. Soc. Fr Dermat. et Syph., 1892, iii., 191; Ann. Derm. et Syph., 1892, iii., 495.
262. Faulkner.—A peculiar growth of hair on the face, N. Y. Med. Jour., 1890, lii., 155.
263. Gottheil, W. S.—Hypertrichosis, Trans. IX. Int. Med. Cong., 1887. iv., 180.
264. Grube, C. H.—Hypertrichosis, or unnatural growth of the hair, Med. World, 1887, v. 424.
265. Hardaway, W. A —A supplemental account of the case of a bearded woman, etc., Med. News, 1888, lii., 490.
266. Hillerz. C.—Electrolysis in removal of superfluous hair, Peoria Med Month., 1886-87, vii., 426.
267. Jamieson, A.—The treatment of hypertrichosis, Pract , 1889, xliii., 1.
268. Joseph, M.—Ein Fall von Schwimmhosen symmetrischen thierfellähnlichen pilifer pigmentosa, Deutsch. med Woch., 1888, xv., 482.
269. Joseph, M.—Ueber Hypertrichosis auf pigmentirter Haut, Berl. klin. Woch., 1892, xxix., 163.
270. Juhel-Renoy, E.—Epilation, Dict. Encyl. d. Sc. med., 1887, xxxv., 111.
271. Lawrence, H.—Removal of superfluous hair by electrolysis, Austral. Med. Jour., 1891. xiii., 504.
272. Leviseur, F. J.—The removal of superfluous hair by electrolysis, Med. Rec., 1892, xli., 209.
273. Leviseur, F. J.—Elektroiyse in der Behandlung von Hautkrankheiten, Montshft. f. prkt. Dermat., 1890, x., 307.
274. McMurray, W.—Notes on some abnormal conditions of the hair, Australas. Med. Gaz., 1891-92, xi., 279.
275. Michelson, P.—Ueber die galvano-chirurgischen Depilations-Methoden, Vrtljhr. Derm. u. Syph., 1887, xiv., 237.
276. Ohmann-Dumesnil, A. H.—Circumscribed hypertrichosis (acquired) in the lumbar region, Weekly Med. Rev. (St. Louis), 1887, xv., 317.
277. Ohmann-Dumesnil, A. H —Circumscribed hypertrichosis (acquired) in the lumbar region, Jour. Cut. and Gen.-Urin. Dis., 1888, vi., 97.

278. Ohmann-Dumesnil, A. H.—Hypertrichosis due to general disease of the nervous system, Alienist and Neurologist, 1887, viii., 483.
279. Ohmann-Dumesmil, A. H.—Hypertrichosis, Progress, Louisville, 1887-88, ii., 402.
280. Oliver, F. W.—Electricity in the removal of superfluous hair, Med. Reg., 1889, v., 273.
281. Ornstein.—Fall eines geschwänzten Menschen, Berl. Geslschft. f. Anthropolog., March, 1885.
282. Ornstein.—Ueber Sakrale Trichosen, Berl. Geslschft. f. Anthropolog., December, 1876; December, 1877.
283. Overton, J. W.—Permanent removal of superfluous hair by electrolysis, Alabama M. and S. Age, 1888-89, i., 75.
284. Prince, M.—Electrolysis: proper and improper methods of using it in the removal of superfluous hair, Amer. Jour. Med. Sci., 1889, xcvii., 479.
285. Regensburger, A. E.—Treatment of hypertrichosis by electrolysis, Occident Med. Times, June, 1892.
286. Rohé, G. H.—Studies in hirsuties, Trans. Int. Med. Cong., Washington, 1887, iv., 180.
287. Rohé, G. H.—Hypertrichosis, Maryland Med. Jour., 1887, xvii., 463.
288. Sack, A.—Ueber radicalepilation auf electrolytischem Wege, Berl. klin. Woch., 1892, xxix., 10:7.
289. Sommer, W.—Ein neuer Fall von Hypertrichosis circumscripta, Arch. path. Anat., 1885, cii., 407.
290. Stierlin.—Spina bifida lumbalis mit Hypertrichosis, Korrespondzbl. f. Schweiz. Arzte, 1892, xxii., 408, 482.
291. Thin, G.—Hypertrichosis, Trans. IX. Int. Med. Cong., 1887, iv., 180.
292. Vanderburg, C. R.—Treatment of hypertrichosis by electrolysis, Columbus Med. Jour., 1887, vi., 245.
293. Woody, S. E.—Permanent removal of hair by electrolysis, Amer. Pract. and News, 1886, ii., 65.

    Anonymous.—Indian Med. Jour., 1886, v., 310.
    Anonymous.—Jour. Anthrop. Soc. Bombay, 1386-87, i., 14.
    Bartels, M.—Ztschrft. f. Ethnol., 1876 1879, 1881, 1883, 1884.
    Bonnet, R.—Anat. Hefte Wiesb., 1891-92, i., 233.
    Cristiani, A.—Arch. di Psichiat., 1892, xiii., 70.
    Ecker, L.—Globus, 1878, xiii.; Arch. f. Anthrop., **1879, xi.**; **1880, xii.**
    Ekama, C.—Album der Natur, 1888.
    Fauvelle.—Bull. Soc. d'Anthrop., 1886, ix., **439.**
    Geyl.—Biolog. Centrlbl., 1888-89. viii., 832
    Giovannini, S.—Gior. d. r. Accad. med. di Torino, 1890. xxxviii., **338.**
    Mansuroff, N.—Klin. Sborn p. dermat. i sif., 1889, page 10.
    Meyer, A. B.—Vrhndlung. d. Brl. Gslshft. f. Anthrop., 1886, page 516.
    Miklucho-Maclay.—Zeitshrft. f Ethnol., 1876 and 1881.
    Neisser.—Jahrb. d. Schles. Gslshft. f. vatrl. Kult., 1885, lxii., 66.
    Ornstein, B.—Zeitschrft. f. Ethnol., 1875, 1876, 1877, 1879, 1880, **and 1884;**
        Archiv f. Anthropol., 1886.
    Parreydt, J.—Deutsche Montshft. f. Zahnheil, 1886, iv., 2.
    Ranke.—Vrhndlng. d. Münch. **Anthrop.** Gslshft., 1888, i., 4.
    Tepljaschin, A.—Med. Obozr., 1888, **xxix**, 39.
    Virchow, R.—Zeitschrft. f. Ethnologie, 1875, **vii.**; **1884, xvi.**
    Zojo, G.—Boll. Scient. Pavia, 1886, xiii., 33.

Sycosis.

294. Bockhardt, M.—Ueber die Aetiologie und Therapie der Impetigo und der Sycosis, Monatshft. f. prk. Dermat., 1887, vi., 450.
295. Brooke, H. G.—The contagious nature of sycosis, Brit. Jour. Dermat., 1888-9, i., 467.

296. Davidson, A. R.—Sycosis non-parasitica, Buf. Med. and Surg. Jour., 1886-7, xxvii., 215.
297. Dubreuilh, W.—Deux cas de sycosis non parasitaire, Annal. de l. Polyclin. Bord., 1889, i., 113.
298. Fabry, J.—Zur Aetiologie der Sycosis simplex, Deutsch. med. Woch., 1891, xvii., 976.
299. Hardaway, W. A.—Inflammation of the hair follicles within the nares, Jour. Cutan. and Ven. Dis., 1886, iv., 360.
300. Hutchinson, J.—Case of severe sycosis of pubes, Archv. Surg., 1889-90, i., 264.
301. Hutchinson, J.—Sycosis; acne of scalp with lichenous and pustular **acne of trunk**, Archv. Surg., 1889-90, i., 371.
302. **Jackson, G.** T —Sycosis: a clinical study, Jour. Cutan. and Gen.-Urin. Dis., 1889, vii., 13.
303. Knott, J.—Tinea sycosis, Lancet, 1890. i., 294.
304. Kromayer, E.—Beitrag zur Therapie der Sycosis, Therap. Monatshft., 1892, vi., 181.
305. Michelson, P.—**Ueber Trichofolliculitis bacterica**, Deutsch. med. Woch., 1889, xv., 586.
306. Ohmann-Dumesnil.—Treatment of sycosis, **St. Louis M.** and S. Jour., 1890, lviii., 137.
307. Rona, S.—(Parasitic sycosis in Hungary) **Pest. med.** chir. Presse, 1887; (abst.) Jour. Cutan. and **Gen.-Urin.** Dis., 1887, v., 351.
308. Rosenthal, O.—Beitrag **zur Aetiologie** und zur **Behandlung der** Sycosis vulgaris, Deutsch. med. **Woch., 1889, xv., 459.**
309. Shoemaker, J. V.—Pathology **and treatment of sycosis**, Jour. Am. Med. Assoc., 1890, **xv., 177.**
310. **Salomon, L. F.**—Sycosis, New **Orleans M.** and S. **Jour., 1887-8, xv., 12.**
311. Tommasoli, B.—Ueber bacillogene Sykosis, Monatshft. f. prkt. **Dermat.**, 1889, viii., 483.
312. **Unna**, P. G.—Hyphogenic, **coccogenic und bacillogenic** Sycosis, St. Louis M. and S. Jour., 1889, lvii., **81.**
313. Unna, **P. G.** - Ueber Ulerythema sykosiforme, Monatshft. f. prkt. Dermt., 1889, **ix., 134.**
314. Williams, **W.—A** localized epidemic of acne sycosis traceable to a barber's shop, Lancet, 1890, i., 346.
315. Zeisler, J.—Epilation; **its range of** usefulness as a dermatotherapeutic **measure**, Jour. Cut. and Gen.-Urin. Dis., 1891, ix., 444; **Trans. Amer. Derm. Assoc., 1891.**

Martin, H.—Arztl. Vereinsbl. f Deutschl., 1890, xvii., **488.**
Nyström, A.—Hygei., Stockholm, 1890.
Rona, S.—Orvosi hetil., 1887, xxxi, 385.
Rona, S.—Pest. med. chir. Presse, 1887, xxiii., 353.
Rosenberg, M.—Arztl. Mitth. a Baden, 1891, xiv., 109.

Trichophytosis Capitis.

316. Allyn, H. B —Treatment of ringworm of the scalp, Med. and Surg. Reprtr., 1887, lvii., 106.
317. Audrain, J.—Contribution à l'étude de la trichophytie tonsurante, Ann. Derm. et Syph., 1892, iii., 1051.
318. Brocq, L.—Traitement local de la trichophytie du cuir chevelu (teigne tondante), Ann Derm. et Syph., 1890, i, 147.
319. Busquet, G. P.—Du traitement antiseptique des teignes, etc., Ann. Derm. et Syph., 1892, iii., 269.
320. Butte, L.—De l'emploi du collodion iodé dans le traitement des

teignes pour remplacer l'épilation, Jour. d. Mal. cutan. et syph., 1892, iv., 459.

321. Cantrell, J. A.—Some experiments with electrolysis in cases of tinea tonsurans, Polyclinic, Phil., 1888-9, vi., 141.

321a. Charon et Gevaert, Jour. de Med. de Brux., 1887, page 673.

322. Dockrell, M.—Hydronapthol as specific in the treatment of tinea tonsurans, Lancet, 1889, ii., 110.

323. Dubreuilh, W.—De quelques formes rares de la trichophytie du cuir chevelu, Jour. Mal. cut. et syph., 1891, iii., 438.

324. Dubrisay.—L'école des teigneux à l'Hôpital St. Louis, Rev. de Hyg. Par., 1887, ix., 296.

325. Duhring, L. A.—Experiences in the treatment of chronic ringworm in an institution, Am. J. Med. Sc., 1892, ciii., 109.

326. Eddowes, A.—The treatment of ringworm of the scalp, Brit. Med. Jour., 1893, i., 785.

327. Eloy, C.—Un traitement nouveau de la teigne tondante, Gaz. hebdom. de Méd., 1889, xxvi., 475.

328. Fournier, H.—Les hôpitaux et les écoles de teigneux, J. Mal. derm. et syph., 1891, iii., 643.

329. Furthmann, W., und Neebe, C. H.—Vier Trichophytonarten, Monatshft. f. prkt. Dermat., 1891, xiii., 477.

330. Harrison, A. J.—Further researches on the treatment of tinea tonsurans, Brit. M. Jour., 1889, i., 465.

331. Hutchison, J —Tincture Siegesbeckia orientalis in ringworm, Brit. M. Jour., 1887, i., 1384.

332. Hutchinson, J.—Notes on cure of ringworm, Arch. Surg., 1889-90, i., 276.

333. Hutchinson, J.—Herpes tonsurans which travelled from scalp to hand, Arch. Surg., 1889-90, i., 367.

334. Juhel-Rénoy, E.—Teignes : nature et traitement, Arch. gén. de Méd., 1887, ii., 84.

335. Kerley, C. G.—A report of thirty-one cases of ringworm of the scalp treated successfully with bichloride, kerosene, and iodine, N. Y. Med. Jour., 1891, liv., 396.

336. Leviseur, F. J.—Irritation and the treatment of ringworm of the scalp, Med. Rec., 18-9, xxxv., 594.

337. Leviseur. F. J.—The prophylaxis of ringworm of the scalp, Post-Grad., 1889-90, v., 36 ; N. Y. Med. Jour., 1889, l., 43.

338. Marianelli, A.—Sulla cura della tigna tonsurante del capillizio, Giorn. ital. d. Mal. ven., 1890, xxv., 359.

339. Mazza, G —Di una forma rara di tricophitiasis. Giorn. ital. d. Mal. ven., 1889, xxiv., 168.

340. Mégnin, P.—Différence spécifique entre le champignon de la teigne des poules et celui de favus, Compt. rend. Soc. biol.,1890, ii., 151.

341. Morris, M.—Ringworm in the elementary schools, Lancet. 1891, ii., 348.

342. Newman, W.—Short notes on ringworm transmission from cows, Brit. Med. Jour., 1889, ii., 1276.

343. Noyes, A. W. F —The artificial cultivation of trichophyton fungus (ringworm) of the skin and hair, Australian Med. Jour., 1891, xiii., 473.

344. Oberndorfer, J.—Die differential Diagnose und Therapie der Tinea trichophytina, N. Y. med. Presse, 1886-7, iii., 143

345. Ohmann-Dumesnil, A. H.—Case of ringworm of the scalp complicated by pustular eczema, Cincin. Med. News, 1890, xix.. 376 ; Brit. J. Dermat., 1888-9, i., 264.

346. Pelizzari.—Trichophyton tonsurans, abst. Monatshft. f prkt. Dermat., 1887, vi., 1049.
347. Purdon, H. S.—Note on the treatment of tinea tonsurans, Dub. J. M. Sc., 1889, lxxxviii , 299.
348. Quincke, H.—Ueber Herpes tonsurans, Monatshft. f. prkt Derm., 1887, vi , 987.
349. Quinquaud.—Les teignes · la teigne faveuse, la trichophytie, la pélade, Gaz. hôp., 1890, lxiii., 990.
350. Quinquaud.—Prophylaxie et traitement de la trichophytie, Union méd., 1890, l., 49.
351. Quinquaud et Butte.—Note sur les résultats obtenus dans le traitement de la trichophytie, Bul. Soc. Franç. d. Dermat. 1891, ii. 152.
352. Reynolds, H. J.—A new method of treating the vegetable parasitic diseases of the skin, Trans. Inter. Med. Cong., 1887, iv., 189.
353. Roberts, H. L.—Observations on the artificial cultivation of the ringworm fungus, Brit. J. Dermat , 1888-9, i., 359.
354. Sabouraud, R.—Contribution à l'étude de la trichophytie humaine, Ann. Derm. et Syph., 1892, iii., 1061 et seq.
355. Sabouraud, R.—Nouvelles recherches sur la mycologie du trichophyton; des espèces de trichophyton à grosses spores, Bul de la Soc. Franç. de Dermat. et Syph , 1893, iv., 59.
356. Smith, A.—Ringworm of the head and its treatment, Lancet, 1886, i., 418.
357. Tenneson et Berdal.—Trichophytie disseminée des régions glabres et du cuir chevelu à l'âge adulte, Bul Soc. Franç. Derm. et Syph., 1882, iii., 314 ; Ann. Derm. et Syph., 1892. iii , 709.
358. Thin, G.—Pathology and treatment of ringworm, Practitioner, 1857, xxxviii., 241.
359. Thin, G.—Experimental researches concerning trichophytina tonsurans, Brit. M. Jour., 1889, i., 397.
360. Thin, G.—Pathology and treatment of ringworm, Wood's M. and S. Monog., 1889, iv., 721.
361. Trichophytosis.—Cong. Internat. de Derm. et Syph., 1890, p. 191.

Arnaud, F.—Marseilles méd., 1888, xxv., 534.
Bertarelli, A.—Boll. di Poliambul. di Milan, 1890, iii., 1.
Bertrand, L. E.—Arch. d. méd. Nav., 1891, lv., 471.
Butte, L.—Assistance, Par., 1891, i , 204.
Cramoisy.—Compt. rend. Cong. inter. d. homœop., 1889, p. 199.
Déclaux, E.—Compt. rend. Soc. de Biol., Paris.
Laennec, T —Gaz. méd. de Nantes, 1891-2, x., 53.
Mazza, G.—Boll. d. r. Accad. med. d. Genoa, 1891, vi., 1.
Mégnin, P.—Bull. soc. centr. d. Méd. vet , 1890, viii., 183.
Monteverdi, I.—Bol. d. Comit. med. Cremona, 1885, v., 209.
Peroni, G.—Giorn. d. r. Accad. med. d. Torino, 1891, xxxix., 33.
Simon, R. M.—Illust. Med. News, 1890. vi., 100.
Simon, R. M.—Rev. méd. de l'Est, 1890, xxii., 493.
Taussig.—Bol. d. r. Accad. med. d. Roma, 1890, xvii., 54.
Trabut.—Alger. med., 1888, xvi., 220.

Trichophytosis Barbæ.

362. Besnier, E.—Traitement du sycosis (parasitaire), Jour. d. Med. et Chir., 1887, lviii., 248.
363. Besnier, E.—Trichophytie érythémateuse circinée, etc., Ann. Derm. et Syph., 1889, x., 111.
364. Feye.—Epidémie de trichophytie, Arch. med. Belges, 1886, xxx., 17.
365. Gottheil, W. S.—Barber's itch, Med. News, 1892, lxi., 342.
366. Hallopeau.—Trichophytie de la barbe, Ann. Derm. et Syph., 1889, x., 327.

367. Hutchinson, J.—Case of ringworm of the beard, Arch. Surg., 1889, i., 55.

FAVUS.

368. Boer, O.—Zur Biologie des Favus, Vrtljhr. f. Derm. u. Syph., 1887, xiv., 429.
369. Busquet, G. P.—Du traitement antiseptique des teignes et en particulier de la pélade, Ann. Derm. et Syph., 1892, iii., 269.
370. Busquet, G. P.—De l'origine muridienne du favus, Ann. Derm. et Syph., 1892, iii., 916.
371. Elsenberg, A.—Ueber den Favuspilz, Arch. Derm. u. Syph., 1889, xxi., 179, and 1890, xxii, 71; Gaz lek. Warsz., 1889, ix., 170, and 1890, x., 208.
372. Esteves, J.—Tratamiento del favus por la resorcina, Annal. d. l. Assistancia publica, Buenos Ayres, 1891; abst. Ann. Derm. et Syph., 1891, ii., 798.
373. Evans, S. G.—Favus and its treatment; results in 169 cases, Med. Rec., 1892, xli., 490.
374. Fabry, J.—Klinisches und Aetiologisches ueber Favus, Arch. Derm. u. Syph., 1889, xxi., 461.
375. Feulard, H.—La favus et la pélade en France, Ann. Derm. et Syph., 1892, iii., 1118
376. Frank L F—Favus, Monatshft. f. prkt. Dermat., 1891, xii., 254.
377. Hutchinson, J.—A very severe case of favus, Arch. Surg., 1890, i., 9.
378. Jadassohn, J.—Demonstration von Favusculturen, Vrhndl. d. deutsch. dermat. Gslschft., 1889, i., 77.
379. Jadassohn, J.—Bemerkung zu der Arbeit Elsenberg's "Ueber den Favuspilz bei Favus herpeticus," Arch. Derm. u. Syph., 1890, xxii., 451.
380. Jamieson, W. A.—Favus occurring under somewhat peculiar circumstances as to origin, Brit. Jour. Dermat., 1893, v., 140.
381. Kaposi, M.—Zur Pathologie und Therapie des Favus, Intrnt. klin. Rundschau, 1891, v., 503.
382. Kral, F.—Ueber den Favuserreger, Wien. med. Woch., 1890, xl., 1441; Trans. X. Int. Med. Cong., 1891, iv., 13.
383. Kral, F.—Untersuchungen über Favus, Beiträge z. Derm. u. Syph., 1891, i. 79
384. Mibelli, V.—Di alcuni casi di tigna favosa curata con l'oleato di rame senza depilazione, Bol d. cult. d. Sc. med., 1888.
385. Mibelli, V.—Di alcuni casi di tigna favosa curata con l'oleato di rame senza depilazione, Giorn. ital. Mal. ven., 1888, xxiii., 329.
386. Mibelli, V.—Sul fungo del favo, Riforma medica, 1891, p. 817, and 1892; Ann. Derm. et Syph., 1892, iii., 228.
387. Morris, M.—An extensive case of favus, Brit. Jour. Dermat., 1891, iii, 101.
388. Munnich, A. J.—Beitrag zur Kenntniss des Favuspilzes, Arch. f. Hygiene, 1888, viii., 246.
389. Neumann, S.—Identité du favus des poules et du favus de l'homme, Compt. rend. Soc. biol. Par., 1886, iii, 173.
390. Peroni, G.—Di un nuovo metodo pratico per curare la tigna favosa, Gior. d. r. Accad. d. Med. Torino, 1891; abst. Ann. Derm. et Syph., 1891, ii., 797.
391. Pick, F. J.—Experimenteller Beitrag zur Kenntniss des Favuserregers, Wien. klin. Woch., 1890, iii., 642.

392. Pick, F. J.—Untersuchungen über Favus, Zeitschr. f. Heilk., 1891, xii., 153; also Arch. Derm. u. Syph. Ergnzhft., 1891; also Beitrag z. Dermat. u. Syph., 1891, i., 57.
393. Plaut, H. L.—Beiträge zur Favusfrage, Cntrlbl. f. Bakteriol. u. Parasitk., 1892, xi., 357.
394. Quincke, H.—Ueber Favuspilze, Archv. expermnt. Path. u. Pharm., 1886-7, xxii., 62; also Vrhndl. d. Cong. f. Int. Med., 1886, v., 467; also Mntshft. f. prkt. Dermat., 1887, vi., 981.
395. Quincke, H.—Doppelinfektion mit Favus vulgaris und Favus herpeticus, Mntshft. f. prkt. Dermat., 1889, viii., 49.
396. Quinquaud.—Les teignes: la teigne faveuse, la trichophytie, la pélade, Gaz. hôp., 1880, lxiii., 990.
397. Reynolds, H. J.—Favus, Med. Age, 1889, vii., 270.
398. Rossi, A.—La tigna favosa della facia, La Riforma medica, **1891**, p. 87.
399. Schuster.—Ueber Favusbehandlung, Mntshft. f. prkt. Dermat., 1889, ix., 1.
400. Schwengers.—Ueber Einwirkung von Medikamenten auf Kulturen von Favus und Trichophyton, Mntshft. f. prkt. Dermat., 1890, xi., 155.
401. Unna, P. G.—Drei Favusarten, Montshft. f. prkt. Dermat., 1892, xiv., 1; Brit. Jour. Derm., 1892, iv., 139.
402. Unna, P. G.—Die bisher bekannten neun Favusarten, Montshft. f. prkt. Dermat., 1893, xvi., 17.

Ambrosi, A.—Raccoglitore med. Forli., 1888, vi., 282.
Andronico, C.—Bol. d. Sc. med. d. Bologna, 1886, xvii., 377.
Dahl, J.—Norsk. Mag f Laegevidensk., 1888, iii., 392.
Desville, L.—Jour. d. Sc méd. d. Lille, 1892, ii., 49.
Desmet, E.—La Clinique, Brux., 1889, iii., 341.
Hiorth, W.—Norsk. Mag. f. Laegevidensk., 1888, iii., 105.
Kovalevski, I.—Archv. vet. rauk. St. Peters., 1887, xvii., 47.
Krasin, A A.—Vrach., 1890, xi., 207.
Mibelli, V—Riforma med., 1891, vii., 817.
Rossi, A.—Riforma med., 1891, vii., 87.
v. Walsen, A. J.—Nederl. mil. geneesk. **Arch.**, 1889, xiii., 335.

PEDICULOSIS.

403. Besnier.—Déstruction des pediculi pubis, Jour. de Méd. et de Chirurg., 1887, lviii., 248.
404. Cantrell, J. A.—Pediculosis capillitii, Ann. Gynec. and Pediat., 1889-90, iii., 485.
405. Frazer, W.—Pediculi: their treatment by parasiticides, Med. Press and Circ., **1885**. ii., 550.
406. Goldenberg, H.—Ueber Pediculosis, Berl. klin. Woch., 1887, xxiv., 866.
407. Greenough, F. B.—Clinical notes on pediculosis, Boston M. and S. Jour., 1887, cxvii., 469.
408. Grellety.—Traitement de la phthiriase, Jour. de Mal. cut. et syph., 1890-91, ii., 20.
409. Heiser, I.—Pediculi pubis auf der behaarten Kopfhaut, Arch. Derm. u. Syph., 1892, xxiv., 589.
410. Jamieson, W. A.—On some of the rarer effects of pediculosis, Brit. J. Derm., 1888-9, i., 321.
411. Mathews, P. W. P.—Notes on phthiriasis, Canad. M. and S. **Jour.**, 1886-7, xv., 45.
412. Payne, J. F.—Maculæ cœruleæ and other symptoms produced

by pediculosis pubis, Brit. J. Derm., 1890, ii., 209; also Mntshft. f. prkt. Derm., 1890, xi., 388.
413. De Schweinitz and Randall.—Phtheiriasis palpebrarum, Univ. M. Mag., 1891-2, iv., 137.
414. Trouessart.—Sur une phthiriase du cuir chevelu causée par la phthirius inguinalis, Compt. rend. Acad. Sc., Par., 1891, cxiii., 1067.

> Grellety.—Actualité méd., 1890, ii., 161.
> Pjatnitski.—Med. Obozr. Mosk., 1886, xxvi., 288.

Parasitic Diseases.

415. Behrend, G.—Ueber Trichomycosis nodosa, Berl. klin. Woch., 1890, xxvii., 464.
416. Juhel-Rénoy, E.—Recherches histo-biologiques et histologiques sur la trichomycosis nodulaire, Ann. Derm. et Syph., 1890, i., 765.
417. Juhel-Rénoy, E.—Note pour servir de l'histoire de la piedra, Compt. rend. Soc. biolog. Paris, 1888, v., 827.
418. Patteson, R. G.—Trichomycosis nodosa, a bacillary disease of the hair, Brit. M. Jour., 1889, i., 1166 ; also Tr. Roy. Acad. Med. Ir., 1889, vii., 85.
419. Patteson, R. G.—Trichomycosis nodosa, a note on its character, etc., Dub. J. M. Sc., 1889, lxxxviii., 207 ; Brit. J. Derm., 1890, ii., 101.

> Oudemans and Pekelharing.—Arch. neerl. d. Sc. exact., 1886, xx., 404.

Dandruff.

420. Gamberini.—Pitiriasi del capo, Giorn. ital. d. Mal. ven., 1891 ; Archv. f. Derm. u. Syph., 1893, xxv., 307.
420a. Jackson, G. T.—Seborrhœa, Gaillard's Med. Monthl., 1890.
421. Mannino, L.—Il microsporon dispar di Vidal nella seborrea, Giorn. ital. d. Mal. ven., 1886, xxi., 84.
422. Unna, P. G.—Was wissen wir von der Seborrhœa, Mntshft. f. prkt. Derm., 1897, vi., 698.
423. Van Harlingen, A.—The pathology of seborrhœa, Arch. Derm., 1878, iv., 97.

Keratosis Pilaris.

424. Brocq, L.—Notes pour servir à l'histoire de la kératose pilaire, Ann. Derm et Syph., 1890, i., 25.
425. Hutchinson, J.—Cacotrophia folliculorum, Annal. Surg., 1892-3, iv., 45.

Plica Polonica.

426. De Amicis.—Du trichoma vrai, Ann. Derm. et Syph., 1892, iii., 1182.
427. Jarochevski, S.—Un cas de tricoma (plique polonaise) aigu, abst. Jour. Mal. cutan., 1892, iv., 533

Nævus Pilosus.

428. Chiari, H.—Ueber Hypertrichosis beim Menschen, Prag. med. Wochsch., 1890, xv., 495.

> Ornstein.—Vrhndlg. der Gesellschft. f. Anthrop., Ethnol., etc., 1884.

INDEX.

	PAGE
Abscess	242
Achor	314
barbatus	329
Achorion Schönleinii	260
Acne	190, 332
decalvante	198
indurata	251
keloidique	344
mentagra	182
pilaire cicatricielle depilante	198
sycosis	182
Adenotrichie	182
Alopecia	80, 221, 269
Alopecia adnata	81
symptoms	81
etiology	81
pathology	81
prognosis	82
treatment	82
Alopecia areata	115, 220
symptoms	116
etiology	118
pathology	124
diagnosis	120
prognosis	131
treatment	138
Alopecia circumscripta	115
follicularis	96
furfuracea	92
occidentalis	115
pityrodes	92
Alopecia prematura idiopathica	85
symptoms	85
etiology	85
pathology	88
prognosis	89
treatment	90
Alopecia prematura symptomatica	91, 129
symptoms	91
etiology	96
pathology	99
prognosis	100

	PAGE
Alopecia prematura symptomatica—diagnosis	101
treatment	101
Alopecia senilis	83, 129
symptoms	83
etiology	83
pathology	84
prognosis	84
treatment	84
Alopecia syphilitica	94, 130
Alopécie cicatricielle innominée	198, 199
Aplasia pilorum intermittens	153
Aplasia pilorum propria	153
symptoms	154
etiology	154
pathology	155
treatment	156
Area Celsi	115
occidentalis diffluens	115
occidentalis serpens	115
occidentalis tyria	115
Arrectores pilorum	31
Athrix depilis	80
Atrichia	80
Atrophia pilorum propria	140
Bacterium decalvans	126
Baldness	80
circumscribed	115
congenital	81
premature	85
senile	84
Barber's itch	182, 245, 329
Bartfinne	182
parasitische	245
Bartflechte	182
Bearded women	164
Bed-hair	39
Beigel's disease	290
Bibliography	365
appendix	391
Black hair	78
Blepharitis ciliaris	334

INDEX.

	PAGE
Blood-vessels	30
Blue hair	78
Brown hair	78
Brushes, selection of	54
Brushing	54
Cacotrophia folliculorum	103
Calotte	271
Calvezza	80
Calvities	80
Canities	63
symptoms	63
etiology	68
pathology	68
treatment	71
Canities, acquired	63
congenital	63
sudden	66
Cheveux moniliformes	153
Clastothrix	144
Coccogenous sycosis	182
Color altered by chemicals	79
change after death	78
Combing	55
Combs, selection of	55
Cortex, the	23
Crusta lactea	314
Cuticle, the	24
Cutis anserina	311
Dandruff, symptoms	299
etiology	301
pathology	302
diagnosis	303
treatment	304
Dartre furfuracée arrondie	205
pustuleuse mentagra	182
Dasyma	158
Defluvium capillorum	95
Depilatio	80
Depilatories	176
Dermatitis	322
Dermatitis papillaris capillitii	198, 344
symptoms	344
etiology	348
pathology	348
diagnosis	348
prognosis	349
treatment	349
Dermatitis papillomatosa capillitii	344
Dermatomykosis barbæ nodosa	245
favosa	256
palmellina	295

	PAGE
Dermatomykosis tonsurans	205
trichophytina	205
Distichiasis	181
Duhring's parasitic disease	296
Dyes for hair	72
Ecthyma	359
Eczema	189, 219, 267, 303, 312, 359
Eczema barbæ	250, 329
symptoms	329
etiology	330
diagnosis	331
prognosis	332
treatment	332
Eczema capitis	314
symptoms	314
etiology	317
diagnosis	320
treatment	323
prognosis	329
Eczema narium	337
Eczema palpebrarum	334
symptoms	334
etiology	335
treatment	335
Eczema seborrhoicum	299, 301
Electrolysis	170
Embryonal hair change	34
End atrophy	156
Epithelioma	251, 332, 360
Erbgrind	256
Erysipelas	321
Erythema ichorosum	314
Favus	129, 218, 256, 323
symptoms	256
etiology	260
pathology	261
diagnosis	266
prognosis	270
treatment	270
Fikosis	182
Flechte, scherende	205
Fluxus pilorum	80
Fœtal hair	32
Folliculite epilante	198, 199
Folliculitis barbæ	182
Folliculitis decalvans	197
symptoms	197
etiology	200
pathology	200
diagnosis	200
treatment	200
prognosis	200

INDEX. 411

Folliculitis pilorum............182
Fragilitas crinium symptoma-
	tica..............140
	etiology.... 141
	treatment..........141
Fragilitas crinium idiopathica.141
	pathology.........142
	etiology..... 143
	treatment..........143
Frambœsia....................344
Fuchsräude 80

Gangræna alopecia............ 80
Gourme 314
Granuloma trichophyticum ...238
Grayness..................... 63
Green hair........ 76

Hair, anatomy of............. 21
	appearances of......... 41
	bed................ 39
	centres 36
	change at puberty...... 35
	chemical constitution .. 48
	color of 43
	curly 47
	cutting.. 57
	development of......... 32
	diameter of......... 46
	discoloration of....... 74
	dressing.... 56
	dyes 72
	elasticity of............ 48
	electricity of 48
	embryonal change in.... 34
	follicle 26
	general description of... 21
	growth of........35, 45, 46
	hygiene of............. 51
	length of...........45, 46
	muscles of... 50
	number of............. 46
	papilla............... 29
	physiology of........... 33
	pigment............... 24
	racial differences....... 47
	regeneration of 40
	root.................. 24
	shedding of............ 38
	superfluous..........158
	technique............. 43
	transplantation of...... 46
	uses of....... 49
Hair-cone, primitive.......... 33

Hairiness....................158
Hats........................ 57
Henle's layer................ 28
Herpes circinatus............205
	circiné parasitaire 205
	pustulosus mentagra..182
	squamosus 205
	tonsurans............205
	tonsurans barbæ......245
	tonsurante...........205
Hirsuties 158
Huxley's layer........ ... 28
Hyperkeratosis pilaris........310
Hypertrichosis...............158
	symptoms. 158
	etiology........166
	treatment......169
Hypertrichosis acquisita par-
	tialis..........163
	congenita par-
	tialis 162
	congenita uni-
	versalis......158
	pilorum..........158
	transitory. ... 165

Ichthyosis....................312
Impetigo.....................322
	figurata.............314
	lactantia...........314
	muciflua 314
	mucosa 314
	sycosiforme... 329

Kahlheit...... 80
	kreisfleckige......... 115
Keratosis pilaris.... 310
	symptoms 310
	etiology.......... . 311
	pathology 311
	diagnosis...... 311
	treatment..........312
	prognosis..........313
Kerion 239, 256, 360
	symptoms 239
	etiology..............241
	pathology 241
	diagnosis..... 242
	prognosis..........243
	treatment..........243
Kerion Celsi 239
Koltun. 339
Kopskurv......... 256

INDEX.

	PAGE
Lanugo hair	25
Lapsus pilorum	81
Lausesucht	270
Lepothrix	293
Lichen menti	182
pilaris	310
planus	312
scrofulosorum	312
Lipsotrichia	80
Loss of hair—condition of subjects of	108
occupation of subjects of	108
age at beginning	108
part of scalp affected	109
complicating diseases	109
diseases of scalp and hair	109
heredity of	110
Lousiness	275
Lupoid sycosis	198
Lupus erythematosus	209, 268, 304, 321, 361
Lupus vulgaris	190, 359, 361
Lymphatics	30
Maculæ ceruleæ	284
Maladie pediculaire	275
Malis pediculi	275
Medulla, the	22
Melitagra	314
Mentagra	182, 245
Microsporon Audouini	125
Milk crust	314
Monilethrix	153
Moniliform hairs	153
Morbus pedicularis	275
Muscles of the hair	31
Mycosis frambœsiodes	344
fungoïde	349
Nævus pilosus	350
symptoms	350
etiology	350
pathology	352
diagnosis	352
treatment	352
Nerves	30
Nodositas crinium	141
Noduli laqueati	156
Oligotrichia	80

	PAGE
Ophiasis	80, 115
Paschkiss's soap	53
Paxton's disease	293
Pedicularia	275
Pediculosis capillitii	275
Pediculosis capitis	275, 320
symptoms	275
etiology	276
pathology	277
diagnosis	279
prognosis	279
treatment	279
Pediculosis palpebrarum	289
Pediculosis pubis	283
symptoms	283
etiology	285
pathology	285
diagnosis	287
prognosis	288
treatment	288
Pelada	80
Pelade	115
acromatosa	115
decalvante	115
ofiasica	115
Phagmesis	156
Phalacrotes	80
Phthiriasis	275
Phytoalopecia	115
tonsurans	205
Pian ruboide	344
Piedra	291
symptoms	291
etiology	292
pathology	292
diagnosis	292
Pigment	43, 44
Pityriasis	320
capitis	300
pilaris	310
Plica polonica	339
symptoms	339
etiology	341
treatment	342
Plique polonaise	339
Poils accidentels	158
Poliotes	63
Poliothrix	63
Polish ringworm	339
Polytrichia	158
Pomades	59
Porrigine tonsurante	205
Porrigo	314

INDEX. 413

	PAGE
Porrigo decalvans	115
favosa	256
furfurans	205
lavalis	256
lupinosa	256
phyta	256
scutalata	256
tonsoria	205
Psilosis	80
Psoriasis	218, 268, 303, 322, 360
Pubic hair	35
Rhizophyto-alopecia	205
Ringed hair	67
Ringskurv	205
Ringworm	266, 304, 321
of the beard	245
crusted	256
honeycomb	256
of scalp	205
Root-sheath	27
Scall, vesicular	314
Scalled head	256, 314
Scissura pilorum	140
Sebaceous glands, anatomy of,	31
glands, physiology of	49
Seborrhœa	268
capitis	217
sicca	322
sicca capitis	299
Shampoo	52
Shaving	58
Sicosi parasitaria	245
Spilosis poliosis	63
Squarus tondens	205
Superfluous hair	158
Sweat concretions	295
Sycosis	182, 250, 331, 348
symptoms	182
etiology	185
pathology	186
diagnosis	188
treatment	190
prognosis	197
Sycosis barbæ	182
capillitii	344
chronique	198
contagiosa	182
framboesia	344
menti	182, 245
non-parasitica	182
parasitica	245

	PAGE
Syphilis	190, 251, 311, 323, 332, 354
symptoms	354
diagnosis	358
treatment	360
Tâches ombrees	284
Teigne annulaire	205
du pauvre	256
faveuse	256
herpetique furfuracée	205
mentagra	245
pelade	115
tondante	205
tonsurante	205
Thin's parasitic disease	296
Tinea amiantacea	314
asbestina	314
barbæ	245
decalvans	115
favosa	256
ficosa	256
furfuracea	314
granulata	314
kerion	239
lupinosa	256
maligna	256
nodosa	292
sycosis	245
tondens	205
tonsurans	205
vera	256
Trichauxis	158
Trichiasis	180
Trichoclasia	144
Trichoma	339
Trichomyces tonsurans	205
Trichomykosis	205
barbæ	245
capillitii	239
favosa	256
Trichonosis cana	63
discolor	63
furfuracea	205
poliosis	63
Trichophytie	205
sycosique	245
tonsurante	205
Trichophyton tonsurans	212
Trichophytosis barbæ, 189, 245, 331
symptoms	245
etiology	248
pathology	249
diagnosis	249

Trichophytosis barbæ—treatment	251
prognosis	255
Trichophytosis capitis	129, 205
symptoms	205
etiology	210
pathology	212
diagnosis	217
treatment	222
prognosis	238
Trichorrhexis nodosa	71, 144
symptoms	145
etiology	146
pathology	149
treatment	153
Trichorrhœa	80
Trichosis athrix	80
decolor	74
furfuracea	205
hirsuties	158
Trichosis pityriasica	205
plica	339
tonsurans	205
Trichosyphilis	144
Trichoxerosis	140
Tricolorosi	74
Twin hairs	29
Ulerythema sycosiforme	198
Verrucæ	348
Vespajo del capillizio	239
Vitiligo capitis	115, 130, 263
Vitreous membrane	27
Vulpis morbus	80
Weichselzopf	339
Wigs	56
Yellow hair	78

www.ingramcontent.com/pod-product-compliance
Lightning Source LLC
Chambersburg PA
CBHW030421300426
44112CB00009B/798